Designing an Internet

Information Policy Series

Edited by Sandra Braman

The Information Policy Series publishes research on and analysis of significant problems in the field of information policy, including decisions and practices that enable or constrain information, communication, and culture irrespective of the legal siloes in which they have traditionally been located as well as state-law-society interactions. Defining information policy as all laws, regulations, and decision-making principles that affect any form of information creation, processing, flows, and use, the series includes attention to the formal decisions, decision-making processes, and entities of government; the formal and informal decisions, decision-making processes, and entities of private and public sector agents capable of constitutive effects on the nature of society; and the cultural habits and predispositions of governmentality that support and sustain government and governance. The parametric functions of information policy at the boundaries of social, informational, and technological systems are of global importance because they provide the context for all communications, interactions, and social processes.

Virtual Economies: Design and Analysis, Vili Lehdonvirta and Edward Castronova

Traversing Digital Babel: Information, e-Government, and Exchange, Alon Peled

Chasing the Tape: Information Law and Policy in Capital Markets, Onnig H. Dombalagian

Regulating the Cloud: Policy for Computing Infrastructure, edited by Christopher S. Yoo and Jean-François Blanchette

Privacy on the Ground: Driving Corporate Behavior in the United States and Europe, Kenneth A. Bamberger and Deirdre K. Mulligan

How Not to Network a Nation: The Uneasy History of the Soviet Internet, Benjamin Peters

Hate Spin: The Manufacture of Religious Offense and its Threat to Democracy, Cherian George

Big Data is Not a Monolith, edited by Cassidy R. Sugimoto, Hamid R. Ekbia, and Michael Mattioli

Open Space: The Global Effort for Open Access to Environmental Satellite Data, Mariel Borowitz

Decoding the Social World: Data Science and the Unintended Consequences of Communication, Sandra González-Bailón

Designing an Internet, David D. Clark

Designing an Internet

David D. Clark

The MIT Press
Cambridge, Massachusetts
London, England

This book was set in Stone Serif by Westchester Publishing Services.

Library of Congress Cataloging-in-Publication Data

Names: Clark, David D. (David Dana), 1944– author.
Title: Designing an internet / David D. Clark.
Description: Cambridge, MA : The MIT Press, 2018. | Series: Information policy series | Includes bibliographical references and indexes.
Identifiers: LCCN 2017061377 | ISBN 9780262038607 (hardcover : alk. paper)
ISBN 9780262547703 (paperback)
Subjects: LCSH: Wide area networks (Computer networks) | Internetworking (Telecommunication) | Internet—History.
Classification: LCC TK5105.87 .C63 2018 | DDC 004.67/8—dc23
LC record available at https://lccn.loc.gov/2017061377

To Susan, who has supported this book with tolerance and sympathy.
I promise not to do this again any time soon …

Contents

Series Editor's Introduction

Sandra Braman

It is one thing to acknowledge that the Internet is a sociotechnical system, quite another to fully incorporate both the social and the technical sides into decision-making for the network and its uses. Putting politics to the side, the greatest challenge for those involved with information and communication policy today is bringing the technical and legal communities together in a common conversation. There have been technical decisions made in ignorance or disregard of legal constraints and policy problems that might or would ensue and, conversely, laws put in place by policymakers who so little understood the technologies being regulated that the statutes or regulations were impossible to implement, bearing no relation to how the systems of concern actually work. The challenges are many, for the modes of thinking, language, and types of specific focal problems that must be addressed—and how they are framed—differ across multiple dimensions.

David Clark's *Designing an Internet* is a foundational work for those seeking to think in both social and technical terms when addressing sociotechnical policy problems. His knowledge of the Internet design process spans almost 50 years, beginning with his work on the technical issues involved in the early 1970s through his leadership of a series of recent efforts funded by the U.S. National Science Foundation that explored alternative approaches to network design issues that might be taken up in the future. The book is not only for those involved in the ever-ongoing Internet design process, but also for legal decision-makers, policy advocates and activists, entrepreneurs, and anyone who thinks about the nature of the Internet, what values are in play and how they might be in tension with each other, and where we might go from here. *Designing an Internet* is beautifully and accessibly written, and is so rich that many readers will find it deserving of more than one read and of keeping on the shelf as a reference work.

The book discusses over two dozen proposals that have been put forward as alternatives to various Internet features and for how to improve the

Internet going forward. Clark explains the design considerations that must be taken into account to achieve goals that include sustainability, security, access, economic viability, manageability, and meeting a wide range of social needs. Significant attention is paid to the economic, social, and political factors that provide the context within which networks are built and operate, and that will determine whether any given element of network architecture will actually succeed over time.

Those struggling with contemporary issues, such as network neutrality, will find Clark's explanations of Internet design enlightening; from a design perspective, to stick with this one example, the fundaments of network neutrality lie with the basic question of whether anything is known about what is in a flow or whether "bits are bits are bits." It is with knowledge that the possibility of constraining, preventing, or violating network neutrality is born; without it, segregating knowledge by content or service or vendor would not be possible. Technical matters such as whether a router can see the content of a packet affect not only the potential for surveillance but also economic issues of interest to content providers and third-party intermediaries. Those thinking about ways in which the nature of the state, as well as governance, is changing will also appreciate Clark's unveiling of what has for those who are not computer scientists been hidden in technical language. While the international legal and political system is of course composed of geopolitically recognized states, for the Internet it is autonomous systems (59,000 of them at the time of writing). When faced with matters of scaling, computer scientists will typically think in hierarchical terms, while lawyers, politicians, and political scientists may well not.

Clark's insights into these and other differences between the technical and the social aspects of the Internet suggest the work could usefully be read hand in hand with the legal scholar Frederick Schauer's *Playing by the Rules: A Philosophical Examination of Rule-Based Decision-Making* (1991) and other works examining the most basic elements of governance of structures of all kinds. His conceptualization of the expressive power of packet headers resonates with Ian Bogost's (2007) notion of procedural rhetoric, although it operates at different levels of the sociotechnical ensemble. His discussions of specific "realizations" of the Internet bring to mind Star and Ruhleder's (1996) discussions of the ways in which global infrastructure always and only, ultimately, exists in specific local manifestations. Social scientists would benefit by taking up the distinction between the singular steps of per-hop behaviors and the multiple steps required for actual functions when they study "effects" of the uses of technologies and of flows of information and communication. The learning can go both ways; "getting to know you" processes that generate trust within the network were learned from the social world.

There are a number of theoretical provocations for those trying to take into account the technical in what had previously been understood as a purely social world. It goes further—Clark is thinking about aspects of communication processes or production chains that have not yet been theorized because we were not driven to do so when thinking about communication in social terms alone, such as a theory of network availability.

Among the gifts of *Designing an Internet* is a succinct, and sometimes funny, history of Internet design. Defining architecture as a process, an outcome, and a discipline, Clark provides a fascinating discussion of the nature of the design process itself. Something is treated as "architecture" if agreement on an issue is necessary for the system to function, it is convenient to agree on an issue, the issue defines the basic modularity or functional dependency of the system, or it is important that the issue in question be stable over time. Core decisions about architecture affect matters of great social importance, such as the balance between surveillance and accountability, on the one hand, and anonymous action and privacy, on the other. The author helps us out with other basic concepts as well; a platform, Clark notes, is really just whatever there is that is one layer below what one is looking at.

The Internet is working and in ever-increasing use globally, but there are lots of choices yet to be made going forward. The policy-maker's problem is how to think about interactions between legal developments, network architecture, management of the physical network, social processes, and policy principles. The sociologist Leigh Star (1998) introduced the idea of a "Durkheim test"—as distinct from the Turing test—to evaluate whether a technical decision makes sense for humans. The computer scientist Clark agrees, arguing that we need to "allow for the future in the face of the present" (p. 43), with a "socially robust design" (p. 205).

For computer scientists, *Designing an Internet* is of immediate use on the ground, contextualizing and raising the level of abstraction of issues currently being discussed within the Internet Engineering Task Force and other technical decision-making and design venues. For policy-makers, social scientists, and citizens, the book is enormously valuable as well. Clark has succeeded in providing a vocabulary, a set of concepts, and a way of thinking that can go far toward bringing the legal and technical communities into a common conversation about Internet governance.

1 Introduction

This is a book about how to design an internet. I say *an internet* rather than *the Internet* because the book is about not just the Internet we have today but also possible alternative conceptions of an internet—what we might instead have designed back then or might contemplate in the future. I take the word *internet* to describe a general-purpose, global interconnection of networks designed to facilitate communication among computers and among people using those computers. The book concerns itself with the implications of globality, the implications of generality, and the other requirements that such a network would have to meet, but it does not take the current Internet as a given—it tries to learn from the Internet of today, and from alternative proposals for what an internet might be, to draw some general conclusions and design principles about networks.

I call these design principles *architecture*, so this book is about architecture as well as the specifics of an internet. There are lots of little design decisions that shape today's Internet, but they could have been made differently and we would still have an internet. It is the basic design decisions that define the skeleton of the design, on which subsequent, more specific decisions are based. I am concerned with the question of what the essence of the design is—what defines a successful skeleton, if you will.

This is a very personal book. It is opinionated, and I write without hesitation in the first person. It is a book-length position paper—a point of view about design. I have drawn on lots of insights from lots of people, but those people might well not agree with all of my conclusions. In this respect, the book reflects a reality of engineering, that while engineers hope that they can base their work on sound, scientific principles, engineering is also a design discipline, and design is in part a matter of taste. So what this book talks about is in part matters of taste, and if I can convince the reader about matters of taste, so much the better.

The inspiration for this book arose out of the NSF-sponsored Future Internet Architecture program, and its predecessors, the Future Internet Design (FIND) program and the Network Science and Engineering (NetSE) program. These programs challenged the network research community to envision what an internet of 15 or 20 years from now might be, without being constrained by the Internet of today. I have been involved in this program for its duration, and I have had the chance to listen to several excellent groups of investigators discuss different approaches to designing an internet. These conversations have been very helpful in bringing into focus what is really fundamental about an internet. There have also been similar projects in other parts of the world, in particular Europe, that have contributed to my understanding. Just as one may perhaps come to understand one's language better by studying a foreign language, one may come to understand the Internet better by studying alternative approaches. Chapter 7 provides an introduction to these various projects.

The Internet is deeply embedded in the larger social, political, and cultural context. Assuming that we aspire to build a future global internetwork, we must accept that different parts of the world will present different contexts into which the technology must fit, so this is not a book just about technology. Indeed, technology is often not center stage. Much of the book centers on the larger issues: the economic, social, and political considerations that will determine the success or failure of a system like this, so woven into the larger world. If this book provides some insights into how the technical community can reason about this larger set of design constraints, it will have been a success from my point of view.

My hope is that this book will be useful both to technologists concerned with network design and also a broader set of readers concerned with the character of the Internet. I hope the book will convey a different perspective on what the Internet is, how it works, and how a range of potentially conflicting requirements have shaped its character. The book goes into detail about some critical aspects of the Internet today, including security and economics, but the deeper goal of the book is to convey how developers think about design. Networking is a subdiscipline of the field of computer science, which has its own ways of thinking about design and structuring solutions. I have tried to avoid (or else explain) some of the terms that engineers use when they talk about the Internet, but it is useful to understand some of these terms and concepts, since they are beginning to be used (often incorrectly) in nontechnical conversations about the Internet—conversations about policy, the place of the Internet in society, and the global consequences of connectivity.

My own professional history coincides with that of the Internet. I received my PhD at MIT in 1973, the same year that the two original inventors of the Internet, Robert Kahn and Vinton Cerf, wrote their seminal paper proposing the Internet. I started working on the Internet about two years later, chaired its architecture group in the 1980s, worked on enhancing the mechanisms of the Internet in the 1990s, became more concerned with the larger sociotechnical context of the Internet in the first decade of the new millennium, and continue to do research both on technology and on the larger societal context in which the Internet sits. This book is a partial distillation of what I have learned over those 40 years.

I begin in chapter 2 with a brief introduction to the current Internet and a review of its history from its inception in the 1970s up to today. Because the computer science community has co-opted the word *architecture*, I then discuss what network designers mean when they use this term. Chapter 3 talks about requirements—what it is that a network like the Internet should do. The superficial answer to that question may seem obvious—move data from point to point. However, there are many other considerations that shaped its initial design and have subsequently shaped its evolution.

Those of us who designed the original Internet were working out some basic design approaches as we developed the running system. In 1988, I wrote a paper in which I tried to capture what we understood at the time. As a way to see how the thinking in the field (or at least my thinking) has evolved, in chapter 5 I have reproduced that paper with extensive commentary on how I feel now about what I said almost 30 years ago.

Chapter 6 discusses how networks are designed to carry out their functions. Textbooks concentrate on how the Internet works, but with a focus on detail that can sometimes lose the forest for the trees unless the reader is already technically informed. That level of detail is not necessary to understand the points in this book, and my discussion of how networks function is a little more abstract and applies to a range of network designs, not just the Internet.

With that chapter as background, chapter 7 discusses about 25 alternative proposals for how to design an internet. They differ widely in the set of requirements that motivate them and their design approach. I try to present them in a way that captures the essential point of the design without diving into too many details, which will no doubt disappoint my friends who have lovingly crafted these proposals, but my goal is to capture the essence of the designs so that I can compare the different approaches.

After this review of these alternative approaches to design, the chapters that follow go into detail about each of the important requirements that

I identify for an internet: longevity, security, availability, economic viability, management and control, and meeting the needs of society. In the final chapter, I go out on a limb and distill out of these alternative proposals and diverse requirements how I might go about designing an internet today if we could do it from scratch. I have no illusion that anyone could do that, but my hope is that by conceiving a preferred future, we can push toward it.

The Appendix contains a more in-depth discussion of the core function of a network—how it takes data and delivers it to the intended destination. For readers who want to understand in more detail the history of addressing and forwarding on the Internet, as well as other proposals that have been made since the Internet was standardized, the Appendix might be a place to start. It has a more complete set of citations of prior work, but there is no way I can cite all of the work that the network research community has produced over the last 40 years. I apologize to all those talented people I did not manage to mention.

Onward…

2 The Basics of the Internet

For those with a less technical background, this chapter provides a foundation for the rest of the book. It provides a brief description of the structure and function of the Internet, and a history of its development from the 1960s to the present. The history serves to make the point that the Internet is not a static thing but a system that has evolved over several decades as it has matured and has been shaped by changing requirements.

Basics

The Internet is a communication facility designed to connect computers together so that they can exchange digital information. Data carried across the Internet is organized into *packets*, which are independent units of data, complete with delivery instructions in the first part, or *header*, of the packet. The Internet provides a basic communication service that conveys these packets from a source computer to one or more destination computers. Additionally, the Internet provides supporting services such as naming the attached computers (the domain name system, or DNS). Many applications have been designed and implemented using this basic communication service, including the World Wide Web, email, newsgroups, distribution of audio and video information, games, file transfer, and remote login to distant computers. A core objective of the Internet is to support a wide range of new applications over time.

An application program on a computer that needs to deliver data to another computer invokes software that breaks that data into some number of packets and transmits these packets serially into the Internet. A number of application support services can assist applications by performing this function, most commonly the transmission control protocol (TCP).

Users of the Internet tend to use the term *Internet* to describe the totality of the experience, focusing on the applications, but to a network engineer, the Internet is a packet transport service provided by one set of entities (Internet service providers, or ISPs); the applications run *on top of* that service. Anyone with the skills and inclination can develop a new application for the Internet. It is important to distinguish the Internet as a packet forwarding mechanism from the applications that run on top of that mechanism. It is also important to distinguish the Internet from the technologies that support it, because the Internet is not a specific communications technology such as fiber optics or radio. It makes use of these and other technologies in order to get packets from place to place. A goal for the designers of the Internet was to allow as many communications technologies as possible to be used in the Internet and to incorporate new technologies as they were invented.

The heart of the Internet itself is a simple service model that allows a wide range of applications to exploit the basic packet carriage service of the Internet over a wide range of communications technologies. The common standards that define the packet carriage service separate the details of the applications from the details of the technologies, so that each can evolve independently. The designer of an application does not need to know the details of each technology but only the specification of this basic communication service. The designer of each technology must support this service but need not know about the individual applications. One way to envision this structure is as an hourglass, as in figure 2.1. The breadth of the top of the hourglass signifies the diversity of the applications the Internet can support. The breadth of the bottom signifies the diversity of communications technologies over which it can operate. The narrow waist signifies the point of common agreement that isolates the diversity of the applications from the diversity of the communications technologies.

The Basic Communication Model of the Internet

The service model for packet delivery contains two parts: the addresses that identify the computers attached to the Internet and the delivery contract that describes what the network will do when it receives a packet to deliver. To implement addressing, the Internet has numbers that identify end points, somewhat similar to the telephone system, and the sender identifies the destination of a packet by using these numbers. The delivery contract specifies what the sender can expect when it hands a packet over to the Internet for delivery. The original delivery contract of the Internet is that the Internet will do its best to deliver to the destination all the packets it

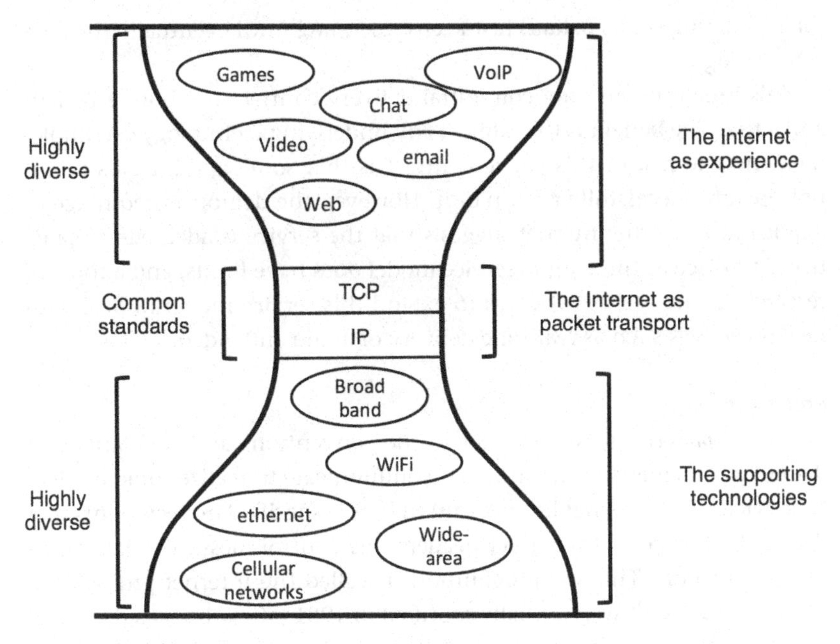

Figure 2.1

The hourglass model of the structure of the Internet, capturing the diversity of applications and technologies, connected through common agreement on the standards for IP (Internet protocol) and TCP. Adapted from *Realizing the Information Future* (National Research Council, 1994, 53).

receives for carriage but makes no commitment as to data rate, delivery delay, or loss rates. This service is called the best-effort delivery model.

Network designers and operators have a well-understood conception of what good service on the Internet is—acceptable levels of loss, delay, and so on—but there is no hard-and-fast specification. The reason for that was clear in the minds of the early designers: a poor service was better than none. Designers and operators should hold themselves to a high standard of network performance, but there are circumstances where best effort is not very good. Application designers must cope with this weak specification and decide how much effort to put into adapting and compensating for circumstances where best effort is not very good. Some applications, such as real-time speech, that depend on low levels of loss and delay may themselves not function, or even attempt to function, when the packet forwarding is functioning poorly. Other applications, such as email, can struggle forward even if most packets are lost in transit. The application (or TCP acting on its behalf) just keeps resending until eventually the data gets through.

This indefinite and noncommittal delivery contract has both a benefit and a risk. The benefit is that almost any underlying technology can implement it. The risk of this vague contract is that some applications might not operate successfully on top of it. However, the demonstrated range of applications on the Internet suggests that the service is adequate in practice. As I discuss, this simple service model does have limits, and a topic of research in the early 1990s was to extend this service model to deal with new objectives such as real-time delivery of audio and video.

Protocols

The word *protocol* is used to refer to the conventions and standards that define how elements in the Internet communicate to realize some particular service.[1] The Internet layer discussed here is specified in documents that define the format of the packet headers, the control messages that can be sent, and so on. This set of definitions is called the Internet protocol, or IP. These standards were initially specified in 1981 (after almost a decade of initial research and experimentation by the early team designing the Internet).[2] The early team of researchers also developed the protocols for early applications such as email. Different groups are responsible for different protocols.

The Role of the Router

The Internet is made up of a series of communication links connected by relay points called routers. There can be many sorts of communication links—the only requirement is that a link be able to transport a packet from

1. The word *protocol* was chosen based on its use in diplomacy, where it describes formal and proscribed modes of interaction. However, the etymology of the word tells another story. The word comes from the Greek *prōtokollon* and means "first page," from *prōtos* "first" + *kolla* "glue." The Greeks glued the table of contents onto the beginning of a scroll once it was complete; we put the header (not literally with glue) onto the front of each packet. Whether or not the Internet researchers who first picked the term *protocol* had a good classical education, I am assured that they did research the etymology of the word.
2. The various standards of the Internet (and other related materials) are published in a series of documents called "Requests for Comment" (RFCs), the title of which captures the idea that the authors of RFCs should be open to suggestions and change. The RFCs can be found at https://tools.ietf.org/html/. The specification for IP is RFC 791.

one router to another. Each router, as it receives a packet, examines the delivery information in the header of the packet and, based on the destination address, determines where to send the packet next. This processing and forwarding of packets is the basic building block of the communication service of the Internet.

Typically, a router is a computer, either general purpose or specially designed for this role, with software and hardware that implements the forwarding functions. A high-performance router used in the interior of the Internet may be an expensive and sophisticated device, while a router used in a small business or at other points near the edge of the network may be a small unit costing less than a hundred dollars. Whatever the price and performance, all routers perform the same basic communication function of forwarding packets.

A reasonable analogy to this process is the handling of mail by the post office or a commercial package handler. Every piece of mail carries a destination address and proceeds in a series of hops using different technologies (e.g., truck, plane, or letter carrier). The address information is on the outside of the envelope, and the contents (the data that the application wishes to send) are inside the envelope. The post office (with some limitations) is indifferent as to what is in the envelope. Each hop examines the address to determine the next hop to take.

In the Internet, when some abnormal situation arises, a router along a path from a sender to a receiver may send a packet with a control message back to the original sender of the packet. Again, by analogy, if the address on a letter is flawed in some way, the post office may write "Undeliverable: return to sender" on the envelope and send it back so the sender can attempt to remedy the situation. To emphasize this analogy between Internet and postal forwarding, the delivery process in the Internet is sometimes called datagram delivery.[3]

Routers do not provide application-specific services. Routers forward packets based on the header, not the application-level data that is inside the packet. Because routers in the Internet do not take into account what application sent the packet, I will use the term *application-agnostic* to describe the router function. This idealized description of how the Internet works was true in the early days but is less true now as routers try to peek at packet content to vary their behavior accordingly. This behavior is called deep packet inspection, or DPI. Advocates of network neutrality decry this

3. Datagram sounds more like a telegram analogy than a postal service analogy. I think the postal analogy is better, but I did not pick the term.

practice; encryption of the packet contents thwarts it. But this level of detail is a matter for later chapters.

Another important aspect of packet forwarding is that routers do not keep track of the packets they forward. The post office also did not traditionally log the letters it forwarded, unless the sender paid a substantial fee for tracking. Now, however, in this era of massive surveillance, both the post office and the Internet are recording and remembering a lot more. The current postal system includes a scheme called the mail isolation and tracking system, which photographs the front of every letter it forwards (Nixon, 2013).

A digression on the word *state* Computer science uses the word *state* in a specialized way to capture the idea that a device, depending on how it is designed, can be in different states based on stored information—information that can reflect what has happened in the past. If a router does not keep any record of packets that it forwards, it is described as *stateless* (or *memoryless*). Given a similar input (e.g., a packet to forward), a system that can be in different states might treat the same input differently. A stateless system will always respond to the same input in the same way. The stored information that defines the state of a system is referred to as its *state variables,* and from time to time throughout this book I will talk about the state variables of a component (or say that a component is stateless or has no state variables) as part of explaining why it can or cannot implement a certain function. In general, stateless systems are simple to build but can implement only simple operations, whereas systems with state are more complex and thus can implement richer functions.

Application Support Services in the End Node

The delivery contract of the Internet is simple: the best-effort service tries its best to deliver all the packets given it by the sender but makes no guarantees—it may lose packets, duplicate them, deliver them out of order, or delay them unpredictably. Many applications find this service difficult to deal with because there are so many kinds of errors to detect and correct. For this reason, the protocols of the Internet include a transport service that runs on top of the basic Internet service and tries to detect and correct all these errors and give the application a much simpler model of network behavior. This transport service is called transmission control protocol, or TCP. It offers a service in which the sending application gives data to TCP, and TCP delivers the data at the receiving end, in order, exactly once. The

TCP software takes the responsibility of breaking the data into packets, numbering the packets to detect losses and reordering, retransmitting lost packets until they eventually get through, and delivering the data in order to the application. This service is often much easier to utilize than the basic Internet communication service.

Division of Responsibility

The router, which implements the relay point between two communication links, has a very different role than the computers or end nodes attached to the Internet. In the Internet, the router is only concerned with forwarding the packets to the next hop toward the destination. The end node has a more complex set of responsibilities related to providing service to the application. In particular, the end node provides the additional services, such as TCP, that make it easier for the application (such as the World Wide Web) to make use of the basic packet transport service of the Internet.

TCP is implemented in the end nodes, but not in the packet forwarding software of the routers. In principle, the routers look only at the IP information, such as the destination address, when forwarding packets. Only the end nodes look at the TCP information in the packets. This division of responsibility is consistent with the design goals of the Internet and is an important example of layered design.

TCP provides a simple service that most high-level applications find easy to use, but some applications, such as real-time streaming, are not well-matched to the service model of TCP. If TCP were implemented in the routers, it would be much harder for the high-level service to bypass it and use a different transport service, so the design principle of the Internet has been to push functions out of the network to the extent possible and implement them only in the end nodes. This design approach is called the end-to-end argument and was described by Saltzer, Reed, and Clark (1984). The ability to create a new application or supporting service without any need to modify the routers is another means by which the Internet can evolve rapidly.

Routing and Forwarding

There are also functions implemented in the routers but not in the end nodes. The router must decide how to forward each packet as it arrives, which requires a forwarding table that specifies, for each destination address (or cluster of addresses), the preferred path toward that destination. In order to construct this table, routers (in addition to forwarding packets) continuously compute the best routes to all the addresses in the network. This process requires that the routers send messages to other routers describing what links in the Internet are currently in operation and which routers they connect. This exchange enables a collective decision, the *routing computation*, to select the best overall routes. Routers perform this task in the background at the same time that they forward packets. If a communication link fails or a new one is installed, this routing computation will construct new routes as appropriate. This adaptation improves the Internet's robustness in the face of failure. A number of routing protocols have been defined and deployed to implement the routing computation.[4]

This decentralized computation of the forwarding tables was the design approach in the original Internet. Recently, there has been a trend toward using a centralized controller within a region of the Internet that computes the correct routing information for the routers in the region and downloads that information into the routers as needed. This approach is called software-defined networking, and I discuss it on page 262.

End nodes do not participate in the routing computation. They know only the identity of a router or routers near them; to send a packet, they just pass it to this first router, which then decides where to send it next, and so on. This division of responsibility makes it possible to replace a routing protocol (which has happened several times in the life of the Internet) without having to change the software in the end nodes, which is effectively impossible to do in a coordinated way for all the millions of end nodes on the Internet.

Regions of the Internet

The Internet is made up of routers, but every router is part of some independently operated *autonomous system*, or AS. Each AS is operated by some entity, which might be a commercial ISP, corporation, university, or another entity. There were about 59,000 ASs across the globe as of 2017. The original Internet used one global routing protocol, but as the Internet got bigger, the designers realized that it needed at least two tiers of routing: schemes that operated inside each AS and provided the routes among the routers and end nodes in that AS, and a routing protocol that connected all the ASs together. The protocol that currently defines this global function is called the border

4. I mentioned earlier, when introducing the concept of *state* and *state variables*, that routers do not have state variables to keep track of the different packets they forward. However, the forwarding table is clearly an example of state in a router.

gateway protocol, or BGP. It is used to tell each AS how to reach the other ASs in the Internet and what destination addresses are inside each AS.

To oversimplify (as I will often do in this book), BGP works as follows. Imagine an AS at the edge of the Internet, for example the AS that represents the portion of the Internet that is at MIT. In order for MIT to get access to the rest of the Internet, it makes a business arrangement with some ISP that offers Internet access, and once that business arrangement (and an interconnection circuit to that provider) is in place, the border router at MIT (hence the name of the protocol) sends a BGP message to that ISP, saying "here is where MIT is connected." That ISP (call it A) now knows where MIT is connected, and it tells its neighbor AS (for example, ISP B), "If you want to get to MIT, just send the packets to A." ISP B now tells its neighbors (for example, ISP C), "If you want to get to MIT, send your packets to B, which will send them to A, which will send them to MIT." These messages (called path vectors) propagate across the Internet until in principle every AS knows where to forward a packet to get it to any other AS on the Internet.

The Internet thus has a nested routing structure. BGP is used to deliver a packet to the correct AS, and an intra-AS routing protocol (of which there are more than one used in the Internet today) is used to deliver the packet to the intended end node within that AS.

The Domain Name System

Internet routers use the destination address in the header of a packet to select the forwarding path, but these addresses are not easy for users to remember or use. Furthermore, if a service moves from one machine to another, users would have to learn a new address for the service. To mitigate these problems, the Internet has a system to provide names for services and end points that are more user-friendly than the destination addresses. The system that keeps track of these names and translates them into addresses on demand is called the domain name system (DNS)—"domain" because the names are typically associated with a naming domain such as MIT. The names in the DNS are hierarchical, with top-level names for organizations of a certain type ("com" for commercial sites, "edu" for universities and the like), and second-level names within each top-level name (like mit.edu) naming individual institutions. The DNS is implemented using servers organized in a hierarchy that reflects the structure of the names: there are top-level, or "root," servers that provide name translation for the top-level names, servers that provide name translation for the next element of the name, and so on. So (for example), to resolve the name "www.mit.edu," the first query would be to the root server to find the addresses of servers that know about the "edu" domain, the next query would be to one of those servers to find addresses of the servers that know about the "mit" domain, and the third query would be to those servers (which would be operated by MIT) to find the address of the machine that has the name "www" within the "mit" domain.

The Design of Applications

The Internet itself, as an entity built of links and routers, is concerned with the delivery of packets. Applications run on top of the basic communication service of the Internet, most often on top of TCP.

The World Wide Web as an Example

The World Wide Web is specified by a set of protocols that allow a Web client (often called a browser) to connect to a Web server.[5] A Web server (a particular kind of end node attached to the Internet) stores Web pages and makes them available for retrieval on request. The pages have names called URLs (uniform resource locators). The first part of a URL is actually a DNS name, and a browser uses the DNS system to translate that name into the address of the intended Web server. The browser then sends a message to that Web server asking for the page. URLs are disseminated in various ways so that potential readers can discover them. They also form the basis of cross-references or links from one Web page to another; when a user positions the mouse over a link and clicks it, the browser uses the matching URL to retrieve the indicated page. There is a longer discussion of URLs in chapter 8.

The operation of the Web depends on a number of protocols; a description of the Web illustrates the layered nature of the Internet design. The protocol called hypertext transfer protocol, or HTTP, provides the rules and format for messages requesting a Web page. At the server, the software implementing HTTP looks at the content of a request and returns the Web page in question. However, this protocol does not specify the format or

5. The protocols that specify the World Wide Web are mostly developed by the World Wide Web Consortium (the W3C) hosted at the Computer Science and Artificial Intelligence Laboratory at MIT. The W3C is intended to be a neutral body that encourages different stakeholders to contribute their points of view about the future of the Web.

meaning of the page itself. The page could be a traditional Web page, an image, music, or something else.

The most common representation of a Web page is HTML, which stands for hypertext markup language. All browsers include software that understands HTML, so that the browser can interpret and display the page on the screen. HTML is not the only page format. Images, for example, are encoded in a number of different ways, indicated by the name of the standard: GIF, JPEG, and others. These are alternative formats that are used to format Web content. Any format is acceptable, so long as the browser includes software that knows how to interpret it.

HTTP uses TCP to move requests and replies. The TCP software takes a unit of data (a file, a request for a Web page, or whatever) and moves it across the Internet as a series of packets. It does not examine the data to determine their meaning; in fact, the data might be encrypted. Finally, the packets that TCP formats are transmitted using IP, which specifies the destination of the packets.

Email

The pattern of communication of the World Wide Web is direct transfer between the browser and the server. Not all applications work this way. Electronic mail, or email, is an example of an alternative application design with a different delivery model. Since many users are not connected full-time, if email were transferred in one step from origin to destination, the transfer could only be successful during those occasional periods when both parties just happened to be connected at the same time. To avoid this problem, almost all email recipients make use of a server to receive and hold their mail until they connect. They then collect all their mail from their server. The concept is that the email server is always attached and available, so anyone can send mail to it at any time, and the receiver can retrieve mail from it at any time. This eliminates the necessity for the sender and the receiver to be attached at the same time. Most mail is actually transferred in three steps, from sending end node to the mail server serving that sender, then to the mail server serving the recipient, and then to the final end node.

Different Applications Are Different

As these two examples illustrate, different applications have different designs. The pattern of mail distribution does not resemble the pattern of Web page retrieval. The delivery of email depends on servers distributed across the Internet, but these servers, in contrast to the application-agnostic routers, are application-aware. They are designed to function as part of a specific application and provide services in support of that application.

There are many applications on the Internet today: email, the Web, computer games, Voice over IP (VoIP, Internet telephone service), video streaming, and so on. The list is almost endless (and in fact there is no list—one does not need to register to invent a new application; one can just do it). Part of the power of the Internet packet transport service is that it is an open, general platform that makes it possible for anyone with the right skills and initiative to design and program a new application.

A Brief Timeline of the Internet

In this book, I will talk about a number of events in the emergence of the Internet, and a brief history of its development may serve to put events in context. I like to think about the history of the Internet in decades, with only a slight necessity to fudge some of the decadal boundaries.[6]

The 1960s

The 1960s was the decade of invention and vision. Three different inventors, with different motivations, independently conceived of the idea of packet switching in the early 1960s. Paul Baran at RAND wanted to build a network that was robust enough to provide launch control over nuclear missiles even after a nuclear attack, so as to assure second-strike capability. He saw this as an important aspect of assured mutual destruction and a contributor to global stability. (His view was that once this scheme was working we should share it with the Russians, in part to prevent an accidental launch.) Donald Davies, at the National Physical Laboratory in England, had the goal of revitalizing the UK computer industry by inventing user-friendly commercial applications. He saw remote data processing, point-of-sale transactions, database queries, remote control of machines, and online betting as potential applications. He is generally credited with coining the term *packet.* Leonard Kleinrock independently conceived the power of packet switching but was less concerned with specific applications than with whether the idea could work in practice. The fear was that the fine-grained interleaving of packets from many sources would lead to

6. This history is brief, and thus necessarily selective, and it does reflect my personal point of view. There are any number of books documenting the various stages of the Internet's development. For the early history, two places to start are Abbate (2000) and Hafner (1998).

an unstable system with intractable fluctuations in traffic load. Kleinrock's early mathematical analysis of packet switching gave confidence that the idea would work in practice, in the process inspiring a large body of work on network queuing theory.

By the late 1960s, the potential of packet switching as a means to hook computers together, and thus to hook people together, mediated by computing, was a powerful vision. The U.S. Department of Defense Advanced Research Projects Agency (ARPA)[7] established a program to build a wide-area packet-switched network called the ARPAnet. J. C. R. Licklider and Robert Taylor (1968) wrote a paper in which they predicted a wide range of applications for a packet-switched network hooking computers and people together.[8] They wrote:

> What will go on inside? Eventually, every informational transaction of sufficient consequence to warrant the cost. Each secretary's typewriter, each data-gathering instrument, conceivably each dictation microphone, will feed into the network.
>
> You will not send a letter or a telegram; you will simply identify the people whose files should be linked to yours and the parts to which they should be linked—and perhaps specify a coefficient of urgency. You will seldom make a telephone call; you will ask the network to link your consoles together.
>
> You will seldom make a purely business trip, because linking consoles will be so much more efficient. When you do visit another person with the object of intellectual communication, you and he will sit at a two-place console and interact as much through it as face to face.

They also saw the network as a basis for widespread access to information. They wrote:

> Available within the network will be functions and services to which you subscribe on a regular basis and others that you call for when you need them. In the former group will be investment guidance, tax counseling, selective dissemination of information in your field of specialization, announcement of cultural, sport, and entertainment events that fit your interests, etc. In the latter group will be dictionaries, encyclopedias, indexes, catalogues, editing programs, teaching programs, testing programs, programming systems, data bases, and—most important—communication, display, and modeling programs.

7. ARPA is now called DARPA, adding the "D" for Defense to the acronym.

8. Both "Lick" and Taylor were instrumental in the development of early packet switching, working at ARPA in different roles to support the development of the ARPAnet. Talyor went on to run the Xerox PARC research lab, which developed perhaps the first real personal computer (the Alto) and the first local area network (Ethernet).

Overall, their assessment of the impact of networked computers was an optimistic one. In their conclusion, they wrote:

> When people do their informational work "at the console" and "through the network," telecommunication will be as natural an extension of individual work as face-to-face communication is now. The impact of that fact, and of the marked facilitation of the communicative process, will be very great—both on the individual and on society.
>
> First, life will be happier for the on-line individual because the people with whom one interacts most strongly will be selected more by commonality of interests and goals than by accidents of proximity. Second, communication will be more effective and productive, and therefore more enjoyable. Third, much communication and interaction will be with programs and programmed models, which will be (a) highly responsive, (b) supplementary to one's own capabilities, rather than competitive, and (c) capable of representing progressively more complex ideas without necessarily displaying all the levels of their structure at the same time—and which will therefore be both challenging and rewarding. And, fourth, there will be plenty of opportunity for everyone (who can afford a console) to find his calling, for the whole world of information, with all its fields and disciplines, will be open to him—with programs ready to guide him or to help him explore.
>
> For the society, the impact will be good or bad, depending mainly on the question: Will "to be on line" be a privilege or a right? If only a favored segment of the population gets a chance to enjoy the advantage of "intelligence amplification," the network may exaggerate the discontinuity in the spectrum of intellectual opportunity.
>
> On the other hand, if the network idea should prove to do for education what a few have envisioned in hope, if not in concrete detailed plan, and if all minds should prove to be responsive, surely the boon to humankind would be beyond measure.
>
> Unemployment would disappear from the face of the earth forever, for consider the magnitude of the task of adapting the network's software to all the new generations of computer, coming closer and closer upon the heels of their predecessors until the entire population of the world is caught up in an infinite crescendo of on-line interactive debugging.

While their last paragraph might reflect a wry understanding of the reality of dealing with computers, the overall vision was positive and forward-looking. They anticipated the possibility of a digital divide, and the movement of work to the place where it could best be done, which might foreshadow offshoring. The article does note that issues of security and privacy were of active concern and were beginning to get the attention they deserve, again a response to a perhaps unwarranted degree of optimism.

In 1969, the year after that report, the first nodes of the ARPAnet became operational, linking UCLA (Kleinrock's lab) to SRI. Packet switching was demonstrated, and a generation of technical folks got on with the task of making it practical. For a moment, the visionaries had done their job, and the funding agencies gave the engineers their marching orders into the next decade.

The 1970s

The 1970s was the decade of engineering and proof of concept. In the early part of the decade, a number of packet-switched networks were built or contemplated, including the ARPAnet (the terrestrial packet-switched network hooking together computers at different ARPA-funded universities and research labs), an international satellite network (SATnet), and a mobile spread-spectrum packet radio network (PRnet). Faced with the challenge of hooking these disparate networks together, Vinton Cerf and Robert Kahn (1974) proposed the core idea of the Internet. While many details about the Internet evolved during the 1970s, the basic design approach in that paper was an enduring foundation for what we have today. ARPA assembled a group of engineers from several institutions, under the direct leadership of Cerf; this group both developed the standards for IP and TCP and implemented the protocols on a range of computers. I was part of that team, implementing the protocols for a large time-sharing system called Multics being developed at MIT. The Internet became operational late in the decade, and the standards for IP and TCP were published in 1981.

By the end of the decade, it became clear to the initial design team that we had misestimated the eventual size to which the Internet would grow. I note with some amusement that in 1978 Danny Cohen and I (Clark and Cohen, 1978) wrote, "We should thus prepare for the day when there are more than 256 networks in the Internet."[9] It was not that the team lacked vision in the 1970s; we hoped to hook all the computers in the world together. It was just that we misunderstood how many computers there might be. The 1970s was the era of the mainframe computer. While Xerox PARC was demonstrating the idea of a personal computer in the 1970s, the importance of that innovation did not really become clear until the advent of the IBM PC in the early 1980s.

9. The initial addressing scheme of the Internet limited the number of networks to 256. In this document, we called for the recognition that the Internet would be composed of regions, with multiple networks within them.

The 1980s

The 1980s was the decade that saw the first substantial use of the Internet: it was a decade of growth and of contention among competing standards to define an internet. With the emergence of the IBM PC early in the decade, and the transformation of computing from mainframe timesharing to personal computing, the Internet design team realized that the Internet should be capable of hooking together millions of computers (a number we now see as billions, or perhaps a trillion).

As the initial design team contemplated the need to scale the Internet to a much larger size than we had first contemplated, it also faced the challenge of its own size. As more organizations became interested in the Internet, the number of people coming to the design meetings grew, and there was a point at which those meetings became ineffective. The original design team, led by Cerf, realized that it needed to create a small steering group, together with working groups to address specific problems. Cerf had the idea that to create a smaller group without having everyone clamor to join, we should give it an unappealing name. We called it the Internet Configuration Control Board (ICCB) to make it sound as boring as possible. This group mutated into the Internet Advisory Board, then the Internet Activities Board, and finally the Internet Architecture Board. (All of these groups shared the acronym IAB.) I chaired this group during the 1980s.

The first working group was the Gateway Algorithms and Data Structures Task Force (GADS), chaired by David Mills. The IAB established it to address the issue of how to scale the routing schemes of the Internet to the size we now contemplated. The approach to routing that the research community developed starting early in this decade (Rosen, 1982) was the hierarchical scheme of autonomous systems connected together with an inter-AS routing scheme (today this is BGP) and a separate routing scheme within each region.

The GADS working group demonstrated the importance of having different groups focus on specific design issues, and the IAB saw that some management structure would be needed as the number of working groups grew. This led to the creation of the Internet Engineering Task Force (IETF), with GADS as its first working group. Today, the IETF has a number of areas, with working groups within each area. The IETF continues as the standards-setting body of the Internet today.

One of the important developments of the 1980s was the DNS, which I described on page 13. Initially, machines on the Internet had simple names like "MIT-1." Jon Postel, at the Information Sciences Institute (ISI) at the University of Southern California, maintained a file that mapped names to

IP addresses, which anyone could download as needed. To register a new name, one sent an email to Jon. This approach would not scale to millions of machines. ISI, under the leadership of Paul Mockapetris and Postel, developed a proposal for the DNS. With the hierarchical structure of the DNS, it was no longer necessary to go to Jon Postel to register a new name but instead the registrar for the relevant domain—to register a name like xyz.mit.edu, for example, one only had to contact a person at MIT. This scheme could scale with much more ease to large numbers of end nodes and is the scheme used in the Internet today, although the management structure for the DNS has become much more complex.[10]

There is a pattern in these stories. Computer scientists, faced with the challenge of scaling, tend to fall back on a hierarchical scheme. In the 1980s, the (now larger) research community designed hierarchical routing, hierarchical naming, and a hierarchical management scheme for itself.

An important event in the 1980s was the transition of support for the deployed Internet from DARPA to the National Science Foundation (NSF). DARPA considered that the research was reaching a conclusion, and it decided to decommission the ARPAnet, which was costly to operate. NSF was interested in hooking its supercomputer centers together and providing access to those centers from other research sites, so it supported the development and deployment of NSFnet, which connected academic sites across the United States using the Internet protocols. As the NSFnet evolved to keep up with demand, there were several versions of technology and communication links; the final version used links with a capacity of 45 mbps (which was fast at that time).

Another feature of the 1980s was the competition among alternative protocol suites as the basis for an internet. There were corporate proposals, such as Xerox Network System (XNS), which proved popular in local area network (LAN) applications, and perhaps most significantly, the effort by the International Organization for Standardization (ISO) to develop their Open Systems Interconnection (OSI) protocol suite. ISO, as a standards body with

10. The functions that Jon Postel initially performed were transferred to an organization called the Internet Corporation for Assigned Names and Numbers (ICANN) in 1998. ICANN created competing registrars for various domains and a large number of top-level domain names, and there is contention over issues ranging from the use of different character sets to trademarks. I am deliberately not dwelling on these issues in this book, as they relate to Internet governance rather than Internet architecture. There are a number of good books on the current tensions over Internet governance, such as DeNardis (2015).

formal representation from national standards bodies, claimed that it had the authority to define a global internetworking standard. Parts of the U.S. government, such as the National Institute of Standards and Technology (NIST), felt the need to stand behind the efforts of the OSI, which split the government between those agencies supporting OSI and those supporting IP. After a contentious period that lasted beyond the end of the 1980s, the market moved toward the IP standards, pushed by DARPA and the NSF, and the OSI effort faded.

A more significant competitor to the Internet protocol suite was Asynchronous Transfer Mode (ATM), an outgrowth of research done at Bell Labs, which reflected a design preference consistent with the history of the telephone system, while at the same time exploiting the idea of switching small data units (called cells) rather than setting up a fixed circuit as the telephone system had previously done. ATM was a serious contender for market share in the 1990s but eventually did not succeed in the marketplace. Some of its ideas were incorporated into technology options (such as multiprotocol label switching, or MPLS) that are used today to provide transport for IP packets. I discuss ATM, MPLS, and related issues in the Appendix.

The application of the 1980s was email, and for users to say that they were "on the Internet" meant that they had an email address. To many users, the Internet was the same as email, an attitude that persisted until the beginning of the next decade.

During most of the 1980s, I chaired the group that started out as the ICCB and ended up as the IAB. I stepped down as chair at the end of the decade and turned to a specific line of research motivated by the emergence of ATM and the realization that the Internet would be used to carry voice and video streams. My collaborators and I set out to propose and develop new protocols that would enhance the original best-effort service of the Internet with new delivery modes that were more suited to the needs of more demanding applications such as real-time voice (Internet telephony, or Voice over the Internet—VoIP) or streaming video.

The 1990s

The 1990s was the decade of commercialization and the World Wide Web. It was also a decade in which there was a recurrence of visions for the future of communications and for the power and utility of computer-mediated access to information and among people. It was also the decade in which fiber optics emerged to drive down the cost of data transmission.

In the beginning of the 1990s, the backbone of the Internet—the long-distance capability that hooked regions together—was the NSFnet. By the middle of the decade, it became clear to commercial communications providers that private-sector provision of Internet connectivity was a viable business prospect, and the NSF, pushed by industry, executed a transition in which the NSFnet was decommissioned, to be replaced by networks from commercial network providers. This transition had the effect, among other things, of removing any residual limitations on the sort of traffic that could be sent across the Internet. The NSF Acceptable Use Policy nominally restricted the use of the NSFnet to the scholarly and research community, broadly defined. While this limitation had not, in practice, prevented the use of email and other applications by a wide range of users, the transition to the commercial backbone opened up the Internet to a much wider range of uses and purposes, including those that were purely commercial and entertainment. In this decade, many users first experienced streaming audio, both "Internet radio" and stored music sites. As a result, the 1990s saw the music industry become anxious about the potential disruption of its somewhat stable business models, an anxiety that was fully justified. Amazon, with its vision of e-commerce, was founded in 1994 and similarly alarmed those with foresight in the retail bookselling business and others. So the Internet itself was commercialized and commerce moved onto it.

The World Wide Web (or just "the Web") was invented in 1990 by Tim Berners-Lee, then at CERN (the European Organization for Nuclear Research), as a tool for collaboration and information sharing among the physics community. In the first part of the decade, the Web was competing with other schemes for information sharing, such as the wide-area information service (WAIS), Gopher, and Archie, which now exist only as memories for a few of us. By 1993 or 1994, the Web had gained a dominant market share (or "mind share"), based in part on the development in 1993 of Mosaic, the first Web browser to allow Web pages to include both graphics and text. Mosaic was developed at the NSF-supported National Center for Supercomputing Applications (NCSA) at the University of Illinois, Urbana-Champaign, and became the browser that defined the Web experience for most users in the 1990s. Its graphics capabilities were essential to a wide range of applications, including the marketing of commercial products.

In 1991, then Senator Al Gore proposed the national information infrastructure (NII), a vision that was sometimes referred to as the information superhighway. It was a high-speed, ubiquitous network that would make available vast amounts of information to both the public and private sectors and provide a wide range of services to citizens. In part motivated by this vision, the Computer Science and Telecommunications Board of the National Research Council released two reports on the Internet: *Realizing the Information Future* (National Research Council, 1994) and the *The Unpredictable Certainty* (National Research Council, 1996). I was privileged to help write both of these reports, and they provide a wonderful time capsule of the thinking of the time. In the first, the members of the committee from the telecommunications industry would not allow the report to use the term "*Internet*" to describe the nation's future network infrastructure. Their view was that a loose band of researchers was not qualified to design the network that the nation actually needed and that the telephone companies would in time design the real thing. They required that the text of the report refer to this future network in a neutral way as a "packet bearer service." By 1996, it was acceptable to discuss the Internet by name, and by the end of the decade it was clear that whatever the NII was, it would be based on the Internet.

Fiber optics emerged as a commercial telecommunications technology in the 1980s. Sprint announced its all-optical network in the mid-1980s, but it was really in the 1990s that the potential of fiber optics to carry huge volumes of traffic at ever lower costs compared to copper wires was realized. The links in the fastest version of the NSFnet had a capacity of 45 mbps. Today, fiber optic links go 100 gbps, over two thousand times faster. It was fiber optics that allowed the Internet to grow at reasonable cost to carry the data facilitated by the Web, as well as streaming content—first audio and in later decades high-quality video. Fiber optics is "just" a low-level technology, but the economic implications of its development and deployment are perhaps the single most important factor in the growth and success of the Internet.

Another activity in the 1990s was a substantial effort by the IETF and the research community to enhance and improve the addressing structure of the Internet. In the 1980s, every end node on the Internet had a distinct address. However, as early as the mid-1980s, it had become clear to the designers of the Internet that eventually there would be more end points on the Internet than could be numbered with an address that was only 32 bits long.[11] To deal with the problem of address exhaustion, the IETF held a retreat at which it discussed the challenges of the future, and a group of us wrote a manifesto titled "Towards the Future Internet Architecture" (Clark

11. A number 32 bits long can identify only about 4 billion end points, and because addresses are allocated in blocks to facilitate routing and to allow for future expansion, it is not practical to use every number. Currently, there are about 1.5 billion addresses in active use, and most of the rest have been allocated to regions of the Internet for expansion.

et al., 1991). The document discussed the tension between the IP and the OSI effort, address exhaustion, security, and a range of other issues. The authors were bold enough to state that the Internet must scale to accommodate one billion networks.[12] To deal with address exhaustion, the group proposed two plans. One was to convert to a new packet format for the Internet with a much larger address field. This proposal led to an effort that resulted in what is now called IPv6.[13] IPv6 is now slowly coming into use on the Internet, 25 years later. The other approach was to specify a new network element called a network address translation, or NAT, device. Using network address translation, a number of end nodes can be clustered behind a single access link connecting to the rest of the Internet. Those nodes are then given addresses that only have local meaning. The NAT device is given one or a small number of globally routable Internet addresses and rewrites the addresses in the packets as they leave the cluster. In this way, a number of end devices can share a single address. Most homes today that have broadband Internet have a home router that is both a firewall (to enhance security) and a NAT device. If every computer in the modern home today had its own global address, the Internet would have run out of addresses long ago. In parts of the world where fewer addresses have been allocated, the ISPs are putting much larger groups of users (perhaps a whole town) behind a NAT device.

Another enhancement to Internet addressing proposed at the beginning of the 1990s was the concept of *multicast*. In the original design of the Internet, an address identified the specific attachment point for an end node to which a packet should be sent. Steve Deering, then a graduate student at Stanford University, proposed a new sort of address that identified a *set* of destinations—if a packet was sent to that address, the Internet would route the packet to *all* of those destinations to the best of its ability (Cheriton and Deering, 1985; Deering and Cheriton, 1990). Deering proposed multicast to provide efficient support for multi-way teleconferences and the transmission of streaming content (such as audio and video) to potentially large audiences. The research community devoted a great deal of effort to the technical aspects of multicast, and it was used for such diverse purposes as teleconferences, remote participation in IETF meetings, and a multicast of a Rolling Stones concert in 1994 (preceded by a multicast by a lesser-known

12. One billion networks, not one billion hosts. We were tired of underestimating the eventual challenge.

13. The current format is actually IPv4—the previous ones were experimental, as was IPv5.

band called Severe Tire Damage in 1993) (Strauss, 1994). Multicast is used in private IP-based networks today to support streaming video, but it was not a success in the public Internet, for reasons of economics, as I will discuss in chapter 12.

By the end of the 1990s, about half of Americans were online (Horrigan, 2000). Of these, about 85 percent were using email, but there was also an explosion of Web-based services, as Horrigan documented. The use of the Web was both expanding and mutating, to the point where his report could say: "Today's Internet user is also different from the typical Web surfer of a few years ago."

The 2000s

This was the decade of consumer broadband, the explosion of the Internet globally, and the expansion of mobile Internet access. Issues of security took center stage, and governments began to consider the need to regulate the Internet.

At the beginning of the decade, most residential users connected to the Internet over dialup modems, which had improved to the point where they could provide 56 kbps throughput. Thousands of large and small ISPs offered dialup access. By the end of the decade, two-thirds of Americans had broadband access to the Internet (Smith, 2010), provided by a small handful of telephone and cable TV providers that owned the physical connections into the home (instead of thousands of small dialup ISPs). Broadband, with its multimegabit speeds, enabled a new explosion in applications such as streaming video, the impact of which was only fully felt in the next decade.

This shift in market structure to a small number of broadband providers (and concentration of potential market power) is perhaps the most important transformation of the Internet ecosystem since the move to commercialization in the mid-1990s. The Internet we have today is under the control of this small set of actors. The concerns about the potential power of these firms to control and distort how Internet traffic is delivered to consumers brought for the first time significant attention from regulatory agencies such as the U.S. Federal Communications Commission (FCC). In 2005, the FCC issued a policy statement of its "Four Principles" (Federal Communications Commission, 2005), which state that a user of the Internet should be able to use applications of their choice, access content of their choice, connect devices of their choice, and select among competitive providers of service. This was the opening round in an ongoing struggle, under the tagline of "network neutrality," to determine whether the government (in particular through the FCC) should have a voice in shaping the future of the Internet, or whether the future of the Internet would be defined by the decisions of the private sector that was building it.

While the Internet expanded outside the United States early in its history, it was during the first decade of the new millennium that major penetration of the Internet to citizens across the globe occured. According to the ITU (International Telecommunications Union, 2016), by 2010 there were over 2 billion users of the Internet, a majority from the developing world.

It is hard to say when the issue of Internet security, which I discuss at length later in this book, first got major attention. The original designers understood from the beginning that the Internet, with its global reach and open access, would raise issues of security. However, two events near the start of the new millennium brought a greatly enhanced awareness of security issues as a result of the release of two *computer viruses*—programs that propagated from computer to computer, doing damage as they reached each machine. The Melissa virus in 1999 and the ILOVEYOU virus in 2000, which caused many parts of the Internet email system to be shut down as a defensive action, were highly visible warnings of the potential of malicious actors to cause harm on the Internet.[14]

I am also not sure in what decade to put another transformative phenomenon, the mobile smartphone. DARPA funded the development of a mobile, wireless packet radio network starting in 1973, although the radios (with fewer capabilities than today's smartphones) were the size of a beer fridge and had to be transported in trucks. In Japan, NTT DoCoMo released a smartphone in 1999 that achieved mass penetration. However, this device was conceived not as giving open access to the Internet but instead a more curated and walled experience. Various personal digital assistants, such as the Blackberry, emerged early in the decade but again were not a means to get open access to the Internet. To me, the release of the first Apple iPhone in 2007 and the first phones using the Android operating system in 2008 mark the point at which there is a device available to any user of the Internet that permits a wide-ranging use of Internet applications in a mobile context.

The growth of new Web-based applications continued during the decade. The growth of e-commerce is an important aspect of the decade. Facebook

14. Computer viruses get their cute names in a number of ways. The name is usually invented by the discoverer of the virus, but it often derives from the contents of the virus itself. The ILOVEYOU virus was named for the email attachment by which it propagated, called Love-letter-for-you.txt. Many people opened the attachment, to their regret.

was started in 2004, but its initial focus on college students masked its eventual scale and impact. Twitter was started in 2006; again, its impact was fully felt in the next decade.

The first decade of the millennium was also the decade in which the origins of this book are found. The NSF, with the encouragement of a number of members of the research community, asked whether the specific design of the Internet (what I am calling its architecture) was a contributor to some of the problems and challenges that became so clear in this decade. They asked if perhaps a different approach to design might lessen some of the security problems, lessen the arguments around network neutrality, or deal better with issues of information retrieval or device mobility. They launched a three-phase program, with initial funding of small, exploratory research projects, followed by funding of four major projects to develop alternative conceptions of what an Internet might be, perhaps 15 years into the future. The groups were encouraged to think about a fresh design approach, not limited by the current Internet. The first meeting of the grant recipients was in 2006, and the program (called the Future Internet Architecture, or FIA, program) has just ended. As part of this program, I was funded by NSF to be a coordinator for the research community, with the task of looking across all the projects, seeing what general lessons had been learned, and providing an integrated view of the insights. One of the outcomes of that work is this book.

The 2010s

In this the decade, the application landscape is again being transformed, this time by streaming video and social media.

The emergence of streaming video on the Internet, as typified by Netflix as a source of more traditional content and YouTube as a source of a wide range of alternative content, from professional to very amateur, is a dominant aspect of this decade. The gating functions in the Netflix transition from renting video by mailing a DVD to streaming over the Internet were the deep deployment of effective consumer broadband access and the falling cost of on-line data transport.[15] When the cost to Netflix of sending a video over the Internet dropped below the postage to mail and return a DVD, the shift to online delivery was almost inevitable. Today, the traffic

15. There were also issues of how content was licensed to Netflix. The rights that Netflix had to mail a DVD did not grant them the right to transmit the content over the Internet, which was a critical business issue but one removed from network architecture.

from just Netflix and YouTube makes up over half of the total volume on the Internet in North America, and all streaming audio and video add up to over 70 percent of Internet traffic (Sandvine, 2016).

To users, their conception of the Internet has always been tied to the experience, not to the technology or the architecture. In the 1980s, the Internet was email. In the 1990s, the Internet was the Web. To a generation of users today, the Internet is Facebook, Netflix, YouTube, and the other social media and content-sharing sites. Again, this reality challenges anyone concerned with the design of the Internet to balance generality versus efficiency. If video traffic dominates the Internet today, should the Internet evolve so that it is specialized to carry this sort of traffic, or should the Internet retain its general character so as to facilitate whatever comes next? This is a recurring question, and I will return to it.

Since I have mentioned the power of visionaries to motivate and inspire, I will close out this discussion with praise for some of my favorite visionaries, the science fiction storytellers. Their futures may be somewhat dark, as opposed to the optimistic visions of Licklider and Taylor, but at their best they are remarkably insightful. We get the term *cyberspace* from a novel by William Gibson, *Neuromancer*, written in 1984. That date, which resonated with some folks for other reasons, is before the Web and while the initial Internet design team was struggling with basic issues of scale. The novel *Shockwave Rider* by John Brunner, published in 1975, included the concept of a "worm," a form of computer virus that moved through the network, collecting secret information and making it public, thus revealing horrible government practices. In the end, there are no secrets and no privacy. Brunner's vision predates the revelations by Edward Snowden by almost 30 years. Feel free to put this book down and go read some good science fiction, if you don't mind a dystopian view of the future.

Onward

With this background, it is time to proceed to the real focus of the book, which is a study of the basic design principles of the Internet (what I call its "architecture"), how those principles relate to the requirements that the Internet should meet (requirements both technical and social), and how alternative conceptions of an Internet might better address these requirements, or perhaps focus on different requirements that today's designers consider more important.

3 Architecture and Design

This is a book about architecture, so, to get off on the right foot, it is important to understand what I mean by that word. It is perhaps overused, and is used in a variety of contexts. Without having a shared understanding between writer and reader, there is a risk of failing to communicate, so what does the word *architecture* mean?

What Is Architecture?

Architecture is a process, an outcome, and a discipline. As a process, it involves putting components and design elements together to make an entity that serves a purpose. As an outcome, it describes a set of entities that are defined by their form. The architectural form we know as Gothic cathedral is characterized by a set of recognized design elements and approaches—the purpose may have been a place of worship, but "Gothic cathedral" implies a lot more. Finally, as a discipline, architecture is what architects are trained to do. The field of computer science borrowed the term from the discipline that designs physical things, such as buildings and cities, where there is a well-understood process of training and accreditation. All three of these faces of architecture apply both to "real architecture" and to computer science.

As a process There are two important parts to my definition: *putting components together* and *for a purpose*.

- Putting components together: This is what computer scientists are doing when they consider issues such as modularity, interfaces, dependency, layering, abstraction, and component reuse. These are design patterns that computer scientists are trained to consider as they contemplate a design challenge.[1]

1. The term *design pattern* was introduced into architecture by Christopher Alexander. Alexander has also influenced the software development community. The

- For a purpose: The process of design must be shaped by the intended purpose of the artifact—a hospital is not a prison, a low-power processor is not a supercomputer, and the network in your car that hooks your brake pedal to the brakes is not the Internet.[2] As part of architecture, the designers must address what the system cannot do (or do well) as well as what it is intended to do. In computer science, there is a peril in system design that is so well-known that it has a name, *second system syndrome*, the tendency, after having first built a system that perhaps does a few things well, to propose a replacement that tries to do everything.

As an outcome In the practice of designing buildings, the design normally results in one copy of the result. There are exceptions, such as tract houses, where one design is constructed many times, but there is only one copy of most buildings. The term *architecture*, when describing an outcome, normally implies a class of design, typified by its most salient features (e.g., flying buttresses). The term is applied to this abstract class, even though the architect has to specify the building down to a very fine level of detail before the construction team takes over.

When computer scientists co-opted the term *architecture*, they redefined it slightly. With respect to the Internet, there have been many different networks built based on the same design: the public global network we call "the Internet," private networks belonging to enterprises, militaries, and the like, and special-use networks such as financial networks. In this context, the word *architecture* only describes part of what is built, and much of the design process for a given instantiation occurs at a later point, perhaps specified by a different group.

As a discipline "Real" architects—those who design buildings—go to school to learn their trade. Looking over the fence at what they do is instructive. Architecture (as opposed to structural engineering) is not a design discipline built on an underlying base of science and engineering principles. Architects do not normally concern themselves with issues such as

---

networking community needs to develop design patterns for application developers, and my use of the term in the networking context is inspired by his work.

2. In my view, they should not be hooked together. Because of a design error adding advanced communications technology to a Jeep Cherokee, security researchers were able to take over control of the car remotely while it was driving down the road (Greenberg, 2015). There are times when it is not a good idea to hook everything in the world together.

structural engineering; they leave that to others. Of course, technical considerations may need to enter the design process early, as the architect deals with such issues as energy efficiency or earthquake resistance, but architects are primarily trained in the process of design. They study not engineering but buildings. They learn by case study—they look at lots of buildings and how fit (or not) they are for purpose. Do they meet the needs of the user? Are they visually attractive? How were the design trade-offs handled? And so on.

In computer science, we tend to hope that we can base our designs on strong engineering foundations, theories that give us limits and preferred design options, and so on, but (at least in the past) most of the business of system architecture has more resembled that of the building architect (e.g., learning from previous designs, asking what worked well and what did not, asking if the design was fit for purpose). We train computer scientists in both theory and practice, but we tend to deprecate the study of prior designs as "not science" or "not based on fundamentals." This book is unapologetically a study of design, not a book centered on a discipline with quantifiable foundations, such as queuing theory or optimization. I am personally excited by attempts to make architecture more rigorous, but we should not deprecate what we do today with phrases like "seat of the pants" design. Ours is a design discipline, just like building architecture, and we should strive to excel at it, not dismiss it.

So if the architecture of the Internet is not the complete specification but just a part of that specification, what is included in the architecture? We can say what is *not* included—we can look at all the different examples of networks based on Internet technology, or different regions of the global Internet, and note all the ways in which they differ. We see differences in performance, degree of resilience, tolerance of mobility, attention to security, and so on. Design decisions at this level build on the core architecture but are not *specified by* the core architecture, so what should we see as being in that core architecture?

Elements of Network Architecture

I identify several criteria that can determine whether a particular issue rises to the level of architecture: whether agreement on the issue is necessary for the system to function, whether it is convenient to agree on an issue, whether the issue defines the basic modularity or functional dependency of the system, or whether it is important that the issue in question be stable over time.

Issues on which we must all agree for the system to function For example, the Internet architecture is based on the use of packets and the assumption that the packet header will have the same format everywhere. (A different design might allow different formats in different regions, in which case the architecture might choose to describe what sort of architectural support is provided for the necessary conversion.)

As another example, when we first designed the Internet, we thought that the design depended on having a single, global address space. It is now clear that this assumption was not necessary—there need not be global agreement on a uniform meaning of addresses. The network address translation device, or "NAT box" (see page 24), allows regions at the edge of the Internet to use a private address space and converts those addresses into globally routed addresses only if the packets exit into the public Internet. It is interesting to note that once the Internet designers realized that they could build a network out of regions with different address spaces, there was no rush to extend the architecture to provide any support or guidance as to how disjoint address spaces are interconnected.[3]

There are a few other points where global agreement is necessary. The autonomous systems of the Internet (the ASs) exchange information about how to route packets among the ASs using the border gateway protocol, or BGP (see page 12). To some extent, all the ASs must agree on the use of BGP, or at least the meaning of the packets exchanged across the inter-AS interface. Even if there were a region of the Internet that did not use BGP for interconnection, it would probably be necessary to agree on the existence and meaning of AS numbers, and within the global address space of the core of the Internet, it is necessary to agree on the meaning of certain address classes, such as multicast (see page 24). It is worth noting that both multicast addresses and AS numbers were not conceptualized as part of the Internet's original design but were designed later. In some sense, they have earned the right to be considered part of the core architecture exactly because a critical mass of designers have agreed to depend on them. Things that the original designers thought were mandatory, such as a global address

3. Of course, many people view this state of affairs as deplorable, since it prevents certain sorts of applications from being deployed easily, and the IETF has recently considered standardizing methods to deal with the implications of network address translation. See, for example, RFC 6887 on port control protocol. On the other hand, architectural purists and those pushing for the deployment of IPv6 have been actively resistant to the idea of NAT and attempts to mitigate its limitations. The better that NAT could be made to work, the lower the incentive would be to convert to IPv6.

space, have turned out not to be mandatory, and other things that they were not contemplating have crept in and acquired the status of "that on which we must all agree."

Issues on which it is convenient to agree There is no requirement that applications use the domain name system (DNS), but since essentially all application designers use it, it has become mandatory as a part of the Internet, even though, as I discussed in chapter 2, the DNS was not part of the initial design. Similarly, although it is not necessary that communicating applications use TCP, so many applications depend on it that it is a mandatory part of the Internet.

The basic modularity of the system Computer science uses the word *module* to describe the subcomponents of a system: a module has some specified *interface* through which it connects to the rest of the system, and the internals of the module, beneath the interface, are hidden and not accessible from outside the module. The designer of a module would normally keep the interface specification constant, since other modules may depend on that interface, but is free to change the internals of the module, since those are private to the module. The specification of the Internet protocol (IP) defines three module interfaces. It defines two layer interfaces, the *service interface,* on top of which higher-level services are built, and the interface to the technologies below the IP layer. It also defines (implicitly and partially) the *AS interface*: the interface among the different ASs of the Internet. The service interface is the *best-effort* packet-level delivery model of the Internet: if an end node sends a packet with a valid destination IP address in the packet, the routers of the Internet will forward the packet to the destination interface defined by that IP address to the extent the network can do so at the moment. The service interface hides all the details of how specific technology is employed to provide communication paths inside the Internet. This service interface thus defines the abstract interface between the network and the end node. The technical details of that interface are dependent on the particular network technology being used to connect to the end node, and will vary depending on the specifics of the technology, so these sorts of details are *not* part of the architectural specification.

Functional dependencies One aspect of architecture is to make clear the *functional dependencies* of the design. I will use the Internet to illustrate what this means. The basic operation of the Internet is simple. Routers compute routing tables in the background so that they know routes to all parts of the

Internet. When they receive a packet, they look up the best route and send the packet on. While there is a lot of stuff going on inside the Internet, at its core that is really all that it does. The proper operation of the Internet depends of necessity on the proper functioning of the routers. But what else is necessary for the Internet to provide service? In fact, the early designers of the Internet tried to limit the number of services or elements that had to be operating in order for packets to flow. An early design goal was stated as follows: "If there are two computers hooked to a network, and each one knows the address of the other, they should be able to communicate. Nothing else should be needed."[4] This design preference could be expressed as the goal of minimal functional dependencies. Some of the proposals for the design of an internet that are discussed in chapter 7 have many more functional dependencies—they depend on more services to be up and running for basic communication to succeed. They are trading functionality for (perhaps) less resilience when things go wrong.

Aspects of the system that are viewed as long-lasting In a system like the Internet, we know that much will change. Indeed, the ability to change, and to upgrade and replace aspects of the system, is a key to successful longevity. (See chapter 9 for an extended discussion of longevity.) But to the extent that there are aspects that seem like durable invariants, specifying them as part of the design can provide stable points around which the rest of the system can evolve.

The Role of Interfaces

Interfaces are the specification of how modules interconnect to make up the overall system. Interfaces become fixed points in the architecture—points that are hard to change, precisely because many modules depend on them. Kirschner and Gerhart (1998) developed the idea of interfaces as "constraints that deconstrain": points of fixed functionality that separate modules so that the modules can evolve independently rather than being intertwined. Their work was in the context of evolutionary biology

4. In 1987, Leslie Lamport, a well-known and thoughtful computer scientist, sent an email in which he offered the following observation: "A distributed system is one in which the failure of a computer you didn't even know existed can render your own computer unusable." That is a condition the designers of the Internet were trying to avoid. See http://research.microsoft.com/en-us/um/people/lamport/pubs/distributed-system.txt.

but seems to apply to man-made systems, whether the designers are clever enough to get the interfaces right from the beginning or whether the interfaces in this case also evolve to reflect points where stability is beneficial and evolution elsewhere is also beneficial.[5]

Layering

Layering is a particular kind of modularity, in which there is an asymmetry of dependence. A system is layered, or more specifically two modules have a layered relationship, if the correct functioning of the lower-layer module does not depend on the correct functioning of the higher-layer module. Operating systems display a layered structure: the design goal is that the system itself will not be harmed or disrupted if an application running on the system crashes. Similarly, the Internet was designed in a layered fashion—the goal was that the applications running on top of the packet forwarding service should not be able to affect it.

The idea of asymmetry of dependency may help with the overall conception of a system, but it is often not quite accurate in practice. One issue is performance—different applications can interact because they compete for resources. In networking, we see extreme examples of this in distributed denial of service (DDoS) attacks, in which a malicious actor tries to send enough traffic so that a host on the network or a region of the network itself does not have sufficient remaining capacity to forward legitimate packets. One response to the reality of DDoS attacks would be to say that the design of a layer, if it is truly a layer with no dependence on what the modules above it do, must include mechanisms to protect itself from malicious applications and to isolate the different applications. Of course, the simple service model of the Internet has no such protections in its architecture. In chapter 7, I discuss potential approaches for mitigation of DDoS attacks.

Summary—Thinking about Architecture

I have sketched a basic conception of what I mean by the word *architecture*. In my view (and, as I warned in the Introduction, this book is a personal point of view), a key principle is *architectural minimality*. In the computer science context, the architecture of a system should not try to specify every aspect of the system. This conception of architecture seems at variance with the architecture of buildings. When the architect of a building hands off the plans to the builder, the specification is complete down to the small

5. I thank John Doyle for bringing this work to my attention.

details—not just the shape and structure but where the power outlets are. But I do not consider most of these decisions to be architecture. As I said earlier, one of the distinctions between the architecture of a building and the architecture of an artifact like the Internet is that there are lots of networks built using the same Internet technology, not just one. There are obvious benefits if it is possible to use Internet technology in different contexts: commercial products are cheaper and likely to be more mature, the relevant software is found in almost all computer systems, and so on. However, these networks may not have exactly the same requirements for security, resilience, and other aspects, so the power of architecture is not that it defines exactly how the network is built (as building plans specify exactly how the building is built) but that it allows these requirements to be met, though perhaps in different ways in different contexts.

I will argue, to paraphrase Einstein, that architecture should be as minimal as possible, but no less. One might argue that the most fundamental aspect of the architecture of the Internet as I characterize it is its preference for minimality. Given that point of view, the scope of what we take as the *architecture* of a network system should include only those aspects that fit within the framework I have laid out here, given the requirements that a given architecture sets out to address.

The next step in understanding how to define the architecture of an internet is to return to the first point of the chapter, that architecture is the putting together of components for a purpose. We must ask what the purpose of an internet is. That is the topic of the next chapter.

4 Requirements

In the previous chapter, I talked abstractly about architecture and about the architecture of an internet. I assumed some common understanding of what an internet actually does. But if we are to be both concrete and precise, we need to start with a specification of that understanding: what an internet should do. In this chapter, I review possible design requirements for the Internet (or for *an* internet), which will set the stage for the chapters to follow.[1]

Fitness for Purpose—What Is a Network For?

The first requirement for an internet is that it provide a useful service. The service model of the original Internet, while perhaps never carefully written down, is simple: deliver a packet (of a certain maximum size) as best it can from any source to a destination specified by an IP address. This specification tolerated failure of the delivery, and indeed it was a rather explicit decision not to include in the specification any bound on the rate of failure. If the network is doing its best, then so be it—the user can decide whether

1. The NewArch project (Clark et al., 2004), which I will discuss in chapter 7, spent a great deal of its effort trying to understand the set of requirements that a successful future architecture would have to address. The list of requirements discussed in the final report included economic viability and industry structure, security, dealing with tussle, supporting nontechnical users (balanced with a desire for user empowerment), the requirements of new applications and new technology, and generality. This list has a lot in common with the set of requirements I will discuss in this chapter, which is not an accident. The NewArch work laid the foundation for much of my subsequent thinking. While the NewArch project did propose some distinctive mechanisms, which I will discuss in chapter 7, the discussion of requirements is perhaps as important a contribution as the exploration of new mechanisms.

this service is better than nothing. The meaning of the IP address is not a part of the specification—it is just a field used as an input to the forwarding algorithm in the routers. The limitations on our ability to design highly scalable forwarding algorithms impose soft constraints on the use of IP addresses—it is highly preferable that they be allocated so that they can be aggregated into blocks that the routing protocols can process, as opposed to having the routing and forwarding mechanisms deal with each address separately.[2] But there is no outright prohibition against having the routing protocols compute a distinct route for a single destination address.

As I will detail in chapter 5, there were good reasons for this rather weak specification of what the Internet was to do. Had the initial designers challenged themselves with a much more constraining specification that set limits on loss rates, throughput, and other parameters, it is possible that the network would never have been built successfully in the beginning. However, as I will discuss in chapter 10, this weak specification, which is silent on what the network should *not* do, opens the door to the malicious behavior we see on the Internet today.

Should the Network Do More?

Part of the appeal of thinking about a new internet is the challenge of devising new services that would make the network more useful. New services might make it easier to design applications, make it possible to serve a broader class of applications, or allow the network to function in a wider range of circumstances. Adding more complex functions to the network might make it easier to deploy new classes of applications but obviously adds complexity to the network itself. There is thus a trade-off between what the network should do and what a service layer on top of the network could do for a class of applications. This trade-off is a recurring one in system design—the early history of operating systems was marked by functions initially being implemented by applications and then migrating into the kernel as their value was proven.[3] So several threads of network research today are exploring the addition of new functionality to the network.

Over time, the Internet research and development community has added new services to the specification of the Internet. An IP address originally

2. Caesar et al. (2006) provides an assessment of the practicality of relaxing this constraint. The conclusions are not optimistic. See chapter 7 for further discussion.
3. The operating system on the IBM 1620 (the first computer I encountered, in the mid-1960s) did not include support for a file system but left disk management to the application. The system would continue to run if the disk was powered down during operation.

referred to a single destination, associated with a network interface on a specific machine. However, IP addresses are now used in different ways. The concept of *anycast* is that multiple destinations can have the same IP address and the routing protocols will direct the packet to the "closest" one. The concept of *multicast*, discussed on page 25, is that multiple destinations can have the same IP address and the routing protocols will direct copies of the packet to *all* of them. Multicast is distinctive in that it requires a different set of routing and forwarding algorithms to be implemented in the system. Another service objective was that the network could tailor the parameters of delivery to the requirements of the application. This concept, which today is commonly called quality of service (QoS), requires more complex scheduling in the forwarding mechanisms and/or more complex routing mechanisms. Without debating here the merits of either multicast or QoS forwarding, I note their implications on overall network design—if different packets receive alternative treatments, there has to be some signal, either in the packet or stored as state in the router, that indicates which treatment each packet gets. With respect to QoS, the original designers of the Internet contemplated such a scheme and defined the Type of Service field in the IP header to trigger different services. With respect to multicast, which was not initially contemplated, a set of distinct IP addresses were later set aside to trigger the desired behavior.[4]

Implicit in the specification of the original Internet was that a router could only forward a packet or drop it. The idea that it might store the packet was hardly even discussed, since memory was scarce in the 1970s, and the unstated assumption was that the goal of the Internet was rapid delivery—an important early application was remote login, followed (in the NSF era) by distributed computation among supercomputers. Storing packets temporarily in the network if they cannot immediately be forwarded adds complexity to the network (should the specification define how long to store packets, and under what circumstances?) and to the behavior that the application sees. However, allowing storage as a part of the network function might make it possible to design a new class of applications directly on top of the network, as opposed to requiring the deployment of storage servers as part of a higher-level application support service.[5]

One of the more innovative approaches for a future internet is to rethink the basic service objective: reject the idea that the service is delivering a

4. This mechanism is similar to the way telephone numbers with the area code 800 are treated differently by the telephone system.

5. The Delay/Disruption Tolerant Network community and the MobilityFirst FIA project represent two examples of this approach; see chapter 7.

packet to a destination specified by an address. One alternative is that the packet should be delivered to a more abstract conception of a destination, a *service*. This proposal is a generalization of the anycast concept I mentioned earlier; for this to be practical, the routing and forwarding schemes must be scaled to deal with a very large number of such addresses (with the current *anycast* mechanism, such addresses are exceptions and are few in number). Another alternative is that the network should deliver to the requester a packet of contents, without the requester knowing anything about the location of the contents. This concept, called information-centric networking (ICN), has profound implications both for the network and for the application. The network must be able to forward packets based on the name of the desired content rather than the address of the destination.

I return to this design question in chapter 6 when I discuss how network designers should reason about the range of services that the network might usefully offer to higher layers. I will discuss providing generality in the packet header (the *syntax* of the network) to trigger a range of behaviors (the *semantics* of the network). In chapter 7, I return to the design of ICNs.

Generality

One of the reasons why the Internet has been successful is that it was designed with the goal of generality. In fact, there are two important aspects of generality represented in the Internet: generality with respect to the applications that run over it and generality with respect to the sorts of communications technologies out of which it can be built.

Generality of Purpose

The Internet is a general-purpose network, suited to email, video, computer games, the Web, and a wide variety of other applications. This generality seems a natural way to structure a network that hooks computers together: many computers are general-purpose devices, and since the Internet hooks computers together, it, too, was intended to be general.[6] In the early days of the Internet, however, this preference for generality was not uniformly accepted. Indeed, this idea was quite alien to the communications engineers

6. Some computing devices today are becoming more specialized, implementing some fixed function such as sensing or actuation. This class of devices is now called the Internet of Things, or IoT. However, while each device may implement a fixed function, there are many different kinds of these devices, so the generality of the Internet is still important.

of the time, who worked for the telephone companies. They asked what to them was an obvious question: how can you design something if you don't know what it is for? The telephone system was designed for a known purpose: to carry voice calls. The requirements implied by that purpose drove all the design decisions of the telephone system, and the engineers from the world of telephone systems were confounded by the idea of designing a system without knowing what its application would be. One can understand the early history of the Internet by noting that it was designed by people who came from a computing background, not a classical networking (telephony) background. Most computers are designed without knowing what they are for, and this mind-set defined the Internet's design.

But this generality has its price. The service a general network delivers is almost certainly not optimal for any particular application. Design for optimal performance does not end up in the same place as design for generality. (There is thus perhaps a tension between design preferences such as generality, optimality, minimality, and the like, to which I will return from time to time.) It may also take more effort to design each application than if the network were tailored to that application. Over the decades of the Internet's evolution, there have been a succession of dominant applications. In the early years of the Internet, that application was email. Email is an undemanding application to support, and if the Internet had drifted too far toward supporting just that application (as was happening to some degree), the Web might not have been able to emerge. But the Web succeeded, and the emergence of this new application reminded people of the value of generality. Now this cycle repeats, and the emergence of streaming audio and video has tested the generality of an Internet that had drifted toward a presumption that the Web, and not email, was "the application." Now that the application that drives the constant reengineering of the Internet is streaming, high-quality video, it is once again easy to assume that we know what the Internet is for and optimize it for streaming video. In my view, the community that designs the Internet should always be alert to protect the generality of the Internet and allow for the future in the face of the present.

Generality of Technology

The other dimension of generality that historically was critical to the Internet's success is that it was structured so that it could work over a wide range of communications technologies. The early Internet interconnected three communications technologies: the original ARPAnet, SATnet (the wideband experimental multipoint Atlantic satellite network), and a spread spectrum mobile packet radio network (PRnet). Because the goal was to operate over as broad a selection of technologies as possible, the architecture made minimal assumptions about what these technologies could do. Had the design targeted a known communications technology, it might have been possible to exploit the particular features of that technology (for example, some wireless systems are inherently broadcast), which might have led to a more efficient outcome. But the decision to architect an Internet that could operate over anything allowed new sorts of technologies to be added as they emerged. For example, Local Area Networks (LANs) emerged after the Internet was designed. We see this tension between generality and optimization repeating today: a network of limited scope, for example a network internal to a car, can use a specific network technology that will allow more sorts of cross-layer optimizations. Some designers of wireless networks have argued that the characteristics of those networks are sufficiently different that they would benefit from an architecture designed for that context.

Longevity

One measure of the Internet's success is how long its design has remained viable. Presumably, any proposal for a network architecture should aspire to durability. One view is that a long-lived network must be evolvable; it must have the adaptability and flexibility to deal with changing requirements while remaining architecturally coherent. The goal of evolution over time is closely linked to the goal of operating in different ways in different regions in response to regional requirements such as security. On the other hand, a factor that can contribute to longevity is the stability of the system: the ability of the system to provide a platform that does not change in disruptive ways. I explore different theories of how to design a long-lived system in chapter 9.

For an architecture like the Internet to survive over time, there are several subsidiary requirements.

Support for tomorrow's computing The Internet arose as a technology to hook computers together, so as the shape of computing evolves, so should the Internet. In 10 years, the most numerous computing device will not be the PC, or even the smartphone or tablet, but most probably the small embedded processor acting as a sensor or actuator that today is called the Internet of Things (IoT). At the same time, high-end processing will continue to grow, with huge server farms, cloud computing, and the like. Any future Internet must somehow take this wide spectrum of computation into account. One point of view is that this wide range of requirements, ranging from high performance to low-cost ubiquitous connectivity, cannot be met by one approach to transport and interconnection. In this case, we will see the emergence of more than one network architecture: the single Internet architecture of today will be replaced by a range of alternatives at this level of the design, each targeted toward a subset of devices and only interconnected at higher levels. On the contrary, it is possible that one set of standards will span this range of requirements just fine.

Utilize tomorrow's networking At least two communications technologies will be basic to tomorrow's networks: wireless and optical. Wireless (and mobile) implies additional sorts of routing (e.g., broadcast), the tolerance of intermittent connectivity, and dealing with high levels of packet loss. Advanced optical networks not only bring huge transmission capacity; they can also offer rapid reconfiguration of the network connectivity graph, which also has large implications for routing and traffic engineering. One point of view about the Internet is that the emergence of wireless networks requires more cross-layer optimization to make effective use of wireless technology, and the architecture of a future internet should allow more variation in how functions are realized. This variation would raise the challenge of interconnection among the different approaches, but the requirement for interoperation does not mean that an internet has to be based on the same design everywhere. Interoperation can be achieved at different layers. One debate about a future architecture will be regarding the extent to which more variation in functional design is required to deal with diversity in technology.

There is an interesting interplay between architecture and technology. In the early days of the Internet, the networks made use of communications technology that had been designed for different purposes (e.g., telephone circuits). One of the early goals of the Internet was to work on top of essentially any communications technology, because that approach seemed to be the only path to rapid, wide deployment. But as the Internet has matured and proven its success, network technology has evolved to provide efficient support for it. Over the long run, technology can be expected to follow the architecture rather than the architecture having to bend itself to accommodate technology designed for other purposes. The tension between short-term deployment and long-term effectiveness is a design challenge for any architecture. Careful design of the architecture can also either facilitate or hinder the emergence of useful sorts of technological heterogeneity.

Support tomorrow's applications Today's Internet has proved versatile and flexible in supporting a range of applications. If there are important applications that are prevented from emerging because of the current Internet, their absence is seldom noted. However, applications of today and tomorrow present requirements that a future internet should take into account. These include a range of security and privacy requirements, support for highly available applications, real-time services, new sorts of naming, and the like.

Security

The Internet of today is marked by a number of serious security issues, including weak defenses against attacks on hosts, attacks that attempt to disrupt communications, attacks on availability (denial of service, or DoS, attacks), and attacks on the proper operation of applications. Ideally, an internet architecture would have a coherent security framework that makes clear what role the network, the application, the end node, and other components have in enabling and maintaining good security. I explore the issue of Internet security, and the relationship between architecture and the resulting security properties, in chapter 10.

Availability and Resilience

These two goals are sometimes lumped into security, but I have listed them separately because of their importance and because availability issues arise in the Internet of today independent of security attacks. Improving availability requires attention to security, good network management and preventing errors by operators, and good fault detection and recovery. What is needed is a *theory* for availability. While the Internet of today deals with specific sorts of faults and component failures (lost packets, links and routers that fail), the research community does not have an architectural view of availability. I return to this topic in chapter 11.

Management

Management has been a weak aspect of the current Internet from the beginning, to a considerable extent because the shape and nature of the management problem was not clear in the early days of the design. It was not clear early on (and is not yet clear) what aspects of network operation would (or should) involve human operators and which would preferably be automated if possible. Also network management plays only a supporting role compared to the core objective of forwarding data, and researchers like to work on the lead problem, not the supporting role. As I will argue in chapter 13, there may not be a single coherent issue that is management, just as there is no single issue that defines security. The key to both security and management is to break the problem into its more fundamental parts and address them without reference to basket words like security and management.

Economic Viability

A fundamental fact of the current Internet is that many of the physical assets from which it is built—the long-distance and residential access links, the large routers in the core of the network, the wireless towers, and the like—are expensive. These assets, often collectively called facilities, come into existence only if some actor chooses to invest in them. Chapter 12 explores the relationship between system design (and core design methods such as system modularity) and industry structure. To argue that a system is viable as a real-world offering, a designer must describe the set of entities (e.g., commercial firms) the architecture implies and make an argument that each will have the incentive to play the role the architecture defines for it. Using the current Internet as an example, there is a tension between a core value of the current Internet—its open platform quality—and the desire of investors to maximize their return on investment. In chapter 6 (see page 84), I introduce the term *tussle* to describe the situation where different actors in an Internet ecosystem do not have aligned incentives or motivations, and I call the working out of this tension between an open architecture and the desire to monetize infrastructure the *fundamental tussle.* Any proposal for a network design will almost certainly end up taking a stance in this space, even if implicitly.

Meeting Needs of Society

A network design will not succeed in the real world unless there is a purpose for which it is useful. The Internet is not just a technical artifact connecting computers but also a social artifact connecting people that is deeply embedded in society. Users do not directly observe the core architecture of the system—they partake of the system using the applications that are designed on top of it. As I noted, one measure of a successful network is that it support a wide range of applications, both today's and tomorrow's. On the other hand, the network design may impose conventions and provide features that cut across applications, and as the network layer supports more functions, the consequences will become more visible to the users. The Internet of today provides a simple service, and one could argue that many variants of an internet would be equally successful. But the core design will influence the outcome of some important social considerations, such as the balance between surveillance and accountability on one hand and anonymous action and privacy on the other. Users want a network where they can do what they please—they have choice in their applications and activities—but where criminals have no ability to undertake their activities effectively. They want a network that is reliable and trustworthy, but they do not want either the private sector or government watching what they (and thus the criminals as well) are doing. Chapter 14 explores some of these socially important trade-offs and considers whether, and to what extent, the core architecture defines the balance or whether the applications built on top of the network itself determine the balance.

Moving beyond Requirements

The topics of the previous sections are high level. They are not actionable as posed; they are desiderata, an *aide-mémoire*, as we contemplate design. It is a big jump from any of these to the design of specific mechanisms, and that is a big issue. It would be good if the design community had a process based on principles and theory, but there are no well-honed design methods to aid in the process of moving from these requirements to mechanisms and architecture.

Several things can happen as we move from high-level requirements to a specific architecture and mechanism. One is that in the attempt to reduce an abstract idea such as security to practice, we discover that lurking inside that requirement are subgoals in tension with each other or with other requirements. Design is not optimization along a single dimension but rather a balancing of different priorities. Some of these may be quantifiable (e.g., a performance requirement), but most will end up as qualitative objectives, which makes the balancing harder. There is a tendency in the computer science community to prefer to optimize factors that can be quantified, such as performance (although performance objectives are not specified in the current Internet), but if an internet is going to be relevant in the real world, we must face the messy challenge of evaluating alternative approaches to security or economic viability.

A further problem is that as we move beyond high-level requirements for a system like the Internet, the resulting design problem will almost certainly grow too large for one team to contemplate holistically, so the design process itself may need to be modularized. As various committees are created to work on different parts of the system, that modularity in the committee structure is highly likely to be reflected in the modularity of the system itself. This reality implies that the fundamental modularity of the system had better be specified before the design process itself is modularized, so that the modularity dictates the design process and not the other way around.

Requirements and Architecture

Several of the subsequent chapters are dedicated to exploring in more depth the requirements I have discussed here and refining them so that they become actionable, but there is a high-level question that cuts across all of these requirements: how do they relate to architecture? Should we look to the architecture of a network to see how these requirements are met? I characterized architecture in chapter 3 in a minimal way: it was those things on which we have to agree, things on which it is highly convenient to agree, the basic modularity of the system, functional dependencies, or aspects of the system that are expected to be long-lasting. Given this preference for architectural minimality, it will turn out that the architecture itself does not directly specify a system that meets these requirements. Rather, what it does is provide a framework within which it is possible to design a system that meets these requirements. In order to make this way of thinking more concrete, in chapter 5 I use the existing Internet as an example and go back to an earlier attempt to list the requirements that the Internet was intended to meet and how its architecture addressed them.

5 The Architecture of the Internet—A Historical Perspective

My introduction to architecture in chapter 3 was abstract. In this chapter, I use the architecture of the current Internet as a more concrete example. In 1988, I wrote a paper titled "The Design Philosophy of the DARPA Internet Protocols" (Clark, 1988), which tried to capture the requirements that have shaped the design of the Internet and the basic design decisions that had been made in meeting these requirements—what I might now call architecture but then called design philosophy. It is now more than 25 years since that paper was published, and looking back at that paper is a way to get started with a less abstract, more concrete example of network architecture.

What follows is that original paper, as first published in 1988, with extensive commentary from the perspective of 2017.

THE DESIGN PHILOSOPHY OF THE DARPA INTERNET PROTOCOLS

David D. Clark

Massachusetts Institute of Technology
Laboratory for Computer Science (now CSAIL)
Cambridge, Mass. 02139

This paper was originally published in 1988 in the ACM Sigcomm '88 symposium proceedings. Original work was supported in part by the Defense Advanced Research Projects Agency (DARPA) under Contract No. NOOOIJ-83-K-0125. It was annotated with extensive commentary in these text boxes in 2017. The original text has been reformatted but is otherwise unchanged from the original except for a few spelling corrections.

Abstract
The Internet protocol suite, TCP/IP, was first proposed fifteen years ago. It was developed by the Defense Advanced Research Projects Agency (DARPA), and has been used widely in military and commercial systems. While there have been papers and specifications that describe how the protocols work, it is sometimes difficult to deduce from these why the protocol is as it is. For example, the Internet protocol is based on a connectionless or datagram mode of service. The motivation for this has been greatly misunderstood. This paper attempts to capture some of the early reasoning which shaped the Internet protocols.

Introduction
For the last 15 years [1], the Advanced Research Projects Agency of the U.S. Department of Defense has been developing a suite of protocols for packet switched networking. These protocols, which include the Internet Protocol (IP), and the Transmission Control Protocol (TCP), are now U.S. Department of Defense standards for internetworking, and are in wide use in the commercial networking environment. The ideas developed in this effort have also influenced other protocol suites, most importantly the connectionless configuration of the ISO protocols [2,3,4].

While specific information on the DOD protocols is fairly generally available [5,6,7], it is sometimes difficult to determine the motivation and reasoning which led to the design.

In fact, the design philosophy has evolved considerably from the first proposal to the current standards. For example, the idea of the datagram, or connectionless service, does not receive particular emphasis in the first paper, but has come to be the defining characteristic of the protocol. Another example is the layering of the architecture into the IP and TCP layers. This seems basic to the design, but was also not a part of the original proposal. These changes in the Internet design arose through the repeated pattern of implementation and testing that occurred before the standards were set.

The Internet architecture is still evolving. Sometimes a new extension challenges one of the design principles, but in any case an understanding of the history of the design provides a necessary context for current design extensions. The connectionless configuration of ISO protocols has also been colored by the history of the Internet suite, so an understanding of the Internet design philosophy may be helpful to those working with ISO.

This paper catalogs one view of the original objectives of the Internet architecture, and discusses the relation between these goals and the important features of the protocols.

This paper makes a distinction between the architecture of the Internet and a specific realization of a running network. Today, as discussed later, I would distinguish three ideas:[1]

1. The core principles and basic design decisions of the architecture.
2. The second level of mechanism design, which fleshes out the architecture and makes it into a complete implementation.
3. The set of decisions related to deployment (e.g., the degree of diversity in paths) that lead to an operational network.

Fundamental Goal

The top level goal for the DARPA Internet Architecture was to develop an effective technique for multiplexed utilization of existing interconnected networks. Some elaboration is appropriate to make clear the meaning of that goal. The components of the Internet were networks, which were to be interconnected to provide some larger service. The original goal was to connect together the original ARPANET[8] with the ARPA packet radio network[9,10], in order to give users on the packet radio network access to the large service machines on the ARPANET. At the time it was assumed that there would be other sorts of networks to interconnect, although the local area network had not yet emerged.

This paragraph hints at but does not state clearly that the Internet builds on and extends the fundamental goal of the ARPAnet, which was to provide useful interconnection among heterogeneous machines. Perhaps even by 1988 this point was so well understood that it did not seem to require stating.

There is also an implicit assumption that the end points of network connections were machines. This assumption seemed obvious at the time but is now being questioned, with architectural proposals that addresses refer to services or information objects.

An alternative to interconnecting existing networks would have been to design a unified system which incorporated a variety of different transmission media, a multi-media network.

1. I am indebted to John Wroclawski, both for the suggestion that led to this revision and for the insight that there are three concepts to be distinguished, not two.

> Perhaps the term *multi-media* was not well-defined in 1988. It now has a different meaning, of course.

While this might have permitted a higher degree of integration, and thus better performance, it was felt that it was necessary to incorporate the then existing network architectures if Internet was to be useful in a practical sense. Further, networks represent administrative boundaries of control, and it was an ambition of this project to come to grips with the problem of integrating a number of separately administrated entities into a common utility.

> This last statement is actually a goal, and probably should have been listed as such, although it could be seen as an aspect of goal 4 that follows.

The technique selected for multiplexing was packet switching.

> Effective multiplexing of expensive resources (e.g., communication links) is another high-level goal that is not mentioned explicitly but was very important and well-understood at the time.

An alternative such as circuit switching could have been considered, but the applications being supported, such as remote login, were naturally served by the packet switching paradigm, and the networks which were to be integrated together in this project were packet switching networks. So packet switching was accepted as a fundamental component of the Internet architecture. The final aspect of this fundamental goal was the assumption of the particular technique for interconnecting these networks. Since the technique of store and forward packet switching, as demonstrated in the previous DARPA project, the ARPANET, was well understood, the top level assumption was that networks would be interconnected by a layer of Internet packet switches, which were called gateways.

From these assumptions comes the fundamental structure of the Internet: a packet switched communications facility in which a number of distinguishable networks are connected together using packet communications processors called gateways which implement a store and forward packet forwarding algorithm.

In retrospect, the previous section could have been clearer. It discussed both goals and basic architectural responses to them without teasing these ideas apart. Gateways are not a goal but rather a design response to a goal.

We could have taken a different approach to internetworking; for example, providing interoperation at a higher level—perhaps at the transport protocol layer, or a higher service/naming layer. It would be an interesting exercise to look at such a proposal and evaluate it relative to these criteria.

Second Level Goals

The top level goal stated in the previous section contains the word "effective," without offering any definition of what an effective interconnection must achieve. The following list summarizes a more detailed set of goals which were established for the Internet architecture.

1. Internet communication must continue despite loss of networks or gateways.
2. The Internet must support multiple types of communications service.
3. The Internet architecture must accommodate a variety of networks.
4. The Internet architecture must permit distributed management of its resources.
5. The Internet architecture must be cost effective.
6. The Internet architecture must permit host attachment with a low level of effort.
7. The resources used in the Internet architecture must be accountable.

This set of goals might seem to be nothing more than a checklist of all the desirable network features. It is important to understand that these goals are in order of importance, and an entirely different network architecture would result if the order were changed. For example, since this network was designed to operate in a military context, which implied the possibility of a hostile environment, survivability was put as a first goal, and accountability as a last goal. During wartime, one is less concerned with detailed accounting of resources used than with mustering whatever resources are available and rapidly deploying them in an operational manner. While the architects of the Internet were mindful of accountability, the problem received very little attention during the early stages of the design, and is only now being considered. An architecture primarily for commercial deployment would clearly place these goals at the opposite end of the list.

Similarly, the goal that the architecture be cost effective is clearly on the list, but below certain other goals, such as distributed management, or support of a wide variety of networks. Other protocol suites, including some of the more popular commercial architectures, have been optimized to a particular kind of network, for example a long haul store and forward network built of medium speed telephone lines, and deliver a very cost effective solution in this context, in exchange for dealing somewhat poorly with other kinds of nets, such as local area nets.

The reader should consider carefully the above list of goals, and recognize that this is not a "motherhood" list, but a set of priorities which strongly colored the design decisions within the Internet architecture. The following sections discuss the relationship between this list and the features of the Internet.

At the beginning of the NSF Future Internet Design (FIND) project, around 2008, I proposed a list of requirements that a new architecture might take into account. Here, for comparison with the early list from the 1988 paper, is the one I posed in 2008:

2008

1. Security
2. Availability and resilience
3. Economic viability
4. Better management
5. Meet society's needs
6. Longevity
7. Support for tomorrow's computing
8. Exploit tomorrow's networking
9. Support tomorrow's applications
10. Fit for purpose (it works?)

The list from 1988 does not mention the word security. The first 1988 requirement, that the network continue operation despite loss of networks or gateways, could be seen as a specific subcase of security, but the text in the next section of the original paper does not even hint that the failures might result from malicious actions. In retrospect, it is difficult to reconstruct what our mind-set was when this paper was written (which was in the years immediately prior to 1988). By the early 1990s, security was an important if unresolved objective. It seems somewhat odd that the word did not even come up in this paper.

The modern list calls out availability and resilience as distinct from the general category of security, a distinction that was motivated by my sense that this set of goals in particular was important enough that it should not be buried inside the broader category. So there is some correspondence between goal 1 in the 1988 list and goal 2 in the 2008 list.

The 2008 list has economic viability as its third objective, although it does not explicitly mention accounting for the use of resources. As I noted earlier, the 1988 paper discussed "the problem of integrating a number of separately administrated entities into a common utility," which seems like a specific manifestation of the recognition that the network is built out of parts. But an implication of the paper is that the design community of that time did not have a good understanding of the issue of economic viability.

Survivability in the Face of Failure

The most important goal on the list is that the Internet should continue to supply communications service, even though networks and gateways are failing. In particular, this goal was interpreted to mean that if two entities are communicating over the Internet and some failure causes the Internet to be temporarily disrupted and reconfigured to reconstitute the service, then the entities communicating should be able to continue without having to reestablish or reset the high level state of their conversation. More concretely, at the service interface of the transport layer, this architecture provides no facility to communicate to the client of the transport service that the synchronization between the sender and the receiver may have been lost. It was an assumption in this architecture that synchronization would never be lost unless there was no physical path over which any sort of communication could be achieved. In other words, at the top of transport, there is only one failure, and it is total partition. The architecture was to mask completely any transient failure.

In retrospect, this last sentence seems a bit unrealistic or perhaps poorly put. The architecture does not mask transient failures at all. That was not the goal, and it seems like an unrealizable one. The rest of the paragraph makes the actual point—if transient failures do occur, the application may be disrupted for the duration of the failure, but once the network has been reconstituted, the application (or, specifically, TCP) can take up where it left off. The rest of the section discusses the architectural approach to make this possible.

Again in retrospect, it would seem that an important sub-goal would be that transients are repaired as quickly as possible, but I don't think there was any understanding then, and perhaps there isn't now, of an architectural element that could facilitate that subgoal. So it is just left to the second-level mechanisms.

To achieve this goal, the state information which describes the on-going conversation must be protected. Specific examples of state information would be the number of packets transmitted, the number of packets acknowledged, or the number of outstanding flow control permissions. If the lower layers of the architecture lose this information, they will not be able to tell if data has been lost, and the application layer will have to cope with the loss of synchrony. This architecture insisted that this disruption not occur, which meant that the state information must be protected from loss.

In some network architectures, this state is stored in the intermediate packet switching nodes of the network. In this case, to protect the information from loss, it must be replicated. Because of the distributed nature of the replication, algorithms to ensure robust replication are themselves difficult to build, and few networks with distributed state information provide any sort of protection against failure. The alternative, which this architecture chose, is to take this information and gather it at the endpoint of the net, at the entity which is utilizing the service of the network. I call this approach to reliability "fate-sharing." The fate-sharing model suggests that it is acceptable to lose the state information associated with an entity if, at the same time, the entity itself is lost. Specifically, information about transport level synchronization is stored in the host which is attached to the net and using its communication service.

There are two important advantages to fate-sharing over replication. First, fate-sharing protects against any number of intermediate failures, whereas replication can only protect against a certain number (less than the number of replicated copies). Second, fate-sharing is much easier to engineer than replication.

There are two consequences to the fate-sharing approach to survivability. First, the intermediate packet switching nodes, or gateways, must not have any essential state information about on-going connections. Instead, they are stateless packet switches, a class of network design sometimes called a "datagram" network. Secondly, rather more trust is placed in the host machine than in an architecture where the network ensures the reliable delivery of data. If the host resident algorithms that ensure the sequencing and acknowledgment of data fail, applications on that machine are prevented from operation.

> See the later discussion about where failures should be detected, the consequences of misbehaving hosts, and the role of trust.

Despite the fact that survivability is the first goal in the list, it is still second to the top level goal of interconnection of existing networks. A more survivable technology might have resulted from a single multimedia network design. For example, the Internet makes very weak assumptions about the ability of a network to report that it has failed. Internet is thus forced to detect network failures using Internet level mechanisms, with the potential for a slower and less specific error detection.

Types of Service

The second goal of the Internet architecture is that it should support, at the transport service level, a variety of types of service. Different types of service are distinguished by differing requirements for such things as speed, latency and reliability. The traditional type of service is the bidirectional reliable delivery of data. This service, which is sometimes called a "virtual circuit" service, is appropriate for such applications as remote login or file transfer. It was the first service provided in the Internet architecture, using the Transmission Control Protocol (TCP)[11]. It was early recognized that even this service had multiple variants, because remote login required a service with low delay in delivery, but low requirements for bandwidth, while file transfer was less concerned with delay, but very concerned with high throughput. TCP attempted to provide both these types of service.

The initial concept of TCP was that it could be general enough to support any needed type of service. However, as the full range of needed services became clear, it seemed too difficult to build support for all of them into one protocol.

The first example of a service outside the range of TCP was support for XNET[12], the cross-Internet debugger. TCP did not seem a suitable transport for XNET for several reasons. First, a debugger protocol should not be reliable. This conclusion may seem odd, but under conditions of stress or failure (which may be exactly when a debugger is needed) asking for reliable communications may prevent any communications at all. It is much better to build a service which can deal with whatever gets through, rather than insisting that every byte sent be delivered in order. Second, if TCP is general enough to deal with a broad range of clients, it is presumably somewhat complex. Again, it seemed wrong to expect support for this complexity in a debugging environment, which may lack even basic services expected in an operating system (e.g. support for timers). So XNET was designed to run directly on top of the datagram service provided by Internet.

Another service which did not fit TCP was real time delivery of digitized speech, which was needed to support the teleconferencing aspect of command and control applications. In real time digital speech, the primary requirement is not a reliable service, but a service which minimizes and smooths the delay in the delivery of packets. The application layer is digitizing the analog speech, packetizing the resulting bits, and sending them out across the network on a regular basis. They must arrive at the receiver at a regular basis in order to be converted back to the analog signal. If packets do not arrive when expected, it is impossible to reassemble the signal in real time. A surprising observation about the control of variation in delay is that the most serious source of delay in networks is the mechanism to provide reliable delivery. A typical reliable transport protocol responds to a missing packet by requesting a retransmission and delaying the delivery of any subsequent packets until the lost packet has been retransmitted. It then delivers that packet and all remaining ones in sequence. The delay while this occurs can be many times the round trip delivery time of the net, and may completely disrupt the speech reassembly algorithm. In contrast, it is very easy to cope with an occasional missing packet. The missing speech can simply be replaced by a short period of silence, which in most cases does not impair the intelligibility of the speech to the listening human. If it does, high level error correction can occur, and the listener can ask the speaker to repeat the damaged phrase.

It was thus decided, fairly early in the development of the Internet architecture, that more than one transport service would be required, and the architecture must be prepared to tolerate simultaneously transports which wish to constrain reliability, delay, or bandwidth, at a minimum.

This goal caused TCP and IP, which originally had been a single protocol in the architecture, to be separated into two layers. TCP provided one particular type of service, the reliable sequenced data stream, while IP attempted to provide a basic building block out of which a variety of types of service could be built. This building block was the datagram, which had also been adopted to support survivability. Since the reliability associated with the delivery of a datagram was not guaranteed, but "best effort," it was possible to build out of the datagram a service that was reliable (by acknowledging and retransmitting at a higher level), or a service which traded reliability for the primitive delay characteristics of the underlying network substrate. The User Datagram Protocol (UDP)[13] was created to provide an application-level interface to the basic datagram service of Internet.

The architecture did not wish to assume that the underlying networks themselves support multiple types of services, because this would violate the goal of using existing networks. Instead, the hope was that multiple types of service could be constructed out of the basic datagram building block using algorithms within the host and the gateway.

I am surprised that I wrote those last two sentences. They did not properly reflect our thinking at that time. RFC 791 (Postel, 1981b) states:

> The Type of Service provides an indication of the abstract parameters of the quality of service desired. These parameters are to be used to guide the selection of the actual service parameters when transmitting a datagram through a particular network. Several networks offer service precedence, which somehow treats high precedence traffic as more important than other traffic (generally by accepting only traffic above a certain precedence at time of high load).
>
> ...
>
> Example mappings of the internet type of service to the actual service provided on networks such as AUTODIN II, ARPANET, SATNET, and PRNET is given in "Service Mappings" (Jon Postel, 1981).

At the time this RFC was written (around 1981), the group clearly had in mind that different sorts of networks might have different tools for managing different service qualities, and the gateway (what we now call the router) should map the abstract ToS field to the network-specific service indicators.

For example, (although this is not done in most current implementations) it is possible to take datagrams which are associated with a controlled delay but unreliable service and place them at the head of the transmission queues unless their lifetime has expired, in which case they would be discarded, while packets associated with reliable streams would be placed at the back of the queues, but never discarded, no matter how long they had been in the net.

This section of the paper may reflect my own long-standing preference for QoS in the network. However, the discussion is about a much more basic set of service types and an architectural decision (splitting IP and TCP) that gives the end node and application some control over the type of service. There is no mention in this paper of the ToS bits in the IP header, which were the first attempt to add a core feature that would facilitate any sort of QoS in the network. Discussions about QoS at the IETF did not start for another several years, but this section does suggest that the idea of queue management as a means of improving application behavior was understood even in the 1980s, and the ToS bits (or something like them) would be needed to drive that sort of scheduling. I think, looking back, that we really did not understand this set of issues, even in 1988.

It proved more difficult than first hoped to provide multiple types of service without explicit support from the underlying networks. The most serious problem was that networks designed with one particular type of service in mind were not flexible enough to support other services. Most commonly, a network will have been designed under the assumption that it should deliver reliable service, and will inject delays as a part of producing reliable service, whether or not this reliability is desired. The interface behavior defined by X.25, for example, implies reliable delivery, and there is no way to turn this feature off. Therefore, although Internet operates successfully over X.25 networks it cannot deliver the desired variability of type service in that context. Other networks which have an intrinsic datagram service are much more flexible in the type of service they will permit, but these networks are much less common, especially in the long-haul context.

Even though this paper came about five years after the articulation of the end-to-end arguments (Saltzer et al., 1984), there is no mention of that paper or its concepts here. Perhaps this was because this paper was a retrospective of the early thinking, which predated the emergence of end-to-end as a named concept. The concept is lurking in much of what I wrote in this section, but perhaps in 1988 it was not yet clear that the end-to-end description as presented in Saltzer et al. (1984) would survive as the accepted framing.

Varieties of Networks

It was very important for the success of the Internet architecture that it be able to incorporate and utilize a wide variety of network technologies, including military and commercial facilities. The Internet architecture has been very successful in meeting this goal: it is operated over a wide variety of networks, including long haul nets (the ARPANET itself and various X.25 networks), local area nets (Ethernet, ringnet, etc.), broadcast satellite nets (the DARPA Atlantic Satellite Network[14,15] operating at 64 kilobits per second and the DARPA Experimental Wideband Satellite Net[16] operating within the United States at 3 megabits per second), packet radio networks (the DARPA packet radio network, as well as an experimental British packet radio net and a network developed by amateur radio operators), a variety of serial links, ranging from 1200 bit per second asynchronous connections to TI links, and a variety of other ad hoc facilities, including intercomputer busses and the transport service provided by the higher layers of other network suites, such as IBM's HASP.

The Internet architecture achieves this flexibility by making a minimum set of assumptions about the function which the net will provide. The basic assumption is that the network can transport a packet or datagram. The packet must be of reasonable size, perhaps 100 bytes minimum, and should be delivered with reasonable but not perfect reliability. The network must have some suitable form of addressing if it is more than a point to point link.

There are a number of services which are explicitly not assumed from the network. These include reliable or sequenced delivery, network level broadcast or multicast, priority ranking of transmitted packet, multiple types of service, and internal knowledge of failures, speeds, or delays. If these services had been required, then in order to accommodate a network within the Internet, it would be necessary either that the network support these services directly, or that the network interface software provide enhancements to simulate these services at the endpoint of the network. It was felt that this was an undesirable approach, because these services would have to be re-engineered and reimplemented for every single network and every single host interface to every network. By engineering these services at the transport, for example reliable delivery via TCP, the engineering must be done only once, and the implementation must be done only once for each host. After that, the implementation of interface software for a new network is usually very simple.

Other Goals

The three goals discussed so far were those which had the most profound impact on the design of the architecture. The remaining goals, because they were lower in importance, were perhaps less effectively met, or not so completely engineered. The goal of permitting distributed management of the Internet has certainly been met in certain respects. For example, not all of the gateways in the Internet are implemented and managed by the same agency. There are several different management centers within the deployed Internet, each operating a subset of the gateways, and there is a two-tiered routing algorithm which permits gateways from different administrations to exchange routing tables, even though they do not completely trust each other, and a variety of private routing algorithms used among the gateways in a single administration. Similarly, the various organizations which manage the gateways are not necessarily the same organizations that manage the networks to which the gateways are attached.

> Even in 1988 we understood that the issue of trust (e.g., trust among gateways) was an important consideration.

On the other hand, some of the most significant problems with the Internet today relate to lack of sufficient tools for distributed management, especially in the area of routing. In the large Internet being currently operated, routing decisions need to be constrained by policies for resource usage. Today this can be done only in a very limited way, which requires manual setting of tables. This is error-prone and at the same time not sufficiently powerful. The most important change in the Internet architecture over the next few years will probably be the development of a new generation of tools for management of resources in the context of multiple administrations.

> It is interesting that we understood the limitations of manual configuration of routing policy in 1988, and we are not yet really beyond that stage almost 30 years later. It is not clear even now whether our persistent lack of progress in this area is results from poor architectural choices or just the intrinsic difficulty of the task. Certainly, in the 1970s and 1980s we did not know how to think about network management. We understood how to "manage a box," but we had no accepted view on system-level management.

It is clear that in certain circumstances, the Internet architecture does not produce as cost effective a utilization of expensive communication resources as a more tailored architecture would. The headers of Internet packets are fairly long (a typical header is 40 bytes), and if short packets are sent, this overhead is apparent. The worst case, of course, is the single character remote login packets, which carry 40 bytes of header and one byte of data. Actually, it is very difficult for any protocol suite to claim that these sorts of interchanges are carried out with reasonable efficiency. At the other extreme, large packets for file transfer, with perhaps 1,000 bytes of data, have an overhead for the header of only four percent.

Another possible source of inefficiency is retransmission of lost packets. Since Internet does not insist that lost packets be recovered at the network level, it may be necessary to retransmit a lost packet from one end of the Internet to the other. This means that the retransmitted packet may cross several intervening nets a second time, whereas recovery at the network level would not generate this repeat traffic. This is an example of the tradeoff resulting from the decision, discussed above, of providing services from the end-points. The network interface code is much simpler, but the overall efficiency is potentially less. However, if the retransmission rate is low enough (for example, 1%) then the incremental cost is tolerable. As a rough rule of thumb for networks incorporated into the architecture, a loss of one packet in a hundred is quite reasonable, but a loss of one packet in ten suggests that reliability enhancements be added to the network if that type of service is required.

Again, this 1988 paper provides a nice "time capsule" as to what we were worrying about 25 years ago. Now we seem to have accepted the cost of packet headers and the cost of end-to-end retransmission. The paper does not mention efficient link loading as an issue, or the question of achieving good end-to-end performance.

The cost of attaching a host to the Internet is perhaps somewhat higher than in other architectures, because all of the mechanisms to provide the desired types of service, such as acknowledgments and retransmission strategies, must be implemented in the host rather than in the network. Initially, to programmers who were not familiar with protocol implementation, the effort of doing this seemed somewhat daunting. Implementers tried such things as moving the transport protocols to a front end processor, with the idea that the protocols would be implemented only once, rather than again for every type of host. However, this required the invention of a host to front end protocol which some thought almost as complicated to implement as the original transport protocol. As experience with protocols increases, the anxieties associated with implementing a protocol suite within the host seem to be decreasing, and implementations are now available for a wide variety of machines, including personal computers and other machines with very limited computing resources.

A related problem arising from the use of host-resident mechanisms is that poor implementation of the mechanism may hurt the network as well as the host. This problem was tolerated, because the initial experiments involved a limited number of host implementations which could be controlled. However, as the use of Internet has grown, this problem has occasionally surfaced in a serious way. In this respect, the goal of robustness, which led to the method of fate-sharing, which led to host-resident algorithms, contributes to a loss of robustness if the host misbehaves.

> This paragraph brings out a contradiction in the architectural principles that I might have stated more clearly. The principle of minimal state in routers and the movement of function to the end points implies a need to trust those end points to operate correctly, but the architecture does not have any approach for dealing with hosts that misbehave. Without state in the network to validate what the hosts are doing, there are few ways to discipline a host. In 1988, we anticipated the problem but clearly had no view as to how to think about it.

The last goal was accountability. In fact, accounting was discussed in the first paper by Cerf and Kahn as an important function of the protocols and gateways. However, at the present time, the Internet architecture contains few tools for accounting for packet flows. This problem is only now being studied, as the scope of the architecture is being expanded to include non-military consumers who are seriously concerned with understanding and monitoring the usage of the resources within the Internet.

> Again, a deeper discussion here might have brought out some contradictions among goals: without any flow state in the network (or knowledge of what constitutes an "accountable entity"), it seems hard to do accounting. The architecture does not preclude what we now call "middleboxes," but the architecture also does not discuss the idea that there might be information in the packets to aid in accounting. I think in 1988 we just did not know how to think about this, and it was not a high priority.

Architecture and Implementation

The previous discussion clearly suggests that one of the goals of the Internet architecture was to provide wide flexibility in the service offered. Different transport protocols could be used to provide different types of service, and different networks could be incorporated. Put another way, the architecture tried very hard not to constrain the range of service which the Internet could be engineered to provide. This, in turn, means that to understand the service which can be offered by a particular implementation of an Internet, one must look not to the architecture, but to the actual engineering of the software within the particular hosts and gateways, and to the particular networks which have been incorporated. I will use the term "realization" to describe a particular set of networks, gateways and hosts which have been connected together in the context of the Internet architecture. Realizations can differ by orders of magnitude in the service which they offer. Realizations have been built out of 1200 bit per second phone lines, and out of networks only with speeds greater than 1 megabit per second. Clearly, the throughput expectations which one can have of these realizations differ by orders of magnitude. Similarly, some Internet realizations have delays measured in tens of milliseconds, where others have delays measured in seconds. Certain applications such as real time speech work fundamentally differently across these two realizations. Some Internets have been engineered so that there is great redundancy in the gateways and paths. These Internets are survivable, because resources exist which can be reconfigured after failure. Other Internet realizations, to reduce cost, have single points of connectivity through the realization, so that a failure may partition the Internet into two halves.

As I said earlier, today I believe that there should be three distinctions:

1. The core principles and basic design decisions of the architecture.
2. The second level of mechanism design, which fleshes out the architecture and makes it into a complete implementation.
3. The set of decisions related to deployment (e.g., degree of redundancy in paths) that lead to an operational network.

The word realization seems to map to the third set of decisions, and the second set is somewhat missing from this paper. One could argue that the omission was intentional: the paper was about the architecture, and what this text is saying is that one of the goals of the architecture was to permit many realizations, a point that I might have listed as another goal. But it is equally important to say that a goal of the architecture was to allow many different alternatives for mechanism design as well—the design decisions of the architecture should permit a range of mechanism choices, not embed those decisions into the architecture itself. I believe that in 1988 the Internet designers saw, but perhaps did not articulate clearly, that there is a benefit to architectural minimality—that is, to specify as little as possible, consistent with making it possible for subsequent mechanisms to meet the goals of the architecture. Were I writing the paper now, I would add a new section, which draws from the previous sections the set of core principles of the architecture, linking them back to the goals they enable.

Core architectural principles

Packet switching.

Gateways (what we call routers today)

- Minimal assumptions about what the networks would do.
- No flow state in routers, which implies no flow setup, and thus the "pure" datagram model.
- Implies strict separation of IP from TCP, with no knowledge of TCP in routers.

Collocation of flow state with end points of flows (fate-sharing).

No mechanisms to report network failures to end points.

Trust in the end node.

Minimal assumptions about service functions and performance.

Totally missing from this paper is any discussion of packet headers, addressing, and so on. In fact, much earlier than 1988, we understood that we had to agree on some format for addresses but that the specific decision should not influence our ability to address the goals in the list. Early in the design process (in the mid-1970s), the design team proposed a design based on variable-length addresses, which would have served us much better with respect to the goal of longevity. The people who were building the first routers rejected that idea because, at the time, the difficulty of building routers that could operate at line speeds (e.g., 1.5 mbs) made parsing of variable-length fields in the header a challenge. In my 1988 list, "longevity" is missing, probably a significant oversight, but in the 1970s we made a design choice that favored the pragmatics of implementation over flexibility, and we still viewed the project as research.

The packet header also embodied other design choices that we thought we had to make in order to facilitate or enable the design of the second-level mechanisms that flesh out the architecture into a complete implementation.

- The idea of packet fragmentation supported the goal that we be able to exploit preexisting networks. Today, the Internet is the dominant architecture, and we can assume that issues like network technology with small packet sizes will not arise.
- The use of a TTL or hop count was an architectural decision that tried to allow more generality in how routing was done—we wanted to tolerate transient routing inconsistency. The architecture did not specify the details of how routing was done (the paper notes the emergence of the two-level routing hierarchy), and indeed it was a goal that different routing schemes could be deployed in different parts of the network.

The Internet architecture tolerates this variety of realization by design. However, it leaves the designer of a particular realization with a great deal of engineering to do. One of the major struggles of this architectural development was to understand how to give guidance to the designer of a realization, guidance which would relate the engineering of the realization to the types of service which would result. For example, the designer must answer the following sort of question. What sort of bandwidths must be in the underlying networks, if the overall service is to deliver a throughput of a certain rate? Given a certain model of possible failures within this realization, what sorts of redundancy ought to be engineered into the realization?

Most of the known network design aids did not seem helpful in answering these sorts of questions. Protocol verifiers, for example, assist in confirming that protocols meet specifications. However, these tools almost never deal with performance issues, which are essential to the idea of the type of service. Instead, they deal with the much more restricted idea of logical correctness of the protocol with respect to specification. While tools to verify logical correctness are useful, both at the specification and implementation stage, they do not help with the severe problems that often arise related to performance. A typical implementation experience is that even after logical correctness has been demonstrated, design faults are discovered that may cause a performance degradation of an order of magnitude. Exploration of this problem has led to the conclusion that the difficulty usually arises, not in the protocol itself, but in the operating system on which the protocol runs. This being the case, it is difficult to address the problem within the context of the architectural specification. However, we still strongly feel the need to give the implementer guidance. We continue to struggle with this problem today.

This paragraph reflects an issue that I could have explored more clearly. The goal of continued operation in the face of failures (resilience) motivated us to design good mechanisms to recover from problems. These mechanisms were in fact good enough that they would also mask implementation errors, and the only signal of the problem was poor performance. What is missing from the Internet, whether in the architecture or as an expectation of the second-level mechanisms, is some requirement to report when error detection and recovery mechanisms are triggered. But without a good architecture for network management, it is not surprising that these reporting mechanisms are missing, because it is not clear what entity the report would go to. The user at the end node is not a trained network professional, so telling that person about a failure is generally not useful, and there is no other management entity defined as part of the architecture that can receive such a notification.

The other class of design aid is the simulator, which takes a particular realization and explores the service which it can deliver under a variety of loadings. No one has yet attempted to construct a simulator which takes into account the wide variability of the gateway implementation, the host implementation, and the network performance which one sees within possible Internet realizations. It is thus the case that the analysis of most Internet realizations is done on the back of an envelope. It is a comment on the goal structure of the Internet architecture that a back of the envelope analysis, if done by a sufficiently knowledgeable person, is usually sufficient. The designer of a particular Internet realization is usually less concerned with obtaining the last five percent possible in line utilization than knowing whether the desired type of service can be achieved at all given the resources at hand at the moment.

The relationship between architecture and performance is an extremely challenging one. The designers of the Internet architecture felt very strongly that it was a serious mistake to attend only to logical correctness and ignore the issue of performance. However, they experienced great difficulty in formalizing any aspect of performance constraint within the architecture. These difficulties arose both because the goal of the architecture was not to constrain performance, but to permit variability, and secondly (and perhaps more fundamentally), because there seemed to be no useful formal tools for describing performance.

From the perspective of 2017, this paragraph is telling. For some goals such as routing, we had mechanisms (e.g., the TTL field) that we could incorporate in the architecture to support the objective. For performance, we simply did not know. (We proposed an ICMP message called Source Quench, which never proved useful and may have just been a bad idea. It is deprecated.) At the time this paper was written, our problems with congestion were so bad that we were at peril of failing 2017 goal 10: "It works." However, there is no mention of congestion and its control in this paper. Arguably, the design community still does not know what the architecture should specify about congestion and other aspects of performance. The community seems to have some agreement on the ECN bit, but not enough enthusiasm to get the mechanism actually deployed. There are also many alternative proposals, such as such as XCP (Katabi et al., 2002) and RCP (Dukkipati, 2008), that would imply a different packet header. The debate seems to continue as to what to put in the packet (e.g., specify as part of the architectural interfaces) in order to allow a useful range of mechanisms to be designed to deal with congestion and other aspects of performance. The design community also still debates the degree to which the end node should be trusted to respond properly to indications of congestion, as opposed to adding a mechanism to the network to police the behavior of the hosts (Briscoe et al., 2005).

This problem was particularly aggravating because the goal of the Internet project was to produce specification documents which were to become military standards. It is a well known problem with government contracting that one cannot expect a contractor to meet any criteria which are not a part of the procurement standard. If the Internet is concerned about performance, therefore, it was mandatory that performance requirements be put into the procurement specification. It was trivial to invent specifications which constrained the performance, for example to specify that the implementation must be capable of passing 1,000 packets a second. However, this sort of constraint could not be part of the architecture, and it was therefore up to the individual performing the procurement to recognize that these performance constraints must be added to the specification, and to specify them properly to achieve a realization which provides the required types of service. We do not have a good idea how to offer guidance in the architecture for the person performing this task.

Datagrams

The fundamental architectural feature of the Internet is the use of datagrams as the entity which is transported across the underlying networks. As this paper has suggested, there are several reasons why datagrams are important within the architecture. First, they eliminate the need for connection state within the intermediate switching nodes, which means that the Internet can be reconstituted after a failure without concern about state. Secondly, the datagram provides a basic building block out of which a variety of types of service can be implemented. In contrast to the virtual circuit, which usually implies a fixed type of service, the datagram provides a more elemental service which the endpoints can combine as appropriate to build the type of service needed. Third, the datagram represents the minimum network service assumption, which has permitted a wide variety of networks to be incorporated into various Internet realizations. The decision to use the datagram was an extremely successful one, which allowed the Internet to meet its most important goals very successfully. There is a mistaken assumption often associated with datagrams, which is that the motivation for datagrams is the support of a higher level service which is essentially equivalent to the datagram. In other words, it has sometimes been suggested that the datagram is provided because the transport service which the application requires is a datagram service. In fact, this is seldom the case. While some applications in the Internet, such as simple queries of date servers or name servers, use an access method based on an unreliable datagram, most services within the Internet would like a more sophisticated transport model than simple datagram. Some services would like the reliability enhanced, some would like the delay smoothed and buffered, but almost all have some expectation more complex than a datagram. It is important to understand that the role of the datagram in this respect is as a building block, and not as a service in itself.

This discussion of the datagram seems reasonable from the perspective of 2017, but as I said earlier, were I to write the paper now, I would give similar treatment to some of the other design decisions we made.

TCP

There were several interesting and controversial design decisions in the development of TCP, and TCP itself went through several major versions before it became a reasonably stable standard. Some of these design decisions, such as window management and the nature of the port address structure, are discussed in a series of implementation notes published as part of the TCP protocol handbook [17,18]. But again the motivation for the decision is sometimes lacking. In this section, I attempt to capture some of the early reasoning that went into parts of TCP. This section is of necessity incomplete; a complete review of the history of TCP itself would require another paper of this length.

The original ARPANET host-to host protocol provided flow control based on both bytes and packets. This seemed overly complex, and the designers of TCP felt that only one form of regulation would be sufficient. The choice was to regulate the delivery of bytes, rather than packets. Flow control and acknowledgment in TCP is thus based on byte number rather than packet number. Indeed, in TCP there is no significance to the packetization of the data.

This decision was motivated by several considerations, some of which became irrelevant and others of which were more important than anticipated. One reason to acknowledge bytes was to permit the insertion of control information into the sequence space of the bytes, so that control as well as data could be acknowledged. That use of the sequence space was dropped, in favor of ad hoc techniques for dealing with each control message. While the original idea has appealing generality, it caused complexity in practice.

A second reason for the byte stream was to permit the TCP packet to be broken up into smaller packets if necessary in order to fit through a net with a small packet size. But this function was moved to the IP layer when IP was split from TCP, and IP was forced to invent a different method of fragmentation.

A third reason for acknowledging bytes rather than packets was to permit a number of small packets to be gathered together into one larger packet in the sending host if retransmission of the data was necessary. It was not clear if this advantage would be important; it turned out to be critical. Systems such as UNIX which have an internal communication model based on single character interactions often send many packets with one byte of data in them. (One might argue from a network perspective that this behavior is silly, but it was a reality, and a necessity for interactive remote login.) It was often observed that such a host could produce a flood of packets with one byte of data, which would arrive much faster than a slow host could process them. The result is lost packets and retransmission.

If the retransmission was of the original packets, the same problem would repeat on every retransmission, with a performance impact so intolerable as to prevent operation. But since the bytes were gathered into one packet for retransmission, the retransmission occurred in a much more effective way which permitted practical operation.

On the other hand, the acknowledgment of bytes could be seen as creating this problem in the first place. If the basis of flow control had been packets rather than bytes, then this flood might never have occurred. Control at the packet level has the effect, however, of providing a severe limit on the throughput if small packets are sent. If the receiving host specifies a number of packets to receive, without any knowledge of the number of bytes in each, the actual amount of data received could vary by a factor of 1000, depending on whether the sending host puts one or one thousand bytes in each packet.

In retrospect, the correct design decision may have been that if TCP is to provide effective support of a variety of services, both packets and bytes must be regulated, as was done in the original ARPANET protocols.

Another design decision related to the byte stream was the End-Of-Letter flag, or EOL. This has now vanished from the protocol, replaced by the push flag, or PSH. The original idea of EOL was to break the byte stream into records. It was implemented by putting data from separate records into separate packets, which was not compatible with the idea of combining packets on retransmission. So the semantics of EOL was changed to a weaker form, meaning only that the data up to this point in the stream was one or more complete application-level elements, which should occasion a flush of any internal buffering in TCP or the network. By saying "one or more" rather than "exactly one," it became possible to combine several together and preserve the goal of compacting data in reassembly. But the weaker semantics meant that various applications had to invent an ad hoc mechanism for delimiting records on top of the data stream.

Several features of TCP, including EOL and the reliable close, have turned out to be of almost no use to applications today. The story of the design and evolution of TCP provides another view into the process of trying to figure out in advance what should be in and what should be out of a general mechanism that is intended to last for a long time (the goal of longevity).

In this evolution of EOL semantics, there was a little known intermediate form, which generated great debate. Depending on the buffering strategy of the host, the byte stream model of TCP can cause great problems in one improbable case. Consider a host in which the incoming data is put in a sequence of fixed size buffers. A buffer is returned to the user either when it is full, or an EOL is received. Now consider the case of the arrival of an out-of-order packet which is so far out of order to be beyond the current buffer. Now further consider that after receiving this out-of-order packet, a packet with an EOL causes the current buffer to be returned to the user only partially full. This particular sequence of actions has the effect of causing the out of order data in the next buffer to be in the wrong place, because of the empty bytes in the buffer returned to the user. Coping with this generated book-keeping problems in the host which seemed unnecessary.

To cope with this it was proposed that the EOL should "use up" all the sequence space up to the next value which was zero mod the buffer size. In other words, it was proposed that EOL should be a tool for mapping the byte stream to the buffer management of the host. This idea was not well received at the time, as it seemed much too ad hoc, and only one host seemed to have this problem.[2] In retrospect, it may have been the correct idea to incorporate into TCP some means

2. This use of EOL was properly called Rubber EOL but its detractors quickly called it "rubber baby buffer bumpers" in an attempt to ridicule the idea. Credit must go to the creator of the idea, Bill Plummer, for sticking to his guns in the face of detractors saying the above to him ten times fast.

of relating the sequence space and the buffer management algorithm of the host. At the time, the designers simply lacked the insight to see how that might be done in a sufficiently general manner.

Conclusion

In the context of its priorities, the Internet architecture has been very successful. The protocols are widely used in the commercial and military environment, and have spawned a number of similar architectures. At the same time, its success has made clear that in certain situations, the priorities of the designers do not match the needs of the actual users. More attention to such things as accounting, resource management and operation of regions with separate administrations are needed.

While the datagram has served very well in solving the most important goals of the Internet, it has not served so well when we attempt to address some of the goals which were further down the priority list. For example, the goals of resource management and accountability have proved difficult to achieve in the context of datagrams. As the previous section discussed, most datagrams are a part of some sequence of packets from source to destination, rather than isolated units at the application level. However, the gateway cannot directly see the existence of this sequence, because it is forced to deal with each packet in isolation. Therefore, resource management decisions or accounting must be done on each packet separately. Imposing the datagram model on the Internet layer has deprived that layer of an important source of information which it could use in achieving these goals.

This suggests that there may be a better building block than the datagram for the next generation of architecture. The general characteristic of this building block is that it would identify a sequence of packets traveling from the source to the destination, without assuming any particular type of service with that service. I have used the word "flow" to characterize this building block. It would be necessary for the gateways to have flow state in order to remember the nature of the flows which are passing through them, but the state information would not be critical in maintaining the desired type of service associated with the flow. Instead, that type of service would be enforced by the end points, which would periodically send messages to ensure that the proper type of service was being associated with the flow. In this way, the state information associated with the flow could be lost in a crash without permanent disruption of the service features being used. I call this concept "soft state," and it may very well permit us to achieve our primary goals of survivability and flexibility, while at the same time doing a better job of dealing with the issue of resource management and accountability. Exploration of alternative building blocks constitutes one of the current directions for research within the DARPA Internet program.

Acknowledgments—A Historical Perspective

It would be impossible to acknowledge all the contributors to the Internet project; there have literally been hundreds over the 15 years of development: designers, implementers, writers and critics. Indeed, an important topic, which probably deserves a paper in itself, is the process by which this project was managed. The participants came from universities, research laboratories and corporations, and they united (to some extent) to achieve this common goal.

The original vision for TCP came from Robert Kahn and Vinton Cerf, who saw very clearly, back in 1973, how a protocol with suitable features might be the glue that would pull together the various emerging network technologies. From their position at DARPA, they guided the project in its early days to the point where TCP and IP became standards for the DOD.

The author of this paper joined the project in the mid-70s, and took over architectural responsibility for TCP/IP in 1981. He would like to thank all those who have worked with him, and particularly those who took the time to reconstruct some of the lost history in this paper.

References

1. V. Cerf, and R. Kahn, "A Protocol for Packet Network intercommunication," *IEEE Transactions Communications*, Vol. 22, No. 5, May 1974, pp. 637–648.

2. ISO, "Transport Protocol Specification," Tech. report IS-8073, International Organization for Standardization, September 1984.

3. ISO, "Protocol for Providing the Connectionless-Mode Network Service," Tech. report DIS8473, International Organization for Standardization, 1986.

4. R. Callon, "Internetwork Protocol," *Proceedings of the IEEE*, Vol. 71, No. 12, December 1983, pp. 1388–1392.

5. Jonathan B. Postel, "Internetwork Protocol Approaches," *IEEE Transactions on Communications*, Vol. Com 28, No. 4, April 1980, pp. 605–611.

6. Jonathan B. Postel, Carl A. Sunshine, Danny Cohen, "The ARPA Internet Protocol," *Computer Networks*, Vol. 5, No. 4, July 1981, pp. 261–271.

7. Alan Sheltzer, Robert Hinden, and Mike Brescia, "Connecting Different Types of Networks with Gateways," *Data Communications*, August 1982.

8. J. McQuillan and D. Walden, "The ARPA Network Design Decisions," *Computer Networks*, Vol. 1, No. 5, August 1977, pp. 243–289.

9. R. E. Kahn, S. A. Gronemeyer, J. Burdifiel, E. V. Hoversten, "Advances in Packet Radio Technology," *Proceedings of the IEEE*, Vol. 66, No. 11, November 1978, pp. 1408–1496.

10. B. M. Leiner, D. L. Nelson, F. A. Tobagi, "Issues in Packet Radio Design," *Proceedings of the IEEE*, Vol. 75, No. 1, January 1987, pp. 6–20.

11. "Transmission Control Protocol RFC-793," *DDN Protocol Handbook*, Vol. 2, September 1981, pp. 2.179–2.198.

12. Jack Haverty, "XNET Formats for Internet Protocol Version 4 IEN 158," *DDN Protocol Handbook*, Vol. 2, October 1980, pp. 2-345-2-348.

13. Jonathan Postel, "User Datagram Protocol NIC RFC-768," *DDN Protocol Handbook*, Vol. 2, August 1980, pp. 2.175–2.177.

14. I. Jacobs, R. Binder, and E. Hoversten, "General Purpose Packet Satellite Networks," *Proceedings of the IEEE*, Vol. 66, No. 11, November 1978, pp. 1448–1467.

15. C. Topolcic and J. Kaiser, "The SATNET Monitoring System," *Proceedings of the IEEEMILCOM*, Boston, MA, October 1985, pp. 26.1.1–26.1.9.

16. W. Edmond, S. Blumenthal, A. Echenique, S. Storch, T. Calderwood, and T. Rees, "The Butterfly Satellite IMP for the Wideband Packet Satellite Network," *Proceedings of the ACM SIGCOMM '86*, ACM, Stowe, VT, August 1986, pp. 194–203.

17. David D. Clark, "Window and Acknowledgment Strategy in TCP NIC RFC-813," *DDN Protocol Handbook*, Vol. 3, July 1982, pp. 3-5 to 3-26.

18. David D. Clark, "Name, Addresses, Ports, and Routes NIC-RFC-814," *DDN Protocol Handbook*, Vol. 3, July 1982, pp. 3-27 to 3-40.

The Relation of Architecture to Function

The architecture of the Internet as I have defined it here (and as I did in 1988) clearly illustrates that architecture does not directly specify how the network will meet its functional requirements.

It is worth looking at the various requirements I laid out in chapter 4 and considering how the architecture of the Internet relates to meeting those requirements.

Fit for purpose (it works?) Arguably, the Internet is a success. Its design led to a network that has passed the tests of utility and longevity. The basic ideas of packet switching, datagrams (no per-flow state in the routers), and the like were well-crafted. Those of us who designed the original Internet are so pleased (and perhaps surprised) that it works as well as it does that we feel justified in turning a blind eye to the aspects that don't work so well. If the Internet of today is not quite as reliable as the phone system, and routing takes rather long to converge on new routes after a failure, we say that after all routing is just one of those "second-level" mechanisms and not a part of the architecture, and who said that "5 nines reliability" is the right idea for the Internet?[3] But overall, I think it is fair to argue that the architecture of the Internet produced a system that is fit for purpose.

Security In chapter 10, I will argue that the Internet itself (the packet carriage layer as opposed to the larger definition that includes the applications and technology) can only solve part of the security problem. Securing the network itself, which seems to call for secure versions of routing protocols, among other things, was relegated to that second stage of mechanism design that turns the architecture into a complete implementation. This approach was probably reasonable, since different circumstances call for different degrees of security, but there is an open question as to whether there are architectural decisions that could make this task easier. Protecting the packet transport layer from abuse by the applications (most obviously in the context of denial of service attacks) is an area that the architecture needs to address, but the early designers did not consider this issue. Overall, the link between the key architectural features of the Internet and the requirements for security seems a bit fragmentary and weak.

Availability and resilience In the 1980s we did not understand how to think about availability in general. We understood that packets might get lost, so we designed TCP to retransmit them. We understood that links and routers might fail, so we needed dynamic routing. The Internet packet header provides a Time to Live (TTL) field to allow for dynamic inconsistency in routing.[4] The TTL field illustrates the point that architecture does

3. The term *5 nines reliability* is shorthand for a system that is up 99.999 percent of the time, which would imply a downtime of 5.26 minutes per year. It is often stated, although I can find no specific citation, that the U.S. telephone system was designed to meet this objective. The Internet, for the most part, certainly does not.

4. The TTL field is designed to deal with a specific problem with routing protocols. During a recomputation of routes after a link has failed or been added, it is possible that different routers will have inconsistent forwarding tables, and two routers may each think that the best route to a destination is through the other. If a packet is caught in this sort of routing loop, it can bounce back and forth between the two routers until the inconsistency is resolved, clogging the network uselessly. To prevent this, as the routers forward a packet through the Internet, each router decrements the TTL field. If the value of the field reaches zero, the router discards the packet. In this way, looping packets can be quickly discarded. The sender will have to retransmit the packet, and with luck the inconsistency will have been resolved by then.

not always define how to meet a requirement but tries to make it possible (or easier) for a system based on that architecture to meet that requirement. Our intuition was that no other architectural support was needed for routing, or for availability more generally. As I will argue in chapter 11, an architecture of a future Internet should take a more integrated view of availability.

Economic viability One way to think about economic viability is that all the actors in the ecosystem created by the architecture must have the incentive to play the role assigned to them by that architecture. In particular, if there is a class of actor that does not find an economic incentive to enter the ecosystem and invest, the design will not thrive. This way of looking at things was roughly understood early on, but we had no tools to reason about it. In fact, the issues have really only become clear in the last decade, with ISPs (which make large capital investments) trying to find ways to increase revenues by "violating" the architecture: peeking into packets, exercising discrimination of various sorts, and so on. There is nothing in the Internet architecture about accounting, billing, money flows, or other issues that relate to economics, although the issue of accounting was identified in the initial Internet design paper (Cerf and Kahn, 1974).

Management The original Internet architecture contained few design elements intended to address the issues of network management. The specification of IP does include control messages (the Internet control message protocol) that allow a router to report back to a sender errors of various sorts that occur as the router attempts to forward the packet, but it is not clear that notifying the sender is a useful component of a good system for network management, since the problem is probably not the fault of the sender but instead some region of the network. We received some criticism from our friends in the telephone industry about the lack of attention to network management; they said that a major part of the design of the telephone system was to address issues of management: fault detection and isolation, performance issues, and the like. Many of the basic data formats used to transport voice across the digital version of the telephone system contain fields related to management, and we were asked why we had not understood that.

Meet society's needs This very general heading captures a range of issues such as privacy (on the one hand), lawful interception (on the other hand), resilience of critical services, control of disruptive or illegal behavior by users, and others. There is little in my 1988 paper that speaks to these issues. It may not have been clear in 1988 that the way Internet addresses are specified and used (for example) has a material influence on the balance between privacy, traffic analysis, lawful interception, and the like. These issues have now emerged as important, but I do not think we have clear ideas even now about how to deal with them, and in particular how to deal with them in a way that leaves the appropriate degree of flexibility as various networks consistent with the architecture are reduced to practice.

One could ask whether the principle of architectural minimality is the correct approach. Perhaps the architecture left too many problems for the designers, who then had to define the second-level mechanisms such as routing. Perhaps a more expansive definition of what we classify as architecture would lead to better outcomes when we deploy the resulting system. Alternatively, perhaps a different approach, with a different conception of what is minimally necessary, might lead to better outcomes. These mechanisms were designed based on our best intuition at the time, but it is reasonable today to rethink these decisions from scratch—what might the architecture do to better support goals such as security and management, which we dealt with poorly if at all in the 1970s. In the next chapter, I develop a framework (one of several in the book) that can be used to compare architectures, and then in chapter 7 I look at some different conceptions of what an internet architecture might be, again mostly with a preference for minimality but a very different view of what it is "on which we must all agree."

6 Architecture and Function

Introduction

Chapter 4, with its long list of requirements, may in fact distract the discussion from what is perhaps most central: that the network has to be fit for purpose—it has to perform a useful set of functions in support of the applications that run over it and the users that employ those applications. So before turning to the question of how the Internet (or a different possible internet with a different design) might address those various requirements, I want to start with the question of how we describe, in architectural terms, what a network does.

Computer scientists often use the word *semantics* to describe the functional capabilities of a system—the range of things it can do. However, what computer networks do is very simple compared (for example) to what an operating system or a database system does. The loose packet carriage model of "what comes out is what came in" is intentionally almost semantics-free. The packets just carry data from source to destination. Packet boundaries can have some limited functional meaning, but not much. The original design presumed some constraints that we might view as semantics, such as global addresses, but the progress of time has violated these, and the Internet keeps working. TCP does impose some modest semantic constraints, but of course TCP is optional and not a mandatory part of the architecture.

What defines the range of behavior that is available in the Internet is the *expressive power* of the packet header, which has to do with its format (what we might call its *syntax*) rather than any semantics. The various fields in the packet header specify how the packet is treated by the routers (and other elements) in the network; defining more or fewer fields changes the expressive power of the packet header. Most fields in the packet header (e.g., packet length) are unremarkable; some, like the Type of Service (TOS) field, have been redefined several times in the history of the Internet; some (like the optional fields) have atrophied; and the IP addresses have had a most interesting history in which the only constants are that they are 32 bits long (in the current version), that whatever value they have at each end must not change for the life of a TCP connection,[1] and that at any locale in the network, they must provide the basis for some router action (e.g., forwarding). They can be rewritten (as in NAT devices) or be turned into logical addresses (as in multicast or anycast). All that really seems to matter is that they be 32 bits long and that at any point they must have at least local meaning to a forwarding process.

The evolution in thinking with respect to IP addresses sheds some light on architectural thinking. The initial idea that addresses were drawn from a single, global address space and mapped uniquely to physical ports on physical machines turned out to be not a necessary constraint but just a simple model to get started. We were initially fearful that if we deviated from this definition, the coherence of the network would fall apart, and we would not be able to ensure that the Internet was correctly connected or debug it when it was not. Indeed, these fears are somewhat real, and it is possible today to mess with addresses in such a way that things stop working. But mostly the Internet continues to work, even with NAT devices and private address spaces, because the consequences of messing with addresses are restricted to regions within which there is agreement to assign a common meaning to those addresses. Those self-consistent regions need not be global; it is the scope of the self-consistent binding from addresses to forwarding tables that defines them.

At a minimum, each region could just be two routers, the sender and the receiver for each hop along the path of the packet. (This would somewhat resemble a scheme based on label rewriting, which is described in the Appendix.) Regions this small would be hard to manage without some sort of overarching framework for state management (and would have other drawbacks, as I discuss in the Appendix), but a single, global region—the starting point for the Internet design—has also proven to have complexities. In practice, the operational Internet has gravitated to regions that represent a rough balance among the issues that arise from big and small regions.

1. The source IP address is used at the receiving end to dispatch the packet to the right process. In addition, TCP computes a checksum over the packet to detect modification of the packet in transit. It incorporates parts of the IP header (called the pseudo-header in the spec) into the checksum. For this reason, the IP address in the packet cannot be changed without taking these limitations into account.

My point is that the format of the packet header is a defining feature of the Internet, in contrast to assertions about the semantics of addresses. It is for this reason that I focus on the *expressive power* of the packet header as a key factor in the specification of a network architecture.

Per-Hop Behaviors

We can generalize from this discussion of addressing and ask more abstractly about the local behavior of routers (and other network elements) and the resulting overall network function. In fact, the network is built up of somewhat independent routers. What applications care about is that the local behavior at a sequence of routers (the *per-hop behavior*, or PHB) can be composed to achieve some desired results end-to-end.[2] If the packets get delivered, which is really the only thing that defines today's properly operating Internet except in the context of defense against attack, then the details of how PHBs are configured (e.g., the routing protocols and the like) are a matter left to the regions to work out. The expectation about forwarding is a core part of the architecture; how routing is done is not. (If the packets do not get delivered, then debugging may be more or less a nightmare, depending on the tools for coordination and analysis, but this is a separate issue, which I address in chapter 13.)

Today, a router has a rather simple set of behaviors. Ignoring QoS and source routes for the moment, a router either picks (one or more) outgoing paths on which to forward a packet or drops the packet. The router can have as much state as inventive people define for it—static and dynamic forwarding tables, complex routing protocols, and static tables that define unacceptable addresses.[3] The router can also rewrite many parts of the packet header. But even today, and certainly looking to the future, not all elements in the network will be simple routers. Once elements receive a packet, they can perform any PHB that does not cause the end-to-end behavior to fail. So when we consider PHBs as the building blocks of network function, we should be careful not to limit ourselves to a model where the only PHB is forwarding.

2. The term *per-hop behavior* was coined as part of the effort in the IETF to standardize the mechanisms that underpin the Differentiated Services or *diffserv* mechanism, which is part of the scheme to implement enhanced QoS on the Internet (Nichols and Carpenter, 1998, section 4).

3. For example, so-called Martian and "bogon" packets, which contain addresses that are not allocated or are reserved.

Perhaps in introducing the term *expressive power* I have not actually said anything new. What is the difference between discussing the expressive power of an architecture and just discussing the architecture? I introduce expressive power to draw attention to and gather together the aspects of architecture that relate to its network function, as opposed to other aspects such as economic viability or longevity. It is equally important to conceptualize expressive power in the context of invoking PHBs. Some PHBs can be designed and deployed without any support from the architecture: we have added firewalls and NAT boxes to the current Internet more or less as extra-architectural after-thoughts. But thinking about expressive power in the context of invoking PHBs is a structured way to reason about both function and security. In the following sections, I will develop a taxonomy for how PHBs can be invoked, taking into account different possible alignments of interest among the various relevant actors. In chapter 10, I will use this taxonomy to reason about the security implications of an architecture.

Tussle and Alignment of Interest

One of the distinctive features of networks and distributed systems is that they are composed of elements whose interests are not necessarily aligned. These actors may contend with each other to shape the system behavior to their advantage. My co-authors and I picked the word *tussle* to describe this process (Clark et al., 2005). Sometimes one of the actors is a clear "bad guy" (e.g., someone wants to infiltrate a computer against the wishes of its owner). This tension leads to devices such as firewalls, which are an example of a PHB that is not simple forwarding but rather forwarding or dropping based on the content of the packet. Firewalls are an attempt by the receiver to overrule the intentions of the sender: a PHB that the receiver wants executed on the packet but the sender does not.

Sometimes the issues are not black and white but instead more nuanced: I want a private conversation, but law enforcement wants to be able to intercept any conversation with proper authorization. I want to send a file privately, but copyright holders want to detect whether I am serving up infringing material. To the extent these tussles are played out in the network (as opposed to in the end nodes or the courts), they will be balanced through the relative abilities of the different actors to exploit the expressive power of the design. So our discussion of expressive power, and the tools that implement it, will be strongly shaped by the reality of tussle. Looking at the balance of power created by a specific feature in the architecture is a way to integrate security considerations into the design process of an architecture.

Expressive Power and Per-Hop Behavior

As I said at the beginning of this chapter, most computer systems, such as operating systems or database systems, are characterized by a rather detailed specification of the functions they can perform—what I called the semantics of the system. However, the functional capabilities of the Internet are not defined by its specification. If (in general) a network element can be programmed to do a wide range of things as its PHB, then the resulting overall capability is the result of the execution of these PHBs in some order, where the execution of the PHBs is driven by the fields in the packet header (the expressive power), and the order of execution is defined by the routing of the packet among these devices. Of course, since the devices themselves can modify the routing, the resulting computational power could be rather complex. Computer scientists are accustomed to thinking about the implications of semantics: the limitations of some semantic construct. We are less accustomed (and less equipped with tools) to thinking about the expressive power of a packet header—what functions are consistent with some format and syntax. It is sort of like asking what ideas can be expressed in sentences of the form "subject, verb, object." The question seems ill-defined and unbounded. Even harder is to catalog what cannot be expressed. But this question is the one that actually captures the limits of what the Internet can and cannot do, so we should try to think about how to think about it.

This view of packet processing has not been seriously explored,[4] because in the Internet of today, the overall function we want to achieve is very simple—the delivery of the packet. If that is the desired overall function, there is not much demand for the complex chaining of arbitrary PHBs within the network, but as we think about wanting to do more complex things as a packet moves from source to destination (many having to do with security), the range of interesting PHBs will grow (see chapter 7 for examples), so it is worth considering what factors define or limit the expressive power of a network.

In this section, I pose a three-dimensional framework to describe PHB execution: *alignment of interests, delivery,* and *parameterization.*

4. With the exception of some of the active network research, which I discuss here and on page 140.

Alignment of Interests

The first dimension of the model is to capture the relationship between the sender of the packet and the owner of the element that implements the PHB. This dimension directly captures the nature of tussle. I will propose two cases: *aligned* and *adverse*.

Aligned In this case, the interests of the sender and the element match. Simple routing, multicast, and QoS, for example, usually fall in this obvious class. The sender sent the packet, the router forwards it, and this is what both parties expected.

Adverse In this case, the PHB performs a function that the sender does not want. A firewall is a good example here, as would other sorts of content filtering, deep packet inspection, and logging.

Delivery

The second dimension of the model is to ask why or how the packet arrives at the element that implements the PHB. There is a simple four-case model that covers most of these circumstances: delivery is either *intentional, contingent, topological,* or *coerced.*

Intentional In this case, the packet arrives at the element because it was specifically sent there. For example, with source routes, the destination address is actually a series of addresses, each of which directs the packet to the next addressed router. As another example, a packet arrives at a NAT box because it was intentionally sent there.

Contingent In this case, the packet may or may not arrive at a given device, but if it happens to arrive, then the device will execute the PHB. This is the basic mode of datagram operation—if a router gets a packet, it forwards it. There are no preestablished paths from source to destination (which would be examples of intentional delivery). Each router computes routes to all known destinations, so it is prepared to respond if a packet happens to arrive.

Topological In this case, there is nothing in the packet that causes it to arrive at a particular device, but instead the topology of the network (physical or logical) is constrained to ensure that the packet does arrive there. Firewalls are a good example of topological delivery. The sender (assuming he is malicious) has no interest in intentionally sending his attack packet to a firewall. He would prefer to route around the firewall if he could. The receiver wants some assurance that the firewall will be in the path. The receiver will not normally be satisfied with contingent protection, so the remaining tool available is to constrain the connectivity or routing graph so that the only path or paths to the receiver pass through the firewall.

Coerced This is a special case of intentional or topological delivery, in which the sender is compelled to subject itself to a PHB even though the interests of the sender and the owner of the PHB are adverse. An attacker attempting to reach a machine behind a NAT box has no choice but to send the packet to that element—there is no other means of reaching beyond it. In this case, we can expect the sender to cheat or lie (in terms of what values are in the packet) if it is possible.

Parameterization

The third dimension of the model is that the packet triggers the execution of a PHB, and the data in the packet is the input to that PHB, like arguments to a subroutine. The values in the packet are the input parameters to the PHB, and if the PHB has the effect of modifying the packet, this is similar to the rewriting of arguments in the invocation of a subroutine. (Using the vocabulary of programming language, the parameters in the packet are passed to the PHB by reference rather than by value.) The element that executes the PHB can have lots of persistent state (which can be modified as a result of the PHB) and can have distributed or more global state if suitable signaling and control protocols are devised.

In this context, I will again offer two cases, *explicit and implicit,* although these define ends of a spectrum more than they do distinct modes:

Explicit While the PHB can in principle look at any data fields in the packet, in common cases there will be specific fields set aside in the header as inputs to specific PHBs. This is the common case for packet forwarding; since packet forwarding is the basic operation of networking, there is an explicit address field used as input to the forwarding lookup. IP was designed to support QoS, so there is an explicit field in the packet (the ToS field) that is the input parameter to the QoS algorithm.

Implicit In other cases, there is no specific field used as input to the PHB; the PHB looks at fields intended for other purposes. Firewalls block packets based on port numbers, some ISPs assign QoS based on port numbers, and packets are sometimes routed based on port numbers (e.g., when a Web query is deflected to a cache or outgoing mail is deflected to a local mail server). If the PHBs have state, they can also base their actions on implicit information such as the arrival rate of packets.

Implicit parameters can be expensive for the router to process. In the worst case (deep packet inspection), the PHB may process the entire contents of the packet as input to its operation. Clearly, a PHB that uses implicit arguments will not be as efficient as one based on a specific header field (e.g., an address field), so implicit arguments must be used sparingly, but in the case of adverse interests, implicit parameters may be the only option.

This model suggests that there is some rough analogy between the expressive power of a network and a programming language of some sort, where the computation is a series of subroutine executions, driven by the input parameters carried by the packet, and where the order of execution is defined by the routing protocols, together with the expressive power of the packet to carry the addresses that drive the forwarding. Of course, the addition of tussle and nodes that are hostile in intent with respect to the sender adds a twist that one does not find in programming languages, and in fact this twist may be one of the most important aspects of what the network actually computes. So the power of an analogy to a programming language remains to be explored.[5]

This taxonomy classifies activity based on the alignment of interests among the senders and the PHBs in the network. Another way to classify activities is to look at the alignment of interests between sender and *receiver*. In the case where the interests of the sender and receiver are aligned, the PHBs would normally be a part of providing the desired service unless they are inserted into the path by an actor with interests adverse to the communicating parties. They are *functional* in that the application being used by the communicants is invoking them as part of the service. (While only the

5. This idea by no means originated with me. In an early paper with the delightful title of "Programming Satan's Computer" (Anderson and Needham, 2004), the authors observe that "a network under the control of an adversary is possibly the most obstructive computer which one could build. It may give answers which are subtly and maliciously wrong at the most inconvenient possible moment." Their focus is on the design of cryptographic systems, but their point is more general: "In most system engineering work, we assume that we have a computer which is more or less good and a program which is probably fairly bad. However, it may also be helpful to consider the case where the computer is thoroughly wicked, particularly when developing fault tolerant systems and when trying to find robust ways to structure program and encapsulate code."

sender can directly control the sending of the packet and its contents, there are certain architectures, which I discuss in chapter 7, where the receiver as well as the sender can directly exercise control over what PHBs are applied to the packet.)

If the interests of the sender and receiver are aligned, then the resulting questions are first whether the architecture is providing support to functional PHBs through some aspect of its expressive power (delivery, parameters, etc.) and (the negative aspect of the analysis) whether the architecture needs to provide support to protect the communicants from the misuse of this expressive power, and whether the architecture needs to provide support for the task of detecting and isolating a faulty or malicious element. (See page 100 on debugging and chapter 11 for a discussion of fault diagnosis.) If there is an adverse PHB in the path, it must be there because a third party (e.g., an ISP or a government authority) has interposed it or because the network itself has previously suffered an attack such that some of its elements have been taken over by an attacker.

If the interests of the sender and receiver are not aligned (in which case the receiver either wants protection during communication or does not want to receive traffic at all), then the PHBs are serving a different purpose: they are deployed to protect the receiver from the sender, a role that creates different potential roles for the architecture. I will return to security analysis in chapter 10.

Pruning the Space of Options

What I just described is a $2 \times 4 \times 2$ design space, but in fact it is less complex than that. The method that helps to sort out this space is *tussle analysis*, which starts with understanding the alignment of interests.

Aligned If the sender wants the PHB to be executed, then intentional delivery and explicit arguments make sense. Contingent delivery may be suitable in some cases (e.g., the basic forwarding function), but explicit arguments (e.g., the address field) still make sense.

Adverse If the sender does not want the PHB to be executed (it represents the interests of the receiver, not the sender), then the receiver cannot expect the sender to provide any explicit arguments to the PHB. In this case, the PHB must depend on implicit arguments. Also, the PHB cannot count on intentional delivery, so coerced delivery is the best option, with contingent or topological delivery as a fallback.

Some Examples

NAT boxes Network address translation devices (NAT boxes) implement a PHB that is not simple forwarding but includes rewriting of the source or destination address field. They are a wonderful example of how one can disrupt two of the most fundamental assumptions of the original Internet and still have enough functions mostly work that we accept the compromised functionality. The assumptions of the original Internet were that there was a single, global address space, and there was no per-flow state in forwarding elements. NAT boxes have per-flow state, but early NAT devices, lacking a protocol to set up and maintain soft state, depended on a trick: they used the first outgoing packet to set up the state, which then persisted to allow incoming packets to be forwarded. This trick does not allow state to be set up for services waiting for an incoming packet that are behind the NAT box. (The IETF has subsequently developed protocols to allow an end node to open a port to a service behind the NAT device.[6])

NAT boxes are an example of intentional delivery with explicit parameters (the addresses and port numbers). If the interests of the end points are aligned, NATs are normally a small nuisance; if the interests are not aligned, they provide a measure of protection, and in that respect fall into the *coerced* category.

Firewalls Firewalls, as I described earlier, are an example of a PHB that is adverse to the interests of the hostile sender (potential attacker) and thus must depend on implicit information. The firewall has the poorly defined task of trying to distinguish good behavior from bad, based on whatever hints can be gleaned from the packets. Normally, all a firewall can do today is a very crude set of discriminations, blocking traffic on certain well-known ports and perhaps certain addresses. The roughness of the discrimination is not necessarily a consequence of the details of the current Internet but perhaps results from the intrinsic limits of making subtle discriminations based only on implicit fields in the packets.

This outcome is not necessarily a bad thing. Sometimes users want the blocking to succeed (when they are being attacked) and sometimes they want it to fail (when some third party, such as a conservative government, is trying to block their access to other sites on the Internet). If we decide to make the job of the firewall easier, we should consider whose interests we have served.

6. The port control protocol (Wing et al., 2013) and the Internet gateway device protocol, part of the UPnP protocols (Open Interconnect Consortium, 2010), allow an end node to set up a new port mapping for a service on the end node.

Tunnels and overlay networks It is possible to take a packet and encapsulate it inside another packet, a technique called packet encapsulation. This PHB causes the packet to be sent to the destination address in the outer packet, not the inner packet. When the packet reaches that destination, a PHB there must either remove the outer packet or reencapsulate the inner packet in a new outer packet. The term *tunnel* is used to describe what happens to the encapsulated packet: the packet reaches the PHB that does the encapsulation (the beginning of the tunnel) and reappears at the second device.

A set of devices across the network can collectively tunnel packets in a way that overrides the default routing computation of the network; the result is called an overlay network. One reason to do this is that the overlay routers may be able to find routes that are more efficient or more resilient than those found by the default routing computation (Andersen et al., 2001).

While tunnels provide a way to control the routing of a packet, more generally they provide a way to interpose an explicit element in the path toward a destination. The encapsulated packet is the explicit information used as input to the end point of the tunnel. Sometimes the starting point of the tunnel is contingent or topological, sometimes it is coincident with the sender, and sometimes it is intentional. For example, The Onion Router (TOR) is an example of nested tunnels, each with explicit information as input to the PHB at each TOR forwarder.[7]

Tussle and Regions

Consider the example discussed earlier of a firewall put in place by the receiver to block attacks by the sender. In this adverse circumstance, the receiver must depend on implicit arguments and topological delivery (or coerced delivery if the architecture permits). For this to work, the region of the network within which the receiver is located must provide enough control over topology (connectivity and routing) to ensure that the firewall is in the path of the packets. The receiver must have sufficient control over

7. TOR, or The Onion Router, is a set of servers scattered across the Internet, with the goal of allowing anonymous communication between parties. The clever use of nested encryption, and the indirect forwarding of a message through a number of TOR nodes between the sender and the receiver, hides the identity of the sender from the receiver. Each TOR node implements a PHB that peels off a layer of encryption (hence the name—an analogy to peeling the layers of an onion). For information on TOR, see https://www.torproject.org/.

this region of the network to make sure that the topology is as desired, and enough trust in the region to be confident that the routers will forward the traffic as requested.

To generalize, what this illustrates is that different actors within the network (the sender, the receiver, the ISPs, other third-party participants) have the right to control certain parts of the network, and within each such region of the network, the PHBs found there will be used to further the intentions of that actor. The factor that will determine the outcome of the tussle (e.g., the balance of power) is not the PHBs (which can be more or less anything) but rather the information in the packet that can serve as the input to the PHB, and the order in which the packet is routed to the PHBs.

The order of processing arises from the nature of packet forwarding: the packet originates in the region of the sender (who thus gets first crack at invoking any desired PHBs), then enters the global network, and finally enters the region of the receiver and the PHBs found there. The information that is in the packet at each stage is a consequence of this ordering. For example, the sender can include data in a packet that is used by the PHBs in the region of the sender and then stripped out so that the other regions cannot see it. While the packet is in the global middle region, some or most of the packet can be encrypted to prevent it from being examined, for example.

But as I have noted, PHBs can do more or less anything that can be derived from the information in the packet, and the routing is under the control of each of these regions. The fixed point in this design is the packet header itself, so when we think about putting more or less expressive power into the header (e.g., a more or less expressive format), we should consider whether the different options shift the balance of power in ways that match our preferences. My preference for architectural minimality and also my concerns over security lead me to conclude that while adding expressive power to the header may be very beneficial, the option should be used sparingly.

The design of applications can also shift the balance of power among the various actors. An application can by design insert higher-level services into the pattern of communication. The design of the email application specifies that mail is sent to a mail forwarding agent, and the packets are addressed to that element—intentional delivery. In this case, especially where the packets of an email are assembled into a larger unit for processing (an application data unit, or ADU), the explicit parameters used by the service are in the body of the packets, not the packet header. That sort of data is not part of the architecture—it is not something on which there has to be agreement; quite the opposite. An ISP might intercept the communication using topological delivery in order to inspect (e.g., deep packet inspection) or modify the contents, an intervention that is thwarted if the data is encrypted. I consider this sort of tussle in chapter 10, but from the point of view of balance of control, I believe that a network operator should have to offer a very strong justification for inserting a service into a communication where neither end point requested or expected it to be there.

Generality

I have been talking about PHBs in a rather abstract and general way. As I have used the term, it could equally apply to a low-level function like forwarding, an overlay service, tunneling, or an application-specific service like content reformatting or detection of malware. The taxonomy of delivery modes, alignment of interests, and parameter modes applies equally to both packet-level PHBs and higher-level services. One could use the term *PHB* generally to mean any service element that is inserted into a data flow or restrict the term to lower-level functions like firewalls or routers. Since my goal is to discuss the role of architecture, I am most concerned with cases where a PHB justifies adding to the expressive power of the packet. In general, application-level services would not fit into this category, but this is a presumption, not a given, as some architectures directly support the insertion of application-level services into the flow of packets and might provide explicit fields in the packet header for these PHBs.

Architectural Alternatives for Expressive Power

Using the lens of expressive power, here are a few architectural concepts that would change (usually enhance) the expressive power of a design. Some of these have been incorporated into the alternative architectures I discuss in the next chapter. I mention some of them briefly here.

Addressing

It is generally recognized that the current approach of using the IP address both as a locator and as an identifier was a poor design choice. Mobility is the obvious justification for this conclusion. In today's Internet, dealing with mobility is complicated by the fact that the IP address is used both for forwarding and for end-node identity. Separating these two concepts into two different data fields in the packet would allow the location field (e.g., the input to the forwarding PHB) to be changed as the mobile host moved.

This division does not solve the two now separate problems: keeping the location information up-to-date and making sure the identity information is not forged. Linking identity to location provides a weak form of security: if two machines have successfully exchanged packets, the location is sufficiently unforgeable that it can stand as a weak identifier, but by separating the two problems, each can be resolved separately and managed differently in different situations as circumstances require.

An alternative design approach might result in two fields, or perhaps three, each serving a distinct purpose.

- Locator: This field is used as input to the forwarding PHB of a router. For example, it may be rewritten (as in a NAT device) or be highly dynamic (in the case of a mobile device).
- End-node identifier (EID): This field is used by each end of the connection to identify itself to the other end(s). There are in general three issues with such a field: how to make sure a malicious sender cannot forge a false identifier, how each end associates meaning with this field (is there some sort of initial exchange of credentials associated with the EID, or do high-level protocols associate some meaning with it once the connection is in place?), and whether elements other than the end nodes (e.g., PHBs in the network) should be allowed to see and exploit this value.
- In-network identifier (INID): If the EID is private to the end nodes of a connection, then an architecture might define some other identifier that can be seen and used by PHBs in the path from the sender to the receivers, perhaps to support accounting or authorization for enhanced services. This possibility in turn raises many subquestions, such as how the INID is obtained, whether there are security issues associated with its use, and the the duration for which it is valid.

So while the general idea of the locator-identity split is well- understood, there is no clear agreement on how to design the system that results. Most of the architectures that I will discuss in chapter 7 implement some sort of location-identity split and illustrate the range of approaches that have been taken to address this issue.

Increasing the Expressive Power of a Design

If there seems to be some value (some increase in function or generality) to the ability to provide richer input data to a PHB, it is worth at least briefly speculating on how this might be done. I have argued that since a PHB can in principle compute almost anything, the expressive power of an architecture will depend on what arguments can be presented to the

PHB—in other words, what data is in the packet header. I will quickly sketch a few options.

Blank "scratch pad" in the packet A simple idea is to leave a fixed, blank area in the packet header, to be used from time to time. One need only look at all the creative ideas for reuse of the fragment offset field in the current IP header to appreciate just how powerful a little extra space can be.[8] One problem with the IP option was that it assumed contingent rather than intentional delivery. The sender would send the packet addressed to the destination, and all routers along the path were to look at the options to see whether they were supposed to process the option. If the design dictated that only intentionally addressed PHBs would look at the field, this would avoid the performance issues that arose with the IP option field. Only elements that have the specific requirement for an input value would parse the field, and the packet would be addressed to them. The drawback of this scheme is that there might be a conflict among different potential uses of the scratch pad, so we might consider a more complex scheme.

Pushdown stack model A more complex model for explicit data in packets is a pushdown stack of explicit data, carried as part of the packet header. A pushdown stack is a common structure in computer science. In a stack, values are "pushed" onto the top of the stack and removed (or "popped") from the top, so a stack provides "last in, first out" behavior, as opposed to a queue, which provides "first in, first out" behavior. In this model, the packet is explicitly directed by the sender to the first element that should perform a PHB. That element (conceptually) pops the first record off the stack of explicit information and uses it as input to the PHB.[9] Then, using either stored PHB state or information in the record that was just popped off the stack, it identifies the next element to which the packet should go. This PHB can push a new record onto the stack or leave the one provided by the original sender, depending on the definition of the intended function.

This sort of mechanism seems to build on the rough analogy between PHB sequencing and some sort of programming language, and packet

8. One feature of the original IP was the ability to fragment the packet into parts to allow them to be carried over networks with small packet sizes. The function of packet fragmentation is no longer used, which left an unused field in the packet header.

9. Issues of performance would suggest that the design would not literally pop a record off a stack, thus shortening the packet and requiring that all the bytes be copied. A scheme involving offset pointers would achieve the desired function.

encapsulation is a rough version of a push-down mechanism, in which the whole header is "pushed" onto the stack by the encapsulating header. A related use of a push-down stack in the header can be found in two of the architectures I will describe in chapter 7, i3 and DOA, which use a pushdown stack to carry the sequence of IDs that order the sequence of PHB executions.

A heap The proposal for Role-Based Architecture (RBA) (Braden et al., 2003) (part of the NewArch project) is perhaps the best example of an architecture that captures the idea of general PHBs and the expressive power of a packet header. In this proposal, PHBs are called roles, and the input data to each node is called the role-specific header, or RSH. The packet header is described as a *heap* of RSHs. The implication of the term *heap* is that the roles are not necessarily executed in a predetermined order, so the idea of push and pop is too constraining. RSHs are an example of an explicit argument. The proposal discusses both intentional and contingent delivery, where the intentional addressing is based either on the ID of a role broadly defined or only at a specific node. The paper does not delve into tussle to any degree or work through the case of roles that are adverse to the interest of the sender, so there is not much attention to implicit arguments or topological delivery. However, the idea of a network as a sequence of computations performed on a packet based on explicit input arguments is the core concept of role-based architecture.

Active networks The concept of active networks, initially proposed by Tennenhouse and Wetherall (1996), was that packets would carry small programs that the routers would execute in order to process the packet. In other words, the packet rather than the router would define the PHB. This idea may well define the maximal end point on the spectrum of expressive power, in contrast to the current Internet, which has only a small number of fixed fields in the packet to express how the packet should be handled. I defer the discussion of active networks to the next chapter.

Per-Flow State

I have described the current Internet as more or less a pure datagram scheme, in which each packet is processed in isolation with no per-flow state in the router, so all the flow-specific parameters to the PHB must come from the packet header. By linking the treatment of different packets in a sequence, per-flow state in the router can enrich the range of PHBs that can be designed. However, the option of per-flow state raises the question of how the state is established and maintained.

Signaling and state setup In the original Internet, the designers avoided any setting up of per-flow state in the routers. There were several reasons for this preference. One was simplicity—if we could do without it, we would avoid yet another thing that could go wrong. In particular, once per-flow state is instantiated in a router, it has to be managed. Questions such as when it should be deleted or what happens if the router crashes have to be answered. The simplicity of the stateless model makes it easier to reason about resilience and robust operation.

Another reason to avoid per-flow state in the routers involves the packet exchanges and resulting latency to set it up. It seems a waste of effort to set up state for an exchange that may involve only one packet. It would be much better to have a system in which the sender can "just send it." But if this works for one packet, why not for all the packets?

However, control messages can be an important aspect of the expressive power of an architecture. Per-flow state might only be needed in specific elements to deal with special cases. We are also now dealing with per-flow state (e.g., in NAT boxes) whether we design for it or not. Some of the architectures in the next chapter depend on per-flow state, so it seems worth revisiting this design decision.

State initiation bit If we are prepared to consider per-flow state as part of the design, we need to consider whether the protocols should include a standard way to establish and maintain this state. The original preference in the Internet design was to avoid an independent control plane as a mandatory component of the network. (Of course, there is no way to prevent parties from attaching controllers to the network if they choose to, but these would not be a part of the architecture.) The original design preference was to carry control information (to the extent that it existed at all) by using fields in the data packets that flowed along the data forwarding path. It is possible to imagine a similar scheme as a standard means for an end node to establish and maintain per-flow state in intermediate elements. Such an idea would enrich the expressive power of the packet header by building the idea of state establishment into the design, which would link the treatment of a sequence of packets.[10]

10. A related activity in the IETF is SPUD, an acronym variously expanded as session protocol underneath datagrams, substrate protocol for user datagrams, or session protocol for user datagrams. Like any protocol that creates a control/communication path between end nodes and the network, SPUD raises security questions, which received attention as a result of the Snowden leak (Chirgwin, 2015).

One can imagine that just as TCP has a state-establishment phase and a connected phase, protocols that establish state in intermediate elements could follow the same pattern. A bit in the header (similar to the SYN indicator in the initial packet of a TCP connection) could signal that the packet contains state-establishment information. This packet might require more processing overhead (and thus represent a vector for DDoS attacks) but in normal circumstances would only be sent at the initiation of a connection. Once the state is established, some much more efficient explicit indication in the packet could link subsequent packets to that stored state. The two sorts of packets could have different formats.

Maintaining state in intermediate elements Per-flow state in routers could either be *hard state* (state that persists until it is explicitly deleted) or *soft state* (state that can be reconstituted if it is lost). Soft state has the advantage that if the mechanisms that are intended to delete the per-flow state fail, the state in the routers goes away after a while as opposed to persisting forever. Assuming that the state is soft-state (a choice that could be debated), the protocol must include a means to reinstate the soft state if it is lost. One could imagine a new sort of error message signaling that some expected state is missing. To recover from lost state, the sender would have to transition back from a fully connected mode into a state-setup mode. The sender could reestablish the state in two ways. First, it could do so from scratch by sending whatever initial information was used. Second, the intermediate node that holds the state could send back to the source a bundle (perhaps encrypted) of state information that could be resent from the source on demand to reestablish the state efficiently. Such a scheme might make sense in the special case of intentionally sending a packet to an anycast address. In this case, the sender is sending to a logical service, but the actual physical machine implementing the service might change. In this case, it might be necessary to reestablish some state in that box.

In-network state associated with receivers The preceding discussion above covered the case of a sender establishing state along a path as part of session initiation, but an equally important case is state set up along a path that comes from the receiver rather than the sender. Setting up and maintaining this state is actually the trickier part of the scheme.

As an illustration of the problems, consider the case where a receiver, as part of protecting itself from attack, outsources connection validation to a set of devices running a PHB. Since a sender (either legitimate or malicious) may connect to any one of these devices (perhaps using an anycast address), every one of these devices must have available the information necessary to validate all acceptable senders or else there must be an authentication protocol for those devices to send off credentials to a back-end service. At a minimum, the protection devices need to be able to find this service. In practice, this pattern sounds more like hard state, somewhat manually set up and torn down, rather than dynamic soft state.

In other cases, soft state may make more sense. A transient service behind a "firewall of the future" may want to open an incoming port temporarily (assuming that a future network has ports, of course), which could be done by dynamic setup of soft state. In this case, mechanisms would need to be provided to make sure the state remains in place for the necessary period, even when no packets are being exchanged.

PHBs and Control of Network Resources

With the exception of architectures that allow for complex PHBs, the objective of the network is very simple—deliver the bits. But a necessary component of delivering the bits is that the network has to manage its resources to that end. These functions, which come under the heading of control and management, are critical but less well researched than the actual forwarding of data. In the early days of the Internet, just getting the packet forwarding right was so challenging that we did not have much attention left over to think about network control. As a result, a key control issue—network congestion and our inability to control it effectively—was a major impediment to good network performance (actually delivering the bits) until the mid-1980s, when Van Jacobson proposed a congestion control scheme that is still in use today (Jacobson, 1988). Since then, there has been a great deal of research on congestion control algorithms, but the relationship between architecture and network control is poorly understood.

An important component of a router's PHB is the manipulation of data related to the management and control of the network. A router performs tasks that are fairly obvious, such as counting the packets and bytes it forwards. The PHBs related to system dynamics, such as congestion control, may be more complex, and are shaped by what data the router retains about the traffic it is forwarding. I will return to this issue, and the relation of architecture to network control and management, in chapter 13.

Debugging

All mechanisms fail, and complex mechanisms fail complexly. If we design a network that permits all sorts of complex routing options and invocation options for PHBs, the potential for failure will certainly go up. Tools to debug and recover from such failures will be critical in meeting the goals of availability and usability.

PHBs that are contingent are the hardest to debug, since the sender did not invoke them intentionally. The idea of trying to diagnose a failure in a box the sender did not even know about is troubling. This fact suggests that when effective diagnosis is desired, the design should prefer intentional invocation of PHBs.

If the interests of all parties are aligned, it would make sense that the tools for debugging would be effective and useful. However, if the interests of the parties are adverse, the situation becomes more complex. If an attacker is being thwarted by a firewall, it may be in the interest of the firewall to prevent any sort of debugging or diagnosis of the failure. The goal (from the point of view of the defender) is to keep the attacker as much as possible in the dark as to what is happening in order to prevent the attacker from sharpening his tools of attack. So while tools and approaches for debugging and diagnosis must be part of any mechanisms to provide expressive power for a future internet, designers of these mechanisms must keep tussle in mind.

Certain classes of failure are easy to debug, even for contingent PHBs. Fail-stop events that cause the element not to function at all can be isolated and routed around just like any other router failure. "Fail-go" events do not require diagnosis. It is the partial or Byzantine failures (see page 215 in chapter 10) of a contingent PHB that may cause diagnosis problems for the sender. This is why intentional invocation of PHBs is preferable unless the goal of the PHB is to confound the sender.

Expressive Power and Evolvability

In this context, the term *evolvability* refers to the ability of the network architecture to survive over time and evolve to meet changing needs while still maintaining its core coherence. Chapter 9 explores this issue in depth, but here I consider the relationship between the expressive power of an architecture and how that architecture may evolve over time. The history of the Internet provides some informative case studies.

I discussed earlier the evolutionary process in which the Internet mutated from having a single, global address space to a number of private address spaces connected using NAT devices. By and large, the Internet has survived the emergence of NAT, and perhaps global addresses did not need to be such a central assumption of the presumed architecture. Perhaps less mourned but more relevant is the atrophy of IP options. They were developed to allow for future evolution of the architecture, and they could have provided a substantial degree of expressive power. However, IP options were costly to process and were deprecated in practice to the point where they are essentially gone. One could speculate about the implications of this fact:

- Perhaps this degree of expressive power was not in fact necessary and made the network too general.
- Perhaps IP options were not well-designed, and required much more processing than a better-designed option.
- Perhaps the loss of IP options represents an unorchestrated decision to favor short-term cost reduction over future evolvability.

However, at the same time that we have seen IP options atrophy, there have been any number of proposals to add some new functionality to the Internet by repurposing underused fields in the IP header, in particular the fields related to fragmentation. This behavior suggests that some additional expressive power in the header would have been of great benefit.

Whatever the mix of actual reasons is, one can learn two lessons from the preceding discussion. First, avoid mechanisms that are costly to maintain unless they are needed. For example, if there are fields in packets that are used to carry extra input values to PHBs, design them so that only the device that actually implements the PHB has to parse those fields or otherwise pay any attention to them. If packets are intentionally addressed to devices, then the processing rule is clear: if the packet is not sent to you, don't look at the extra fields.

Second, any mechanism added to a packet header should have at least one important use from the beginning in order to make sure that the implementation of the mechanism remains current. If designers propose something intended to facilitate evolution but cannot think of a single use for it when it is proposed, perhaps it is overkill and will atrophy over time.

Finally, the addition of tools to promote evolvability may shift the tussle balance, so enthusiasm for rich expressive power may need to be tempered by a realistic assessment of which actors can exploit that power. Indeed, it would seem that the goal of evolution over time is inseparable from the goal of operating in different ways in different regions of the network at the same time, in response to different perceived requirements within those regions.

Making design choices about the potential expressive power of a future internet seems to call for a trade-off between evolvability and flexibility on the one hand, simplicity and understandability on the second hand, and tussle balance on the third hand. However, there is no reason to think that this trade-off is fundamental. Creative thinking might lead to alternative ways of defining packets and routing such that we gain in all three dimensions. To explore this space, it may be helpful to ask ourselves challenging questions of the sort that a clean-slate thought process invites, such as why packets must have addresses in them or why we need routing protocols.

PHBs and Layering

There is a convenient fiction that some Internet architects, including me, like to propagate, which is that there are a limited set of functions that are "in" the network but that most of the elements that we find intermediating communication today (so-called middleboxes) are somehow "on" the network but not "in" it. This fiction lets us continue to argue that what the Internet itself does (and what as architects we might be responsible for) continues to be very simple, and the "middlebox mess" is someone else's problem. It is not hard to argue that complex services such as content caches are not "in" the network, but things like firewalls and NAT boxes are harder to ignore.

One basis for defining a service as "on" or "in" is which actor operates it. ISPs operate the Internet, so if the element is not under the control of an ISP, how can it be "in" the network? An ISP might correctly say that since it cannot be responsible for an element that it does not operate, and since the ISP has the responsibility for making sure that the packet carriage function continues to work even if such services fail, these other services must be at a higher layer. Indeed, sorting different PHBs along an axis of which depend on which is a good design principle. Very few network operators would allow an element (with its PHB) that they do not control to participate in the routing protocol of the region, for the same reason that the early design of the Internet did not anticipate that hosts would participate in the routing protocols. (This view is consistent with the architectural goal of minimal functional dependency, which I discussed in chapter 3.) An ISP that is providing a packet carriage service should work to ensure that the availability of that service does not depend on any element over which it has no control. However, the taxonomy of how PHBs are invoked (modes of delivery, parameters, etc.) is a cleaner way to classify different PHBs than "in" or "on." As I look at the relation between architecture and economic viability in chapter 12, I will argue that design of expressive power will in fact shape which actor is empowered to play one role or another in making an Internet out of its parts, which is a critical factor in the economic viability of an architecture. These design alternatives, which derive from such things as intentional versus contingent delivery, are the important issues, not a vague concept of "in" or "on."

7 Alternative Network Architectures

Introduction

Talking about network architecture in the abstract can seem, in a word, abstract. Chapter 5 used the Internet as one case study, but it is useful to have more than one example to draw on. Having several examples helps to bring into focus which differences are fundamental and which just a secondary consequence of some other design decision.

The motivation for this book arose in the context of the U.S. National Science Foundation's Future Internet Architecture (FIA) project and its predecessors. However, this program was not the first moment when the research community considered alternative network architectures. There have been a number of proposals going back at least 25 years that have looked at different requirements and proposed different architectural approaches. In this chapter, I review a selection of these proposals for a new internet architecture.

In chapter 4, I discussed a range of potential requirements that might shape an internet architecture. In the next section, I organize each of these proposals according to the most important requirement that it sets out to address. Then I look at several important architectural features (separation of identity and location, the service model, how they shape the role of the ISP, functional dependencies within the architecture, dealing with actors that have adverse interests, and expressive power) and compare a selection of the architectures with respect to these features. In subsequent chapters, I will focus on specific requirements and elaborate on these comparisons.

Most of these proposals were put forward as thought experiments within the research community. A few of them, including the FIA proposals, were demonstrated as running prototypes. One way ideas such as this can take hold in the real world is as an overlay on the current Internet. Overlay networks are discussed in chapter 6 on page 90. In most cases, the creators did not expect that their ideas would literally lead to a new Internet; the current Internet is too widely deployed. In a few cases, the proposals have influenced the evolution of specific aspects of the Internet. In my view, this sort of influence is a beneficial outcome. A vision of the future, even if embodied in a specific proposal that is not achievable in practice, can nudge the deployed Internet in a particular direction, so the impact of this work is often indirect, but there is impact.

In many respects, architectural proposals are creatures of their time. Since the Internet has proved quite resilient to changing requirements over time (a point I consider in chapter 9), it is interesting that many of the proposals I discuss here were motivated by a concern that the Internet could not survive some change in requirements. As I have noted, both I and many of the architectural designers discussed here have a preference for architectural minimality, but that minimality is shaped extensively by the set of requirements chosen to address. Indeed, the landscape of requirements has changed over the years, which is reflected in the architectural responses.

A Review of Alternatives

In this section, I briefly summarize a selection of proposals for an alternate internet architecture, looking at the NSF-funded projects that were part of their FIA program, projects funded by the European Commission, and research starting from the 1990s. I have grouped them based on the most important requirement that they were designed to address.

Regional Diversity in Architecture

As of today, the Internet, with its particular format for packets, dominates the world. As I discussed in chapter 2, this outcome was not always certain. In the 1980s and into the 1990s there was much less confidence in the research community that the Internet architecture would prevail as a single, global solution. There were competing architectural proposals, including both the alternative packet-switched OSI protocols from the ISO and the cell-switching Asynchronous Transfer Mode (ATM). The assumption that these different proposals might have to interoperate drove the development of a number of higher-level architectural frameworks that were intended to allow different architectures, each running in a region of an internet, to be hooked together to provide an end-to-end delivery service supporting a general range of applications. I describe four proposals in this category—the Metanet, Plutarch, Sirpent, and Framework for Internet Innovation (FII)—and then compare these schemes with the current Internet.

Metanet About 20 years ago, the Metanet proposal (Wroclawski, 1997) clearly laid out the requirements for a network of heterogeneous regions. Here are some quotations from the Metanet white paper:

> We argue that a new architectural component, the region, should form a central building block of the next generation network.
>
> The region captures the concept of an area of consistent control, state, or knowledge. There can be many sorts of regions at the same time—regions of shared trust, regions of physical proximity (the floor of a building or a community), regions of payment for service (payment zones for stratified cost structures), and administrative regions are examples. Within a region, some particular invariant is assumed to hold, and algorithms and protocols may make use of that assumption. The region structure captures requirements and limitations placed on the network by the real world.
>
> [D]ata need not be carried in the same way in different parts of the network—any infrastructure which meets the user's requirements with high confidence can be used to construct a coherent application. Packets, virtual circuits, analog signals, or other modes, provided they fit into a basic service model, are equally suitable. The overall network may contain several regions, each defined by the use of a specific transmission format.
>
> Paths of communications must thus be established in a mesh of regions, which implies passing through points of connection between the regions. We call these points waypoints.
>
> Three essential aspects of the Metanet are a routing and addressing system designed for region-based networks, end-to-end communication semantics based on logical, rather than physical, common data formats, and abstract models for QoS and congestion management, mapped to specific technologies as required.

While the Metanet white paper lays out these requirements, it does not propose a specific architectural response—this is posed as a research agenda. But the last paragraph of the quotation captures the essence of Wroclawski's thinking. While there is no need to have a common packet format across a global internet if there are rules for conversion, these rules must be clear and must capture what must be in common across the regions—the definition of the end-to-end service. One thing that must be in common is a way to name end points that is globally meaningful. There can be conversion of addresses at region boundaries, but there must be enough common understanding between regions for that conversion to work. If applications are to interoperate across different regions, they must be able to count on common delivery semantics (or at least a minimal baseline of common semantics). In order for this to be so, there must be a common, if abstract, model of key network functions. In the Metanet proposal, these functions are QoS and congestion control. In later proposals of this sort, other designers put forward a different list of global functions.

Plutarch Among those who favor architectural minimalism, a key question is what, if anything, the regions must share in common. Must there be common addresses or names, for example, and if so, at what level in the architecture? Plutarch (Crowcroft et al., 2003) was an experiment in minimality—an attempt to put together a cross-region "glue" architecture that made as few assumptions as possible about common function, common naming, and other shared features. In Plutarch, regions (the Plutarch term is *contexts*) have names, but the names are not globally unique. Within a region, *entities* have addresses, but those addresses are also not unique beyond the scope of a region. Regions are hooked together by interconnection entities (the Plutarch term is *interstitial functions*, or IFs) that have addresses within each of the regions they interconnect.

To deliver data, Plutarch uses an address of the form <entity address, entity address,... entity address>,[1] where each entity address is the address (in a particular region) of the interconnection entity (the interstitial function) that connects to the next region on the path toward the final destination. As a packet is forwarded, each entity address is used in turn to reach the interconnection point that connects to the next region, where the next entity address has meaning, until the final entity address describes the actual destination point.

Plutarch includes a mechanism for state establishment at the region boundaries to deal with the required conversions. In the view of the authors, there might be many regions but perhaps only a few *types* of regions (ten or fewer), so the effort to program the required interstitial functions would be feasible.

There is little discussion in the Plutarch proposal of any global functions such as QoS or congestion. As an exercise in minimalilty, Plutarch might have been just a little too minimal. The key challenge for Plutarch is how the end-point addresses, the sequence of entity addresses that allow a source to identify a destination, are created. Entity addresses are only meaningful in the context of a particular region, where the entity address is well-defined and unique. For one end point to send a packet to another end point, that destination end point must have the Plutarch destination address (the sequence of entity addresses) that is meaningful in the region of

1. As I discuss in the Appendix, this kind of address, which specifies a series of points through which the packet is to flow, is called a source route.

the source. The authors briefly sketch an approach for generating Plutarch addresses, which they describe as *gossip*. Ignoring possible optimizations, the core of a gossip scheme is that an end point announces its presence, and this announcement flows outward until it reaches all parts of the network. As the announcement passes through each point of interconnection into another region, the interstitial function at that point prepends its entity address to the existing Plutarch address of the end point, so that as the announcement flows outward, the source route back to the destination is composed incrementally for each region through which the announcement flows.

This approach to routing is in essence how BGP (see page 12) works in the current Internet. The challenge for Plutarch is scale. BGP provides routes to regions (ASs about 59,000 in the Internet as of 2017), while Plutarch would in principle want to provide routes to end points (billions in the current Internet). Many technologists would want a proof of concept before they would accept that a gossip scheme could work at that scale.

Sirpent Sirpent (Cheriton, 1989) (Source Internetwork Routing Protocol with Extended Network Transfer) was an early proposal to exploit the advantages of source routing to hook together disjoint addressing regions or architectures. Each component in the Sirpent source route was (to simplify) the outgoing interface for the packet at the current router, together with network-specific information adequate to get the packet to the next Sirpent router. This information is similar to the entity address in Plutarch. A clever aspect of Sirpent was that as the packet traversed the series of routers going toward the destination, a source route in the return direction was computed incrementally and carried in the packet, so that when the packet reached the destination, that node had a source address it could use to send a reply.

In order for the sender to obtain a source route to the recipient, the Sirpent architecture assumed a global name space (somewhat like the DNS) that could generate and return the source route from the sender to the receiver. Plutarch did not make this assumption, which is why it depended on the gossip scheme to create the source routes. Routes would in practice be computed by the system that implements the higher-level global name space, and Cheriton's (1989) paper states that having regions of the network compute routes is an optimization that in most cases would be unjustified.

The Sirpent approach reflects the view that giving the user control over the routing is the preferred balance of power between the end node and the network. The Sirpent paper claims that this approach can deal with accounting, congestion control, security, and real-time applications. To deal with access control for regions of the network, the global name-space system, as part of returning a source route, would compute and return authorization tokens (hard to forge, cryptographically constructed values) that confirmed the right of the sender to use the source route. Sirpent mechanisms look to some extent like the source routes in Nebula or the forwarding directives of NewArch. Sirpent is one of the earliest proposals I describe here, and it is worth reading the proposal to get a sense of those times.

Framework for Internet Innovation A key assumption of Plutarch was that the various architectures already existed and had to be taken as given. This assumption drove many of the basic design decisions, since Plutarch could make only the most minimal set of assumptions about the features of each regional architecture. This assumption was reasonable in the era when the Internet was competing with alternatives such as OSI and ATM. However, that motivation has faded over time, replaced by a new concern: that the dominance of the Internet would make it essentially impossible to move to a better solution were it to emerge. The Framework for Internet Innovation (FII) (Koponen et al., 2011) proposed a regional architecture like Plutarch, but in contrast to Plutarch, which assumed that the regional architectures were pre-existing, FII made the assumption that the designers of different regional architectures would create them, taking into account the (minimal) overarching FII design. If the overarching architecture (FII) comes first and the regional architectures are designed to fit into the constraints, the designers of FII can make much stronger assumptions about what the regional architectures must support. At the same time, the designers of FII again strove for minimality—they wished to constrain the different architectures as little as possible while meeting the basic requirements they identified.

The FII proposal identified three critical elements in the architecture:

- The first, similar to Plutarch, is the interconnection function between regions.
- The second is the interface (the API) that the network presents to the application layer. The Metanet proposal used different words for this requirement (consistent end-to-end communication semantics) but had the same requirement. Plutarch does not emphasize this interface, but it is implicit in the design of Plutarch that the end points share a common view of the semantics of the interchange.
- The third critical component of FII is a scheme to mitigate distributed denial of service (DDoS) attacks. The authors of FII argued that the

specific security challenge raised by DDoS attacks (as opposed to other challenges such as viruses and phishing) must be managed at the network layer, so any network architecture must specify how this is done. (DDoS attacks and their mitigation are discussed in chapter 10, which explains in detail why different sorts of security challenges need to be addressed at different layers of the system.) The approach to DDoS mitigation used in the FII proposal, the *shut up message* (SUM), requires that all regions implement a rather complex trusted server mechanism and requires a specification of how to convey certain values consistently across the region boundary.

The central mechanism Koponen et al. (2011) describe at the region interface is an agreed means to implement routing. Their approach is to use pathlets (Godfrey et al., 2009), but they stress that an alternative mechanism might be picked. However, since there must be global agreement on the scheme, it has to be specified as part of the architecture. In fact, there are a number of values that must pass across the region boundaries, which implies that there must be an agreed high-level representation for the values: the destination address, the information necessary to mitigate DDoS attacks, and so on. Any regional architecture that is to be a part of the FII system must comply with the requirement to support these values and the associated mechanisms. This requirement reflects the different design goals of FII and Plutarch. Plutarch is intended to hook together preexisting architectures; FII is intended to allow new regional architectures to emerge over time, within the preexisting constraints imposed by FII. The authors of the FII proposal assert that the constraints are minimal.

There might be an argument as to whether FII is underspecified. For example, the designers take the view that there is no need for any abstract model to deal with congestion or QoS, in contrast to Metanet, which considered congestion to be one of the key problems to be addressed globally. On the other hand, the Metanet and Plutarch proposals make no mention of mitigating denial of service attacks. The FII proposal came more than 15 years after the Metanet proposal—requirements change, and architectures are creatures of their time.

Discussion It is informative to compare the goal of Plutarch or FII with the goal of the original Internet. The original goal of the Internet was to hook together disparate networks: the ARPAnet, a satellite network, and a packet radio network. How does the Internet solution differ from the approach of Plutarch or FII? When dealing with the interconnection of disparate technologies, there are two general approaches: *spanning architectures* and *conversion architectures*. In an architecture based on conversion, such as Plutarch or FII, each of the various interconnected networks must provide, as a native modality, a service that is similar enough that an interconnection point can convert the service of one to the service of the other. Given this approach, what a conversion architecture such as Plutarch or FII must do is define an abstract expression of that service in such a general way that the interconnection point can convert one to another while still making it possible to build useful applications on top of the service.

In contrast, a spanning architecture defines an end-to-end service, perhaps expressed as a common packet format and delivery commitment, and the underlying service of each type of network is used to carry that service over its base service. So, in the case of the Internet, the basic transport service of the ARPAnet was used to carry an Internet packet rather than try to somehow convert an abstract Internet packet into an ARPAnet packet.

The shipping of cargo provides a real-world, if imperfect, illustration of the difference between a conversion network and an overlay network. Packages used to be shipped by packing them into some sort of transportation vehicle and unpacking and repacking them as they were transferred from one sort of vehicle to another. This mode is somewhat similar to a conversion network. Shipping has been transformed by the invention and standardization of the shipping container. Individual items are packed into a shipping container, and that container is moved end-to-end from the sender to the recipient without being repacked. Different sorts of transportation vehicles can be used to move a container, including trucks, railroad cars, and ships. There is a strong analogy between the Internet, with its narrow waist defined by the IP (see figure 2.1), and shipping using containers. The shipper is not concerned with what technologies are used to transport the container, and the transportation vehicle need not be designed taking into account what is in the container (presumably within some limits, such as maximum weight).

Another term for a spanning network is an *overlay network*. The conception of an overlay network has changed over time. As the Internet architecture has gained dominance, the need to deal with different regional architectures has faded. Today, overlay network describes a service that runs on top of the Internet to provide some specialized service, such as content delivery. The possibility that such a specialized service might run over heterogeneous lower-layer network architectures, such as the Internet and "something else," is not currently relevant, but in the beginning, it was the presence of those disparate networks that made the Internet possible.

In this context, FII was designed to solve a particular problem—by abstracting the service model away from the details of how it is implemented (e.g., the packet formats), designers could move from one conception of the embodiment to another over time as new insights about good network architecture design were learned. It is the specification of that abstract service model, and the demonstration that it is really independent of the current embodiment, that is the key challenge for an architecture based on conversion.

As I write this book, the latest technology that might call for a heterogeneous regional architecture is what is currently called Internet of Things (IoT). This technology space, previously called sensor and actuator networks, involves devices that may be very low power, fixed function, and often wireless. Some advocates for IoT assert that the current Internet protocols will not serve these sorts of devices well. Almost any device can be programmed to send an Internet packet—that is not the challenge. The IoT environment raises issues related to management and configuration that the current Internet architecture does not address at all, and some devices will indeed use a network technology that is not compatible with Internet packets—the capacity may be limited, the packet size may be insufficient, and so on. However, my suspicion today is that since many IoT devices are fixed function, for those IoT networks that are based on a different low-level network architecture, the interconnection between the IoT network and the current Internet will happen at the application layer, not at a lower transport layer. IoT interconnection will not be implemented by a general overlay or interconnection architecture.

Performance

Application layer framing A number of architectural proposals address some aspect of performance. For example, architectures that allow a client to find the closest copy of some data are certainly addressing performance, but few of the proposals address the classic, perhaps simplistic, concept of performance, which is getting the protocols to transfer data faster. Application layer framing, or ALF (Clark and Tennenhouse, 1990), did focus on performance, but the aspect of performance addressed in ALF was not *network* performance but end-node *protocol processing* performance. In particular, the functional modularity of ALF was motivated in large part by a desire to reduce the number of memory copies that a processor must make when sending and receiving packets. In principle, ALF allowed the protocol stack to be implemented with as few as two copies of the data (including the application layer processing), which was a key factor in improving end-to-end throughput.[2] This issue seems to have faded as a primary concern in protocol processing, but in fact the overhead of copying data by the various layers of the protocol stack may still be a limiting factor in end-to-end throughput.

ALF also allowed for regional variation in architecture; in particular, the authors were considering IP and ATM as candidate regional architectures. This degree of variation implied that an interconnection device would have to break packets into cells or combine incoming cells into packets at a regional boundary, which in turn implied a lot of per-flow state at a regional boundary. The common element of payload (the *end-to-end communication semantics,* to use the Metanet term) across all the regional architectures was a larger data unit called an application data unit, or ADU. ADUs were sequences of bytes that the interconnection device could fragment or reassemble as desired. The loss of part of an ADU as the result of a lower-layer failure caused the interconnection device to discard the whole ADU, which implied a potentially large performance hit for a lost packet or cell.

Information-centric Networking

The idea behind information centric networking, or ICN, is that users do not normally want to connect across the network to a specific host but rather to a higher-level element such as a service or a piece of information. The information or service might be replicated at many locations across the network, and the motivation for networking is that the choice of which location to use should be a lower (network) level decision, because the choice is not fundamental to the user's goal. The assumption is that the network has access to information (topology, latency, and the like) that can allow it to make a better choice among sources for data than a higher-level service could make. This assumption is arguable—the choice may matter to the user because of considerations such as application performance, availability, security, and data integrity. Different ICN schemes take these considerations into account to different degrees. However, the argument in favor of ICN-style networking is that the service provided by the ICN

2. In order for a computer to perform an operation on data, it must retrieve the data from memory. If the data is transformed in any way, then that transformed data must be written back into memory. Breaking the data into packets, converting from one representation of information to another, and the like all require reading and writing memory. Reading and writing the bytes of a data unit is often the most costly aspect of sending and receiving the data.

network (delivery of data) is fundamentally a better match with the actual goal of the user.

TRIAD TRIAD (Cheriton, 2000) is an ICN that was largely inspired by the design of the Web. The names used for data in TRIAD are URLs (see page 14 and chapter 8), and the goal of TRIAD is to incorporate the mechanism for mapping from a DNS-style name to an IP address into the network architecture. In TRIAD, the user requests data by sending a lookup packet containing a URL as the destination address in the packet. To oversimplify, in TRIAD, routers contain routing tables based on URLs (or prefixes of URLs), so they can forward lookup requests toward a location where the data is stored. The lookup request initiates a traditional TCP connection (it is a modified TCP SYN packet), so once the lookup request reaches the location of the data, TRIAD uses TCP (with lower-level IP-style addresses) to transfer the data.

TRIAD does not require that *all* routers support a URL-based forwarding table. TRIAD assumes a regional structure (a connected set of ASs, similar to today's Internet) and requires that each AS maintain a subset of routers with a URL forwarding table. The routing protocols must be able to forward a setup request to one of those routers when an AS receives it, so most routers could function as they do today, with only an IP-based forwarding table, so long as those routers have the address of a router with a URL forwarding table.

The key challenge with TRIAD was to design a URL-based routing system that scaled to the size of the Internet and the anticipated number of data names in the future. The proposal was a BGP-like (path vector) route announcement scheme, where the route announcement includes the first and second levels of the URL rather than an AS number. If the data names in TRIAD were organized like the DNS names of today, the forwarding table would have to be large enough to hold all the second-level DNS names, which is hundreds of millions but not hundreds of billions. Multiple data sources could announce the same DNS prefix, and the routing protocol would then compute paths to whichever source was closest, so the TRIAD scheme provides a form of name-based anycast. However, forwarding on only the first two elements of a DNS name may not be adequate for effective management of data location.

DONA Data-oriented network architecture (DONA) (Koponen et al., 2007) is similar to TRIAD in many respects. A requesting client sends a lookup packet (called a find request in DONA), which, when it reaches a location with the data, triggers the establishment of a TCP-like connection back to the client to transfer the data. DONA differs from TRIAD in that it uses data names that are derived from the data itself rather than using URLs. Any particular piece of data is created by a principal, an entity defined by a distinct unforgeable identity (a public-private key pair).[3] A name for a piece of mutable data is of the form P:L, where P is the identity of the principal (the *hash* of the principal's public key) and L is a label unique within the name space of P.[4] For immutable objects, L is a hash of the data. Names based on a hash provide no structure to give guidance about location—such naming schemes are sometimes called flat,[5] as opposed to hierarchical. Similar to TRIAD, names propagate in a BGP-like routing system to special routers in each AS that support name-based forwarding. A principal can announce names of the form P:L (which can give limited guidance about location if all the data associated with a given principal is in the same location), or P:*, which provides the location of the principal. Again, the principal can announce these names from multiple locations, providing an anycast-like character to the forwarding process.

Because names are flat in DONA, the number of distinct routing entries will be much higher than with TRIAD. DONA proposes an important scalability assumption. The DONA designers assumed that the network would have a somewhat hierarchical structure, similar to today's Internet, with Tier 1 (wide-area) providers at the core, as well as the option of direct interconnection at any lower level. In today's Internet, a peripheral AS

3. For those not familiar with encryption and the concept of public-private or asymmetric keys, I discuss this on page 195.

4. A hash is a short, fixed-length value that is produced by a function that combines the bits of the input (the public key, the data itself, or whatever) according to some algorithm. DONA uses a *cryptographic hash*, which is a hash function with the property that while it is easy to compute the hash of any given input value, it is impractical to reverse that process and compute some input value given a hash. Cryptographic hashes are called self-certifying because a recipient of an item can confirm that it is the correct item by recomputing the cryptographic hash function and confirming that it generates the correct hash.

5. The term *flat* implies that the addresses have no structure (such as a hierarchical organization) to make the forwarding process more efficient. The current Internet assigns addresses such that topological regions of the network are associated with blocks of addresses; this means that the routers only need to keep track of these blocks (the prefixes of the addresses) rather than every destination in the Internet. Even so, a forwarding table in a core Internet router today has about 680,000 entries. The size of the forwarding table in a router is an important consideration in the design of a forwarding scheme.

need not keep a complete forwarding table but can have a default route toward the core. The central region of the Internet, where the routers must maintain routes to all parts of the address space, is thus sometimes called the default-free region. Similarly, DONA routers outside the core need not store a complete name-based forwarding table. Only in the DONA equivalent of the default-free region would a router store a complete name-based forwarding table.

Named data networking Named data networking, or NDN (Zhang et al., 2014), is one of the projects funded by NSF under the FIA program. NDN takes some of the addressing ideas from TRIAD and pushes them into the data plane. In particular, instead of using a name-based lookup packet to trigger a TCP-like data transfer phase, NDN uses data names rather than lower-level addresses in every packet. One design objective of NDN is to remove from the scheme any concept of routable lower-level addresses. The assumption is that this approach will lead to a simple network architecture with great consistency and clarity of structure. The names, like TRIAD names, are URL-like in format, but now every router must have a name-based forwarding table. Names for data in NDN describe packets, not larger data units.[6]

The NDN architecture is thus distinctly different from many other proposals for addressing and forwarding. In NDN, there are two sorts of packets: *interest* and *data*. Instead of sending a packet to a host, in NDN one sends an interest packet containing the name of the data being requested and gets the information in a return data packet that contains the name of the data, the data itself, and a signature that binds the name to the data, thus confirming the validity of the data. In neither case is there a network-level address in the packet, either as source or destination, only the name of the desired information. A key technical aspect of NDN is that when an interest packet is routed toward the location of the data, information about the packet is recorded in all the routers along the path back to the original requester. In NDN, there is thus per-packet state in each router along the path followed by the interest, which includes the interface on the router from which the interest came and the name of the data being requested.

Names of data are hierarchical and begin with the authoritative owner of the data, followed by the name of the specific data. Each owner/creator of data has a public-private key pair and uses the private key to produce the signature that verifies the data. Thus, anyone with the public key of

6. Interestingly, the DONA proposal also notes that names could describe chunks of data, not the complete data unit, but this is a secondary objective for DONA.

the owner can verify the data packet: in particular, the integrity of the data and the binding to the name. The name of the data also describes its location. When information moves to a new location, there is a variation of a data packet called a link, which encapsulates a data packet and is signed by the operator of the current location. This signature allows anyone with the public key to verify that the location named in the packet is the current source of that packet.

When a router forwards a data packet back toward the original requester, it optionally keeps a cached copy for some period of time. The router can use this cached copy to satisfy a subsequent interest packet rather than fetching the data from the original location. This mechanism allows for the efficient delivery of popular data.

NDN is perhaps the most radical of the FIA proposals for an alternate internet architecture. I will return to the NDN proposal a number of times in the remainder of the book.

PSIRP/PURSUIT The publish/subscribe Internet routing paradigm (PSIRP) and its successor, PURSUIT (Trossen et al., 2008; Trossen and Parisis, 2012), were funded by the European Commission as part of its future Internet research program. This proposal differs in many ways from NDN, though I classify both as examples of ICN. Objects in PURSUIT are higher level, more closely related to what applications, services, or users might require, rather than packets as in NDN. In this respect, PURSUIT objects are similar to DONA or TRIAD objects or to ADUs in ALF. Objects in this scheme have names that are valid within a PURSUIT *scope*. These names are called rendezvous identifiers (RIDs), since publishers and subscribers of an object find each other by using these names. The creator of an object *publishes* it by sending a message to a scope and requesting that it assign an RID to the object. A scope is implemented as a set of *rendezvous servers* that hold a mapping between each RID in that scope and one or more lower-level network addresses by means of which the object can be retrieved. Scopes themselves are named using a form of RID (in other words, scopes themselves do not necessarily have globally unique IDs), so the name of an object is a sequence of RIDs: <top-level RID, second-level RID, ... final RID>. *Subscriptions* express interest in receiving the object. The architecture is explicit about both publish and subscribe in order to create a balance of control between creator and requester. Scopes also contain a topology manager (TM), which is responsible for keeping track of where copies of the data are stored, either in a cache or at the location of the publisher. When a subscribe request is received by the scope, the final rendezvous server contacts the TM, which determines the location from which to deliver the data to the requesting subscriber.

PURSUIT names have the same hierarchical structure as TRIAD names but are in no way modeled on URLs. There is no implication that the nested names in PURSUIT have "meaning" in the same way that DNS names have meaning. The scope names and object names in PURSUIT are just IDs and do nothing but identify which scope is being used to provide the rendezvous service. If there are "meaningful" names in PURSUIT, they will exist in a higher-level naming system.

Network of information (Netinf) Netinf (Dannewitz et al., 2013) was also funded by the European Commission as part of its future Internet program. Netinf, like DONA, uses flat, globally unique names to identify data objects. The security architecture of Netinf, like some other schemes, defines the name of an object as the hash of its contents. Netinf, like NDN, defines a standard format for the data object[7] so that any element in the system, not just the end node, can verify that an object corresponds to the name associated with it. The authors stress the importance of being able to validate a data object independent of where it comes from, since (like many ICN proposals) Netinf includes the management of data caching as part of the architecture.

The Netinf retrieval model is similar to DONA's: a *get* message is routed, using the ID of the object, to a site where the object is stored, at which point the end points open a transport connection to deliver the object. The Netinf designers contemplate a system with regional diversity (like FII or Plutarch) and describe two approaches for realizing this: either state is established at region boundaries as the *get* message is forwarded, so that subsequent transport connections can be linked together through that boundary point to deliver the object, or some sort of return source route is assembled in the get message to allow a response to be returned.

The Netinf specification discusses two modes of ID-based routing. One is a system of name resolution servers (NRSs) organized as a hierarchy, which can forward the request toward a location where the object is stored. This scheme seems similar in general terms to the scheme described in DONA. The specification also discusses direct routing on IDs in some of the routers in the system, perhaps in a local region of the network. The Netinf specification refers to the information stored in an NRS as a *routing hint* and points

7. In this case, a representation based on the Multipurpose Internet Mail Extension, or MIME, standard.

out that the hint need not get the request all the way to the region of the destination but only toward it, where another NRS may have more specific information. As described, Netinf can support *peer caching*—an end node could inform an NRS that it holds a copy of some data or respond directly to a get that is broadcast in some region. With respect to routing, Netinf and NDN have some similarities: both architectures allow for a range of routing and forwarding schemes to get a request to a location of the data. However, Netinf allows for a query to an NRS as well as to a local per-object forwarding table. The paper by Dannewitz et al. includes an analysis arguing that a global NRS of object IDs is practical.

Discussion I noted in chapter 5 that the Internet required but did not specify a routing protocol that could deal with Internet-scale end-node addressing. The Internet of today depends on a degree of topological aggregation of addresses (the Internet routes to blocks of addresses, not individual addresses), but the size and number of blocks are not specified in the architecture. The number of blocks seen in the Internet today is a pragmatic response to the capabilities of current routers. The challenge for schemes such as DONA, Netinf, NDN, and MobilityFirst is whether a routing scheme can be devised that can meet the scale requirements of the architecture—all these schemes require a routing system but do not specify it as part of the architecture. (In each case, the proposals include a sketch of a possible solution to give credence to the scheme but, like the Internet, the routing scheme used in practice would probably evolve over time.)

One difference between these various schemes is the extent to which the data naming is embedded within the actual network architecture. NDN depends entirely on the data names and has no concept of a routable network-level address. TRIAD and DONA require a subset of the routers to understand data names, but once the location of the data has been determined, they depend on network-level addresses to forward the packets that actually transfer the data. The names in PURSUIT are used not to forward packets but only as the basis for a query to a rendezvous server. The PURSUIT architecture could be deployed on top of the existing Internet—it is a higher-level scheme in the same sense that the DNS is.

Architecting for Change

As I will discuss in chapter 9, there are several views as to how to design an architecture so that it survives over time. Different schemes here take different views of this objective.

FII, discussed earlier, assumes that over time new architectures will be designed, and thus the goal of the higher-level FII design is to define a minimal set of constraints on these different architectures so they can interwork, thus providing a path for migration.

Expressive Internet architecture (XIA) The emphasis of the XIA proposal (another of the NSF-funded projects) is on expressive addressing in packets to allow the network to use a variety of means to deliver the packet to the intended destination and to provide a range of services in the network. The rich addressing and forwarding mechanisms in XIA allow a packet to carry several forms of addresses at once. For example, they can carry the content identifier (CID) of a desired piece of data but also the identifier of a service hosting that data (SID), the host where the service is located (HID), or an administrative domain in which the CID is known (AD). This richness is described as expressing *intent*, and the other addresses allow various forms of fallback forwarding. This flexibility allows the end host to select from a richer set of network services. It also should contribute to the longevity of the design, as it would permit a more incremental migration to a new sort of XID than the current migration from IPv4 to IPv6.

Some specific features of the XIA scheme are:

- The various identifiers, collectively called XIDs, are specified to be self-certifying. For example, they may be the hash of a public key or the hash of the data to which they refer. This design allows the end points (e.g., the applications or perhaps the actual human users) to confirm that the action they attempted has completed correctly: that they connected to the host they intended, got the data they intended, and so on. In other words, these mechanisms transform a wide range of attacks into detected failures.
- XIA gives the end-point options for bypassing or routing around points of failure in the network by cleverly using the multiaddress destination field in the packet.
- The scalability, control, and isolation on next-generation networks (SCION) system (part of XIA) provide a structure to break up the overall network into what are called trust domains, which allow a variety of end-point controls over routing.

XIDs provide a means of confirming that the correct action occurred once the end points had the correct XIDs. However, most network operations start with higher-level names such as URLs, email addresses, and the like.

Since different applications may involve different sorts of high-level names, the XIA architecture does not define how these names should be converted to XIDs in a trustworthy way. The XIA architecture gives requirements as to what the application and its supporting services must do but does not dictate how.

Intermittent and High-Latency Connectivity

The Internet and its kin were designed to provide immediate delivery of packets, immediate in the sense that they were suitable for interactive applications. A well-engineered region of the Internet today will often deliver packets within a factor of 2 or 3 of the latency of light.[8] However, not all operating contexts have characteristics that can even attempt to support this class of interactive applications. An interplanetary internet is the extreme form of the challenge—intermittent connectivity (perhaps as satellites come in and out of range) and latencies of seconds if not minutes.

A class of network called delay/disruption tolerant network (DTN) was crafted to deal with these challenging requirements. The objective of DTNs is to provide a useful data delivery service in the context of very long delays and unreliable communication channels. The discussion here is based on a seminal framing paper (Fall, 2003).[9] Because of the long and variable end-to-end delays, the DTN high-level service model must be different from that of the Internet: the forwarding must be based on reliable intermediate storage points rather than direct best-effort end-to-end forwarding. The store-and-forward data unit in a DTN is not the packet (which might not be a uniform standard across the system) but rather a higher-level data unit (similar to an ADU), which is called a *bundle*. The DTN architecture is based on regional variation in underlying architecture (for example, classic Internet on any given planet) and is thus similar in some respects to proposals such as Metanet. (The DTN proposal makes specific reference to Metanet.) A DTN does not attempt to stitch together regional architectures to achieve something end-to-end that resembles the behavior of the individual regions. It is a *spanning* architecture rather than a *conversion* architecture. The DTN architecture depends on devices at region boundaries

8. While the speed of light may seem so fast as to be essentially infinite, it takes about 40 ms for a signal in a fiber to cross the United States and return. Measured round-trip delays across the country on the Internet are typically between 70 and 100 ms.

9. The interested reader should refer to http://ipnsig.org/ for background on the DTN initiative.

to terminate transport connections, receive and reassemble bundles, and potentially store those bundles for long periods of time until onward transfer is possible. Different transport protocols might be used in different regions, depending on whether the latencies are on-planet (where TCP would be suitable) or multiminute. Since the long delays in the system make impractical the classic end-to-end reliability model of the Internet with acknowledgments flowing back to the source, a DTN requires that the interregion store-and-forward elements be reliable enough to assume responsibility for the assured forwarding of bundles. In this respect, the basic communication paradigm of the DTN architecture resembles Internet email, with (presumed reliable) mail transfer agents and no architected end-to-end confirmation of delivery. Also since store-and-forward nodes will have finite storage, DTNs raise issues regarding layered flow control.

The names for destinations in a DTN are of the form <region name, entity name>. The region names are globally meaningful, and the entity names are only routable within the region. In the DTN paper (Fall 2003), each of the names is a variable-length text string rather than any sort of flat, self-certifying ID. However, a DTN could use those sorts of names without disrupting the scheme. Like many of the other schemes I have discussed here, a DTN depends on a routing scheme that is not specified as part of the architecture but must be designed. In this case, the routing scheme is not dealing with routing to objects—it is routing to a potentially small number of regions. The challenge for DTN routing is to deal with the complex and multidimensional parameters of the overall network, which may include satellite links that provide intermittent connectivity on a known schedule or (as a terrestrial example) so-called data mules such as a rural bus or delivery vehicle with a wireless base station that can pick up or drop off bundles whenever it drives by. The DTN paper proposes a framework in which routes are composed of a set of time-dependent *contacts*, which are assembled by a routing algorithm.[10] The resilience of this sort of network will depend crucially on the design of the routing protocol.

Shaping Industry Structure

One of the requirements I listed for an architecture was its economic viability and the industry structure it induced. Few of the academic proposals for architecture directly address this goal, but architects from industry have been quick to understand and respond to this requirement. In 1992, when the government, under the urging of then Senator Al Gore, announced the

10. Perhaps somewhat similar to the way pathlets are composed.

vision of the NII, industry was quick to respond with its critical concern, which was the modularity of the design and the resulting implications for industry structure. It understood that since the private sector was to deploy the NII, the structure of the industry was key to its success. Two groups developed a response to the call for an NII.

CSPP The Computer Systems Policy Project (CSPP), now called the Technology CEO Council, responded to the concept of the NII with a high-level vision document (perhaps consistent with drafting by a group of CEOs) (Computer Systems Policy Project, 1994). It does not talk about architecture but contains a list of requirements that is in some respects similar to what I have discussed here (access, first amendment, privacy, security, confidentiality, affordability, protection of intellectual property, new technologies, interoperability, competition, and freedom from carrier liability).

XIWT The Cross-Industry Working Team (XIWT), convened by Robert Kahn at the Corporation for National Research Initiatives, dug deeper into the potential architecture of an NII (Cross-Industry Working Team, 1994). It described a *functional services* framework and a *reference architecture* model. After its own list of requirements (sharability, ubiquity, integrity, ease of use, cost-effectiveness, standards, and openness), this group focused on the *critical interfaces* that would define the modularity of the NII. It emphasized two interfaces that were not well-defined in the Internet: the network-network interface (the control interface that defines how ISPs interconnect) and the network service control point to network interface, which would allow third parties to control the behavior of the network.

Discussion To my knowledge, these documents did not have much influence on the technical community developing the Internet. They were written to influence potential direction-setting in Washington, and Washington turned out doing little direction-setting. The proposal for a network-network interface was not inconsistent with the current Internet—it just brought a focus on an aspect of the Internet's design that was under-developed at the time. The proposal for a network service control point was less compatible with the current Internet. This interface is somewhat reminiscent of the intelligent network interface being contemplated by the telephone system to allow the control of advanced services and is a hint that the members of the XIWT were not totally committed to the idea of the "dumb, transparent" network. Perhaps in the proposal for a network service control point we see an early glimmer of software-defined networking.

Mobility

Mobile devices have become the norm for many Internet users today, and with the advent of complex mobile systems such as automobiles, it is not just individual devices but local networks with many connected devices that are becoming mobile. The key requirements for an architecture that supports mobility include separation of location from identity; accommodating intermittent connectivity and variable performance; and the ability to track the location of moving devices, networks, and data. Architectures differ in how they deal with these issues.

MobilityFirst (MF) The MF architecture, another of the NSF-funded projects, was motivated by the desire to deal with issues raised by mobile end nodes—in particular, movement of devices from one network to another and transient outages when devices become unreachable. In MF, there are two levels of binding between name and address. At a high level, a number of name services similar to the DNS, which are called naming services (NSs), map from a host, service, sensor, data, or context (a context is a set of things that match some criteria) to a flat ID, a global unique identifier (GUID). At a lower level, there is a service, the Global Name Service (GNS), that maps from a GUID to its current location, which is a network address, or NA.

The essential idea behind the MobilityFirst design is that both the destination GUID and the destination NA are included in the header of a packet, which allows rapid forwarding based on the NA but also allows routers in the network to deal with mobility and redirection by making dynamic queries of the GNS as the data is moving through the network. If the NA included in the packet by the source is not the current location of the destination, routers in the network can attempt to look up a new NA by using the GNS. The routers can also store data in transit at intermediate points if the destination is temporarily unavailable because of wireless connectivity issues. Stored data is identified by its GUID until the router can determine a new destination NA.

To enhance security, both the GUID and the NA are public keys, so anyone in possession of a GUID or NA can confirm that the binding is valid. The name space of NAs is thus flat. The MF design assumption is that today there might be as many NA values as there are routing table entries.

Discussion The GNS is a specific approach to solving a problem that all systems that support mobility must solve—if some element (end point or network) is mobile, there must be some stable service (a service that itself is not mobile) that remembers the current location (the NA in the terminology of MF) of the mobile element. The mobile end point (or mobile network) must report its current location to that service as it moves, and end points that want to contact that mobile point must know how to reach that service. MF takes a very ambitious approach to dealing with mobile end points—it allows for the redirection of a packet to a new point of attachment while the packet is in transit across the network. The mechanism that makes that possible is the GNS, which must be able to map a large set of GUIDs (perhaps 100 billion) to the corresponding NA quickly enough to allow a packet to be redirected in transit.

This ambitious objective could be relaxed—a simpler approach would be that if a GUID is not found at the destination NA, the packet is dropped, an error message is returned to the sender, and the sender has to determine the currently valid NA. In this case, there would still have to be some service that maps a GUID to its current NA, but that service could be designed in very different ways. There is a potential security issue related to the GNS in that since it is a single system in which the location of every GUID is recorded, in principle anyone could query the GNS for a given GUID and track its current location. If it is not necessary for routers to be able to query the service but only end nodes, then there can be more flexibility about how that service is designed. Mobile end nodes that want to be reachable by anyone would still register their location in a public mapping service, but end nodes that have more restricted communication patterns might use a more private mapping service that is only available to selected communicating parties. In some cases, it might not be necessary to provide any mapping from GUID to NA. If there is a higher-level naming system (like the DNS or the naming services of MF), that service might be the repository of the current location of a mobile end point, and the mapping would be from a higher-level name to the NA.

Cross-layer optimization I do not have a specific architectural proposal that illustrates this goal, but designers of some wireless networks, in particular wireless networks designed for highly challenging conditions, such as tactical battlefield networks, have argued that to make these networks practical, it is necessary to abandon the layer abstraction that is the basis for most of the architectures that I have described, which states that the end-to-end service is a technology-independent, general-purpose packet transport service. The alternative point of view is that wireless technology has some powerful technology-specific features, such as broadcast, as well as some technology-specific limitations, such as highly variable signal quality and interference levels. In this view, the higher service layers and applications need to be designed taking into account these specific considerations, exploiting the features to compensate for the limitations. This approach would preclude the inclusion of these networks in either an overlay or conversion internet architecture of the sort I have been describing and would instead require interconnection at the application layer if interconnection to networks based on other technologies was desired.

Minimize the Need for Globally Unique Identifiers

Many of the schemes described here depend on globally unique identifiers, whether for data, services, or end points. Global uniqueness can be assured in a number of ways. One is to make the identifiers hierarchical by creating a few top-level unique prefixes and then delegating to different actors the responsibility for creating subsidiary names that are unique within the context of that prefix. PURSUIT names follow this design; they are of the form <top-level RID, second-level RID,… final RID>, where the top-level RID might be globally unique but the second-level RID is unique only within that top level, and so on. DNS names have the same hierarchical structure.

A second approach is to make the identifiers very large, so that there is a very low probability that two identifiers picked independently will have the same value. Identifiers that are created using a cryptographic hash have this property. MD5, an early cryptographic hash function, produced 16 byte values, which was shown to be much too short for good cryptographic protection. SHA1 produced a hash that was 20 bytes long, but that also proved too short. SHA-3 (a current standard) produces hash values ranging from 28 to 64 bytes. Values of that length are indeed highly likely to be unique, but using these sorts of values as addresses leads to very large packet headers. The packet header of the current Internet (IPv4) is 20 bytes in total.

NewArch An alternative approach to global uniqueness is to try to avoid any need for global identifiers in addressing. The NewArch proposal is somewhat distinctive compared to the others in this chapter in that its abstract forwarding architecture (forwarding, association, and rendezvous architecture or FARA) did not include any sort of global ID of end points. The assumption in FARA was that some sort of ID would be needed so that the sender and the receiver could verify themselves to each other but that these IDs should be a private matter between the end points. The designers of

FARA tried to avoid putting into the architecture (or, more specifically, into the packet header) any ID that would act as a global identifier of the recipient. They assumed that communication occurred between *entities*, a term that FARA applied in a number of ways: to an application on a machine, a machine, a cluster, and others. What the architecture required is that it be possible to construct a *forwarding directive* (FD) that would allow the network (and perhaps also the operating system of the receiving host) to forward the packet to that entity. In most reductions of FARA to practice, it was assumed that the FD would be some sort of source route.

Once the packet had been delivered to the entity, the ID in the packet would identify the *association* to which this packet belonged. A common form of association would be a TCP-like connection, and the association value in the packet would identify the right state variables in the entity for that association.

The FD would contain a locator for the destination entity as one of its elements, but this would have meaning (like many source route elements) only in the context of that step in the forwarding process. For example, it might be a process ID inside a host. One of the goals of FARA was to hide, to the extent possible, exactly what parties were communicating across the network by avoiding IDs with global, long-lasting meaning. Of course, the FD may well contain a machine-level address, which provides considerable information about the communicating parties. But forwarding in FARA might include an indirection step, which could hide the address of the final destination.

The FARA designers presumed, but did not specify, some mechanism by which an entity that wished to receive communication would be able to construct an FD that would cause traffic to reach it. In a network like the Internet, this might just be an IP address followed by a process ID. Different instantiations of FARA might have FDs of a different form. The FARA designers presumed but did not specify that there would be some sort of *rendezvous system* (RS) to allow senders to find receivers. The receiver would install an entry in the RS with some higher-level name that could be used by a sender to query the RS and retrieve the FD. This query operation would take the high-level name as an input and return the FD. The RS also allowed the receiver to pass to the sender a *rendezvous string* that tells the sender how to format the information in the first packet of the communication to the receiver. Again, the meaning of this string was seen as a private matter between the sender and receiver—the mandatory function of the RS was just that an opaque string could be passed through it from receiver to sender. This mechanism could be seen as a limited form of adding expressive power (a "scratch pad" in the packet)—limited since it is only intended for use by the end points and is only used in the rendezvous mechanism.

If the FD turned out to be invalid (the desired entity was not at the location indicated by the FD), the designers of FARA deliberately did not provide any sort of global ID that could be used to find the current location of the entity but used the RS for this purpose. In any global name-resolution scheme that supports mobility, the receiver must update its location information in some fixed location when it moves. In FARA, that information is in the RS and is retrieved by looking up a higher-level name. So in contrast to MobilityFirst, which used globally unique identifiers and the GNS to map IDs to location information, NewArch used its higher-level naming system (the RS) to store the binding of an end point to location and required a mobile host to update the RS (rather than a GNS) if and when it moved. This was seen as having security benefits (packets did not carry any globally identifying long-lasting values) and overhead benefits (the headers could be smaller), but perhaps with slower recovery time when a node moved, since the sender must resend a packet if the location of the receiver has changed.

Conceptual Clarity

The design of the Internet evolved as it was reduced to practice, and its design carries its history in various decisions and the interactions among them. One of the goals of a fresh design effort is to shed that history and find a cleaner and more consistent way to think about architecture. Two similar proposals offer a fresh approach that is based on a reconception of layering.

Recursive network architecture (RNA) and recursive internetwork architecture (RINA) RNA (Touch et al., 2006) and RINA[11] are similar attempts to reconceptualize in a foundational way what networks are and how they work. RINA is less concerned with the match of design to requirements and more concerned with clear thinking about design. RINA is a pure example of a clean-slate design; Day introduces new terms for the concepts in the design, which avoids bringing in the baggage and unstated implication

11. The organizational framework RINA was first proposed by John Day (2008), who was involved in the early days of the Internet and also the ISO effort, and he distilled it out of his experience. His book is fairly long, and for those interested in learning more, I would start with the Wikipedia page on RINA, which also points to some of the research projects based on its framework.

from the reuse of old terms but also requires internalizing a new set of terms in order to understand the concepts. The creation of new terms is the privilege of an inventor, but it sometimes makes comparison of designs difficult. At the risk of not doing justice to RINA, I am going to discuss it without the use of Day's nomenclature.

One way to introduce these proposals is to use the Internet as a counterpoint. The Internet is a network of networks. At the global layer (inter-AS), there is a global routing protocol (BGP) that directs packets to the correct destination AS. Inside each AS there is an internal routing protocol that delivers packets to the correct end points. Conceptually, the internal routing protocol makes a set of routers and links into an abstract entity: the network. BGP routes among these networks to make an internet without knowing about the internals of the network. Networking experts tend to think about these as different layers and give them names, such as the *network layer* and the *internet layer*. The insight that motivates RNA and RINA is that these levels should be seen as different instantiations of the same concept, not different concepts. They do the same thing—establish an association among a set of end points to form an abstract element. These abstractions are then the building blocks for the next layer, which combines them to establish an association. There is no reason why there would be exactly two of these—a network layer and an internet layer. The pattern of combining elements to form an association among the end points can be repeated as many times as appropriate—hence the word *recursive* in the name of the proposals.

RINA uses the concept of *scope* to distinguish the role played by each association. Some associations have local scope (a local area network, or LAN, for example) or the scope of a specific technology (such as a cellular network), some have a scope that relates to the assets of an ISP, and an association with global scope is the RINA equivalent of the Internet. RNA uses the term *layer* and focuses on the allocation of different functions to different layers. The designers of RNA stressed the goal of not repeating (or *recapitulating*) a function like encryption at many layers, stressing instead that the functions of a layer have to be performed in the context of the layers above and below it. This variation in vocabulary raises the obvious but almost philosophical question of how scopes and layers differ. Is scope an attribute of a layer, or are they different concepts? My view is that they are different, but I need to provide more background in order to defend that view. I will return to this view toward the end of the book.

RINA stresses that functions at different scopes should be independent, not entangled. The current Internet, because of its history, entangles its layers in ways that are not optimal. The addresses in the Internet do not include explicit identifiers for ASs.[12] Initially, the 32 bits in the Internet address were divided into two parts: the first 8 bits identified the network and the last 24 bits the end point inside that network. But we quickly realized that 8 bits were not enough for the number of networks that would be built, and a much more complex mechanism was invented (called CIDR, for classless interdomain routing), in which different parts of the Internet use a different number of bits at the beginning of the address to identify the network. Today, the interior routing protocols and the inter-AS routing protocols have to work with the same 32 bit field, dealing with all the complexity created by CIDR. The RINA alternative is that each scope should have its own addresses, which are suited to the needs of that scope.

In this respect, the Internet does indeed reflect its history. Initially, Internet routers were used to connect networks, and different addresses were used inside those networks. The ARPAnet (the first network used as a component network inside the Internet) had its own address structure and its own forwarding devices (called interface message processors, or IMPs). When an Internet packet entered the ARPAnet, there was a binding that mapped from the Internet address to the ARPAnet address, and then the ARPAnet address was used internally. However, when the ARPAnet was decommissioned, the designers of the NSFnet realized that they could use Internet routers to perform the routing function inside the NSFnet as well as between the networks. They could have taken the path of designing a new, network-level addressing and routing function, but it was expedient to reuse the Internet mechanisms. The result is two routing protocols using the same address field, which increases the level of complexity.

In some parts of the Internet, lower-level network schemes for addressing and forwarding have been developed and deployed, for example a scheme called *multiprotocol label switching*, or MPLS;[13] another scheme is *switched Ethernet*. These lower-level technologies have devices (somewhat similar to routers) that receive and forward packets, which are usually called switches or layer 2 switches to distinguish them from the IP routers. My view is that what we have learned from the use of MPLS is that layers need to be independent but not totally. In this respect, I am not convinced by some of the design assertions made by the proponents of RINA; for example, that congestion can be dealt with independently at each scope. I think

12. The Internet does have numbers for ASs, but they are not in the packet header.
13. I discuss MPLS in more detail in the Appendix, in the context of how it was developed.

that congestion has to be handled in an integrated way no matter where it manifests—congestion is a cross-scope (or cross-layer) problem.[14] RNA discusses congestion control as a function that requires cross-layer interfaces. My concern about congestion is just one example of a function that requires coordination between different scopes and layers. I will return to this issue in chapter 13; I think it is the key to the distinction between a scope and a layer. (The work reported in Chiang et al. [2007] could have a direct bearing on RINA. That work similarly assumes a recursive design, and uses mathematical analysis to derive the correct placement of a function within the layers to optimize the design.)

RNA and RINA can exist as an explanatory framework or as running code. As an explanatory framework, the question about recursive layering is not whether it is correct but rather the extent to which this framework leads to a clearer understanding of how the different instantiations at different scopes should be designed and, most importantly, in what respects the different layers can be independent or must have functional interfaces that hook the layers together. Running code is the proof of the framework. RNA in particular stresses that the goal of the scheme is to produce a single code module that implements the range of functions that a layer must perform, with this code being reused at every level by enabling or disabling different functions at different layers. RNA uses the term *meta-protocol* to describe the capabilities of this program.

Both RINA and RNA must deal with the challenge of mapping between the addresses at the different scopes and layers. RINA, more than RNA, draws some specific conclusions about how addresses should be used. I am not convinced that the RINA framework leads to the right conclusions about addressing in all cases. In the Internet, addresses refer to network interfaces on an end node, not the end node itself. If a network has multiple network connections (if it is *multihomed*), it has multiple addresses. The RINA framework seems to lead to the strong conclusion that an address should identify the end node itself, not the interface. The network should make the decision as to which interface to use to reach the end node.

I do not agree with this conclusion, for two reasons. First, an end node with two interfaces may be attached to the Internet at points that are far apart in the topology of the network; for example, a landline connection and a cellular connection. So the decision as to how to route the packet to the end point may have to be made many hops away, which means that

14. The FII proposal also asserted that congestion is a regional matter, so there is clearly some disagreement about this point among designers.

the address of the end node must be meaningful (capable of being mapped to a lower-level forwarding decision) at many points in the Internet. The names of end nodes must not only be globally unique but must be globally routable. (MF has a similar requirement to deal with mobility, which they solve with the GNS, a very demanding design challenge.)

Second, assigning one name to the end node means that the network, rather than the end node, has control over which network is used to deliver traffic to it. I think the end node, and not the network, should have control over this decision. For example, if one network is high-performance but costly and the other is slower but cheap, or one network is more susceptible to censorship or unwelcome traffic discrimination, it should be the decision of the user which to use. To me, the decision as to whether the end node or the network should have control over routing to a multi-homed end point should be resolved through tussle analysis. The addressing scheme described in Yang (2003) uses addresses in a very clever way that allows each party in a communication to control how traffic is forwarded in their region of the network. The proposal deals with tussle and how different actors in the system can express their preferences within their scope.

I think the most exciting result that could come from continued research on RINA and RNA would be the demonstration that the interlayer interfaces can be uniform across all the scopes and layers. Identifying and specifying those interfaces would be a major contribution. Touch et al. (2011) report an interesting follow-on effort, called DRUID, which combines some of the ideas in RNA and RINA.

Expressive Power—Services in the Network

This class of architecture has the goal of invoking services in the network as packets flow from sender to receiver. The proposals in this class directly address the question of the expressive power of an architecture, the balance of control held by the sender and the receiver, and the different sorts of delivery modes and PHB arguments (see chapter 6). Services in the network can be functional (enhancing the overall network service) or adversarial (protecting a receiver from unwanted traffic). I look at six examples of this class of architecture: TRIAD, i3, DOA, Nebula, ChoiceNet, and ANTS (active node transfer system, a network toolkit).

TRIAD TRIAD, which I discussed earlier as an example of an ICN, establishes a connection to an abstract entity (typically an item of information) using a setup packet with a URL in the initial packet. As I described, the provider of data advertises that data by inserting the top-level components of its URL into the TRIAD name-based routing function. This mechanism can be used more generally to route any sort of session initiation request to a location named by the URL, so the scheme is a general form of intentional delivery to a service point. The URL serves as an explicit parameter for whatever PHB that service node implements. However, if applications want to deploy many PHBs in TRIAD, this would imply that the routing protocols would have to keep track of lots of URL-based routes.

Internet indirection infrastructure (i3) The i3 system comprises a set of servers or nodes deployed across a network like the Internet that implement a specialized packet forwarding function. In i3, receivers express an interest in receiving a packet by creating an ID and inserting into the i3 system the pair <ID, addr>, where addr is (for example) an IP address. i3 uses the term *trigger* to describe what the receiver is creating by this announcement. After this, the receiver would normally create an entry in a DNS-like system that mapped some user-friendly name to that ID. The sender would look up that name in this i3 DNS system, which would return the ID to the sender. The sender would then initiate communication by sending a packet addressed to the ID. i3 associates the ID with a node in the i3 system where the trigger is stored. In the simple case, once the packet has been delivered to this node, the trigger provides the actual address of the receiver, and the packet is forwarded onward to the receiver using that address.

i3 actually supports a more complex use of IDs, to allow both the sender and the receiver to forward the packet through a sequence of servers (PHBs) on the way from the sender to the receiver. A receiver can include in a trigger a sequence of items, not just a single address. This *trigger sequence* can include both IDs and addresses. The sender can also include in the packet a sequence of IDs as well as the ID retrieved from the i3 DNS. These IDs identify the services (PHBs) that the sender wishes to have performed on the packet. This results in the packet being sent to a sequence of PHBs, first those selected by the sender and then those selected (by means of the trigger) by the receiver. As the packet is forwarded using each ID in turn, the PHB associated with that ID will be executed, the ID removed from the sequence, and the packet then forwarded again.

A key challenge for i3 is how to design the mechanism by which a sender, knowing only the ID of the trigger, can send the packet to the correct i3 node, since different triggers are stored in different nodes. The mechanism i3 uses to solve this problem is called a distributed hash table, or DHT. The function of a DHT is to take an input (the *key*, the *ID*, or the *hash*, since the input is often the hash of some value) and return some output value.

It works as follows. A DHT system is implemented as a set of nodes distributed around an internet. Part of the definition of a DHT is a set of rules that specify how different subsets of the bindings from the ID to the associated values (in this case, the lower-level network address) are distributed among the nodes. The allocation rule must be dynamic, so that the bindings can be moved from node to node as nodes are added or removed from the DHT. Not all nodes in a DHT know about each other. As part of the rules that define the DHT, each node must keep track of a sufficient subset of the other nodes so that a request to look up a particular ID can be forwarded toward, if not to, the correct node. The forwarding rule should ensure that the request reaches a node that actually stores the binding without too many intermediate hops.

The i3 triggers are mappings from an ID to an address, and the i3 nodes are organized as a DHT to store them. The first packet of a sequence is sent by the originator to a nearby i3 node, which forwards it according to the distributed algorithm of the DHT[15] until the packet reaches the node responsible for that ID, which is where the trigger is stored. After the first packet is processed, i3 returns to the sender the IP address of the i3 node where the trigger is stored, so the sender can transmit the subsequent packets directly to the correct i3 node. However, every packet follows the triangular path from sender, to i3 node, to receiver. If there is a sequence of IDs in the path, the packet will flow through a number of i3 nodes. This process could be highly inefficient, possibly ping-ponging the packet around the globe, depending on which i3 node holds the specified IDs.[16] The performance of i3 will be fundamentally determined by the routing required to send the packets to the nodes in the DHT where the triggers are located.

Delegation-oriented architecture (DOA) DOA (Walfish et al., 2004) has much in common with i3 (including a joint author) but simplifies the forwarding mechanism and makes it more efficient. Like i3, addressing in DOA is based on flat, unique IDs. As with i3, DOA uses a DHT to convert the ID to an output value, which can be either an IP address or one or more IDs. However, in contrast to i3, where all packets are forwarded to the final destination via the node in the DHT hosting the ID, in DOA the sender uses the DHT to retrieve the output value stored there. If the returned data is

15. The prototype used in i3 was the CHORD DHT.

16. The designers suggest an optimization where the receiver creates a session ID (they call this a *private* ID), which has been carefully selected to hash to an i3 node near the receiver.

an IP address, the data packets can then be sent directly to that address. If what is returned is a further sequence of IDs, the lookup process is repeated by the sender until it yields an IP address. The sequence of IDs is included in the header of the packet, so as the packet proceeds through the network from service point to service point, the service points look up those further IDs in order to send the packet on.

What the DHT returns is called an e-record. The security architecture of DOA requires that the ID be the hash of a public key belonging to the entity creating the e-record. The e-record includes the ID; the target (the IP address or sequence of IDs); an IP hint that improves efficiency; a time-to-live, after which the e-record must be discarded; the public key of the creator; and a signature of the e-record made using the private key associated with that public key. So a recipient holding an ID can confirm that the party that holds that private key has created the data. DOA assumes but does not specify some higher-level naming mechanisms that would map a user-friendly name (e.g., a URL) to an ID.

Both the sender and the receiver have some control over how a packet is forwarded in DOA. Similar to the process used in i3, the sender can prepend to the data returned from the DHT a sequence of IDs representing the needs of the sender, so the packet will first flow to services specified by the sender and then to services specified by the receiver. I believe that the expressive power of i3 and DOA to define an execution order for PHBs is equivalent, but there may be creative ways of using the two architectures, perhaps ways not contemplated by their designers, that could create very complex execution orders. Because DOA uses the DHT as a query mechanism rather than as a forwarding mechanism, it is simpler for a sender to discover the execution order of the PHBs. A sender could look up the first ID, look at the returned data, recursively look up the sequence of IDs in that data, and eventually construct the final sequence of service points through which the packet will flow. Of course, an intermediate node could rewrite the header or otherwise misdirect the packet, but since all the services in the path are specified by either the sender or the receiver, if a service mis-directs traffic, that is a signal that the service is not trustworthy. Normal execution would not involve the invocation of a service unless either the sender or the receiver requested it. (In DOA, all delivery is intentional—there is no contingent delivery in DOA.)

DOA must deal with the situation where the interests of the sender and receiver are not aligned. If the sender is an attacker, he might manipulate the data that comes back from the DHT before sending the packet. He might omit some of the IDs and attempt to send a packet directly to the final ID in the sequence, thus bypassing some services specified by the receiver (presumably these might be protection services rather than functional services). DOA does not provide a clean way to map adverse interests into a coerced delivery mode, where the sender must send the data to the IDs in turn. There are two options in DOA that can deal with this situation. If the IDs map to an address space different from the sender's, then the sender cannot address the final receiver directly. (The coercion in this case would be a form of topological delivery, using a DOA node that is an address translator to coerce the sender into traversing it.) DOA also specifies a way for an intermediate node to sign a packet using a key shared between the intermediate node and the receiver, so that the receiver can check that the packet actually passed through that node. This scheme gets complex if multiple intermediate nodes need to sign the packet, but it can ensure that the packet has been properly processed. However, it cannot prevent a malicious sender from sending packets directly to the address associated with the final ID, so DOA cannot prevent DDoS attacks based on simple flooding. The designers note that some other mechanism is required for this purpose.

Nebula Nebula defined a much more robust mechanism to control routing of packets through services in order to prevent (among other things) the problem in DOA of an untrustworthy node that might misdirect a packet, skip a processing step, or make some other error. In Nebula, the sender and receiver can each specify the set of services they would like to see in the path between them, but the DOA query to the DHT to retrieve a set of IDs is replaced in Nebula by a query to the NVENT (virtual and extensible networking techniques) control plane, where there is an agent representing the interest of every such server, and all such servers must consent to serve the packet before it can be sent. NVENT computes, with some clever cryptography, a proof of consent (PoC), which is returned to the sender. The PoC serves as a form of source route, and only by putting it into the packet can the packet be successfully forwarded; Nebula is "deny by default." As the packet traverses the sequence of services, the PoC is transformed into a proof of path (PoP), which provides cryptographic evidence that all the service nodes have been traversed in the correct order.[17] These mechanisms address the issue briefly mentioned in Walfish et al. (2004), where it is suggested that a cryptographic signing be used to confirm that the packet has been processed by an intermediate node.

17. For a detailed explanation of how this mechanism works, see the paper that describes ICING (Naous et al., 2011).

The design of Nebula was motivated by the implications of cloud computing on a future internet architecture. The Nebula architecture gives the applications access to and control over a much richer set of services than are available in the Internet. The range of services is open-ended and is crafted by the user based on the elements available in Nebula. The user might want a path that is HIPAA (Health Insurance Portability and Accountability Act)-compliant, slow but inexpensive, or transforms the data in the packets. The responsibilities of the network are high reliability and availability, predictable service quality, and assuring that the requirements (policies) of all the relevant service providers are taken into account as traffic is routed across the network. The relevant providers include networks themselves and higher-level service elements.

The Nebula proposal, like essentially all architecture proposals, assumes that the network is made up of regions that are separately built and operated, often by private-sector providers (like today's ISPs). These providers will have policies governing which sorts of traffic (e.g., for which classes of senders and receivers) they will carry, among other things. In the current Internet, the only tool that can be used to capture these policies is the expressive power of BGP, which may be limiting. In addition, the application may want to control the routing of traffic by directing it to higher-level service elements. A simple example might be a "packet scrubber" that tries to detect and remove malicious traffic, or a higher-level processing element such as a virus detector in email. A service might wish to assert that it will only receive packets that have first gone through the scrubber, even though the scrubber is not directly adjacent to the service. In Nebula, the network itself can enforce this sort of routing policy by requiring that packets carry a PoC.

Nebula is not a pure datagram network—to send a packet, the NVENT policy mechanisms must first compute and return to the sender a string of information that will authorize the data plane to forward the packet. However, these routes can be computed in advance and cached.

ChoiceNet (CN) In contrast to the other FIA projects, which describe a specific forwarding mechanism (e.g., the data plane), ChoiceNet (Wolf et al., 2014) is focused at a higher level: the control plane and what the authors call the *economy* plane. The assumption is that the data plane (which might be implemented using one of the other FIA proposals, such as Nebula) will provide alternatives among services, such as IPv4, IPv6, paths with different qualities (throughput, enhanced QoS, more secure or less liable to observation), or other services in the network (for example, firewalls, checking for malicious data, format conversion, and the like). For example, a user might choose to pay more to get higher-quality delivery of a movie. The goal is to make these options explicit and allow the user to pick among them.

The assumption is that the network will have a connection setup phase, and the user will express his choices at that time. The setup phase is implemented in the control plane, where service elements can be composed to make the final services. The data plane will then deliver the packets so that the service elements are invoked in the right order.

One model for the economy plane is an app store for services. Different service offerings would be advertised, there would be rating systems, and so on. The user would make simple choices, which would translate into actual services via the composition of elements in the control plane. The term *user* in this discussion might be the actual person using the system, a software agent, or an expert (human) such as a system administrator setting up services for users. One challenge for ChoiceNet is whether the users can realistically make these choices, whether the services offered (in the service app store) will be specified well enough that the user will not be misled, and other related issues. While a sophisticated user might want to compose specific services, this task is probably too complex for the average user. The responsibility might be delegated to the application software, which would then have control over how the services are composed, or alternatively delegated to a human expert, who again would then have control over the resulting service composition. As I will discuss in detail in chapter 10, the decision that a user will actually make in some of these complex schemes is not how to configure them but which other actor the user will trust to do the configuration on his behalf.

Introspection, or verification, is an important part of ChoiceNet. Did the user get what was paid for? Since an overall service may be composed of many parts, they should all get a share of the money but should also get the correct share of the blame for failure. ChoiceNet includes monitoring devices in the network that are included in the service invocation sequence to provide some external verification as part of any service. Service proofs and payment verification are exchanged between data and control plane. However, verification is a challenging problem, since these monitoring boxes need to know what to verify, which seems practical only if the range of service offerings is constrained and predefined.

Discussion While these schemes have the same high-level goal, explicitly configuring services in the path between two end points, they actually differ in character (how robust the scheme is to the presence of actors in the system with adverse interests). i3 tries to force the packets to flow through the intermediate service points by not revealing the address of the final node. It can thus provide a degree of protection to a receiver that wants to avoid traffic from undesirable senders, but if the sender can discover the final address, the scheme has no way to prevent the sender (attacker) from using it. DOA has a weaker form of protection, since the addresses of all the nodes can actually be retrieved by the sender from the DHT. In contrast, Nebula provides a very strong level of assurance, at the cost of a complex management plane and a complex address structure. Nebula seems motivated by the starting assumption that the sender and receiver have aligned interests, while regions of the network are untrustworthy. The Nebula discussion does not focus on the case where the sender and receiver have adverse interests; presumably in this case the attempt to have the NVENT layer create a PoC would fail.

Active Networking

Active networking addresses the requirement of invoking services in the network but represents an entirely different approach to invoking customized PHB and can serve a wider range of objectives, so I put it into a different section of the chapter. Active networking is a concept proposed in Tennenhouse and Wetherall (1996). The general idea of active networks is that the packet carries the code for the PHB within it, and routers execute this code as they receive the packet. The router is not a statically programmed device with functions fixed by the vendor but rather a device in which new functions can be added at run-time. The packet rather than the router specifies the PHB applied to the packet. This is potentially a very powerful way to specify how a packet is treated.

Active networking is a point of view about design, not a specific architecture, and a number of different approaches with different architectural implications have been proposed under the banner of active networks. I describe one approach in detail and compare it to some other proposals.

ANTS In ANTS (Wetherall, 1999), the program executed by the router is not actually carried in the packet; rather, the name of the program (a cryptographic hash of the code) is in the packet, along with initial invocation parameters (what the author calls the type-dependent header fields and I would call the explicit arguments to the PHB). The packet is intentionally delivered to the node by the address in the packet. A packet carrying this information is called a capsule.

The idea of active code in a packet raises a number of thorny challenges related to security, such as protecting the router from malicious code and protecting flows from each other. ANTS takes a pragmatic approach to these problems and requires that the code be audited by a trusted party before being signed and deployed. In addition, the code execution is *sandboxed*—placed in a constraining execution environment within the router that limits what the code can do. The code is distributed to each active router along the path by using a clever trick. Part of the header is the IP address of the previous active node that processed this packet. If this router does not have the code, it asks that previous router to send it. That previous router presumably has the code, because it just processed the packet itself. In this way, the first packet in a flow proceeds from active router to active router, sort of dragging the required code behind it. Code, once retrieved, is cached for some period so subsequent packets can be processed efficiently.[18]

When a capsule is processed at a router, the PHB can modify the type-dependent header fields in order to modify the execution of the program at subsequent nodes. The active code can also store state in the router, which permits a range of sophisticated PHBs. However, the code cannot perform arbitrary transformations on the router state. The code is restricted to calling a set of specified low-level router functions; the Wetherall paper describes the active code as being "glue code" that composes these functions in ways that implement the new service. These low-level router functions (what Wetherall calls the ANTS application program interface, or API) now become part of the ANTS architecture. Ideally, all active routers would provide these same functions, so they end up being an example of "issues on which we must all agree for the system to function," which was one of my criteria in chapter 3 for classifying something as architecture. The ANTS paper (Wetherall, 1999) identifies the router functions, the structure of capsules, the address formats, the code distribution scheme, and the use of a TimeToLive field to limit packet propagation as the elements of the ANTS architecture.

The goal of ANTS was to allow different sorts of forwarding algorithms to be implemented for different classes of packets. In this respect, ANTS has something in common with XIA, which has different routing schemes for different classes of packets. XIA actually has more expressive power, since

18. This caching of active code is an example of per-flow state in a router, and in particular an example of *soft state*. If enough time passes for the code to be discarded from the cache, the router can recover it by again asking the previous node for it, so no special mechanism is needed to determine when to remove unneeded code from the cache—the router can just discard it as necessary.

in ANTS the active code cannot implement a different, long-lived routing protocol among all the nodes—the state variables do not support that kind of distributed algorithm. In XIA, there can be separate routing protocols running to support each class of identifier. But there is a similarity between an XIA identifier of a particular type and an ANTS program name. XIA does not discuss how the code for a new identifier class is deployed—this is left as an exercise in network management. XIA allows a series of identifiers to be in a packet header—it would be interesting to work out whether an ANTS-like scheme could allow more than one program invocation to happen when processing a capsule.

Other examples of active networks can be classified along several dimensions: how the programs are written and delivered, the purpose of the active code, and what set of packets the code is applied to. Depending on the answers to these criteria, the different schemes can be described as more or less "architectural," depending on the degree to which they define and depend on "issues on which we must all agree" for the system to function.

Delivering the code If the code is delivered to the router in a single packet to control the processing of that packet, the code has to be compact. For these reasons, many active network schemes provide a set of primitives that the active code can call, and the active code (several schemes call it *glue code*) composes these primitives to achieve the desired function. The smart packet system (Schwartz et al., 1999) puts a very compact code element into the packet that is executed when the packet arrives at a given node. The PAN proposal (Nygren et al., 1999) uses a language that permits high-performance execution in order to argue that this style of active networking is practical. The ANTS scheme did not put the program in the packet but had each router retrieve the code as needed. This removed the requirement that the code fit into the packet.

The function of the active code Different schemes focus on different functional objectives and accordingly provide different primitives for the active code to invoke. The ANTS scheme had the objective of flexible forwarding of packets, and the primitives support that goal. In contrast, the smart packet scheme has the goal of supporting sophisticated and flexible network management and thus provides primitives to read management data from the router. The PLANet system (Hicks et al., 1999) includes both flexible forwarding and network diagnosis as functional objectives. The paper describes how PLAN active code can be used to implement diagnostics similar to *ping* and *traceroute*.

Program isolation and security For a variety of reasons, most active network proposals carefully limit what the active code can do. One reason is security—code that can make arbitrary modifications to the router state would seem to pose an unacceptable risk of malicious perversion of the router function. Schemes that allow the dynamic installation of highly functional code have to restrict the ability to install such code to trusted network managers. Most schemes, like ANTS, attempt to limit the impact of the active code to the packet that carried it. A more general active network capability (with more power and much more potential for harm) would allow the code to modify or extend the function of the router for *all* subsequent packets, not just the ones in a given flow.

The PLANet scheme (Hicks et al., 1999) supported both programmability of the packet and the ability to extend router functionality. It thus exploits two different languages, Packet Language for Active Networks (PLAN), which is carried in the active packets, and OCaml for router extensions. The router plugin scheme (Decasper et al., 1998) focused only on router extensibility. The active bridge project (Alexander et al., 1997) describes the downloading over the network of active code (which the authors call *switchlets*) that bootstraps a box with Ethernet bridge functionality, starting from a machine that only supports loading dynamic code. This code, once loaded, is used to handle *all* packets passing through the switch, so clearly the ability to install this sort of code must be restricted to the trusted network manager. The active bridge system also uses a strongly typed language (Caml) to provide isolation among different modules, so the system achieves security both through restricting which actors can download code and by using strong, language-based mechanisms to protect and isolate different code elements from each other.

Different active network schemes have different architectural implications. Schemes that carry executable code in packets must standardize a number of conventions that will become part of the architecture. They have to standardize the programming language, how programs are embedded in packets and how they are executed, and what functions that code can call. They may depend on some underlying forwarding architecture, such as IP. These schemes will specify many aspects of the packet header, including the parameters that the packet carries that are handed to the code when it is invoked.

Network Virtualization

The active network concept is that routers can be dynamically augmented with enhanced functionality, and different packet streams can potentially receive different treatments based on the execution of different functionality that has been downloaded in one way or another. A different way of thinking about dynamic code is that a router is just a specialized computer that supports *virtual routers* (by analogy to other sorts of device virtualization). It permits different versions of router software to be loaded dynamically into different *slices* or shares of the physical device.[19] This allows one physical device to support several network architectures (of the sort I have been discussing in this chapter) at the same time. To make a virtual network, the physical links as well must be divided into *virtual links*. Multiple virtual routers connected together with virtual links and running code supporting the same network architecture would then be a *virtual network*. Using virtualization, different architectures can coexist on the same infrastructure.

Virtual networks have something in common with active networks, but there is a difference in emphasis. The idea of virtual networks has emerged in large part from a desire to allow multiple network architectures to coexist on a common physical infrastructure, as opposed to the goal of having different packet streams within a given architecture receive different treatments or perform advanced network diagnostics. Virtual networks require the installation on the physical router of code that performs all of the network functions associated with an architecture (forwarding, routing, network management and control, and so on), which means that virtualization code is a complete protocol implementation, not "glue code" that sits on top of functional building blocks associated with a given architecture. For this reason, the isolation among the different code modules has to occur at a lower level, and the installation of the code is almost certainly much more tightly controlled than if a packet carries its own processing code with it. So active networking tends to focus on the specification of a language for defining functional enhancements, clever schemes for code dissemination, and the expressive power created by different code modules and packet headers, while network virtualization tends to focus on low-level

19. The concept of virtualization illustrates a very general aspect of digital processing. A particular function can be coded in software that runs on a general-purpose computer or directly implemented in specialized hardware. A hardware implementation will generally run faster, and thus give better performance, but is harder to change. Routers today often implement packet forwarding in specialized hardware to get sufficient performance, which implies that if the router is to be virtualized, that specialized hardware must be sufficiently general to support the appropriate range of variation. Designing hardware with this degree of generality may prove difficult in practice, which may limit the utility of virtualization to situations where very high throughput is not a requirement.

schemes for code dissemination and performance isolation. The NetScript system (Yemeni and da Silva, 1996) allows the network operator to install very low-level code (different architectures) but emphasizes the use of a crafted programming language to enhance secure execution. Network virtualization also depends on schemes to disseminate code, but they tend to have a very different character than some of the active network schemes, since they must depend on some basic forwarding architecture to distribute code that then instantiates a different architecture.

The earliest virtualization scheme that I know of is Tempest (van der Merwe et al., 1998), which focuses not on diverse network architectures (all of the virtual networks in this scheme are based on the ATM architecture) but on diverse control schemes for the same data forwarding mechanism. van der Merwe et al. (1998) introduced the term *virtual network*, perhaps for the first time. More recent proposals include VINI (virtual network infrastructure) (Bavier et al., 2006), which puts an emphasis on the use of a virtual network as a research tool to test new ideas, and CABO (Concurrent Architectures Are Better than One) (Feamster et al., 2007), which emphasizes the use of virtual networks in a production context, where multiple Internet *service* providers share a common set of routers and links provided by an Internet *infrastructure* provider. The CABO proposal depends on an assumption about economics: that an Internet Infrastructure Provider can obtain a sufficient and predictable return on investment in infrastructure such that it is willing to make the required major capital commitment. Creative business arrangements may be as central to the success of this scheme as creative technology—see chapter 12.

In fact, commercial routers today are in some respects virtualized in that they can run multiple copies of the forwarding software at the same time, although it is the same software. In other words, today's routers will allow an operator to build multiple networks out of one set of physical elements, but all run IP. Building multiple IP-based networks is an important capability in the commercial world in that it allows ISPs to provision private IP networks for customers (for example, large corporations) that run their own private networks or to offer different services to their own customers.

In the context of the FIA program, a virtualized router was proposed early on as a way to allow the research community to deploy multiple experimental architectures on one common platform (Anderson et al., 2005). This idea, put forward in the early stages of the FIA program, was in part the motivation for the development of the Global Environment for Network Innovations (GENI) project, an NSF-funded experimental network platform.[20]

20. For information on GENI, see https://www.geni.net/.

Whatever the specifics of a virtualization scheme, this approach shifts the architectural common point on which we must all agree to a different locus, which relates not to packet forwarding but to the programming environment in the router, the signaling and management tools to set up and operate the virtual network, and other aspects. Network virtualization schemes try to minimize what is standardized (or made a part of the architecture). They have to specify how code is written and installed into routers, but the only requirement they impose on the packet header is some simple identifier that the virtualized router can use to dispatch the packet to the correct virtual router (or *slice*, or *switchlet*, depending on which scheme is being discussed).

Some Architectural Comparisons

I close this chapter with some comparisons among a few of the architectures I have discussed (in particular, the NSF-funded FIA projects) using some criteria that I developed in the earlier chapters.

Identity and Location

In general, all the schemes that specify a data plane share a feature that distinguishes them from the current Internet: a two-step rather than one-step name-to-address resolution scheme. In the current Internet, a high-level name (e.g., a URL) is mapped to an IP address by the DNS. This happens in one step. The IP address is then used as both an identifier for the end point and its location. All of the FIA schemes separate the concepts of identity and location, except for NDN, which has effectively eliminated any concept of a routable location. Most of the schemes have a way to assign an identity to things other than physical end nodes, including services and data.

At the same time that these schemes separate identity from location, most of them attempt, in various degrees, to perform forwarding based on the identity information. The question of how the identity is bound to the location is a major point of divergence in these schemes.

- NDN: Packets are forwarded based on the identity information in the packet. Location is implicit, and not actually expressed in the packet.
- XIA: XIA supports a range of routing and forwarding schemes (tied to types of XIDs) from traditional Internet-style delivery based on an end-node address to forwarding based on a content ID.
- Nebula: In Nebula, there is a strong distinction between the location/forwarding component (which is related to the PoC/PoP mechanism) and any end-to-end identity components.

- MobilityFirst: Packets are delivered based on a location *hint* in the packet, the NA. However, MF includes the function of binding identity to location, the GNS.

With respect to routing, all these schemes (except perhaps NDN) take the view that the network is built of regions (like the ASs of today) that are separately managed and deployed. Nebula and ChoiceNet give the end points control over routing at the level of picking the series of ASs to be used but give each AS control over its internal routing. XIA and MF assume a rather traditional two-level routing scheme, but routing in NDN has a very different flavor, since it is identifiers and not locations that drive the forwarding decisions.

Service Model

As I have mentioned a number of times, the service model of the current Internet is best-effort delivery, although the packet header includes the ToS field, which allows a packet to signal that it wants a different sort of service. While this feature is not utilized in the public Internet, it is a part of the architecture. The various FIA projects differ substantially with respect to the range of services that they provide to the higher layers.

- Nebula and ChoiceNet: These proposals assume that service building blocks in the network can be composed to present a rich selection of end-to-end services to the applications.
- XIA and MF: These designs provide a small number of service classes, corresponding to different classes of IDs—for example, data, services, and hosts. Each of these classes would correspond to a forwarding behavior in each router. MF also allows additional functions to be installed on routers in the path. In the initial specification of MF there is no discussion of per-flow QoS, but the function could probably be easily added.
- NDN: This design implements a single general service that returns a packet associated with a name. It allows for variation in service quality (e.g., QoS) by using a field in the packet similar to the ToS field of today. The only way that an application could embed a service element into the path from a client to the data would be to create a URL in an interest packet that names an intermediate service as its target and then embeds the name of the desired data into the body of the URL.[21]

21. One would have to consider whether this "trick" of putting one URL inside another would confound the security architecture of NDN, where the names are used to link keys to data. It is not clear clear how this approach would work in the case where the requester and the provider have adverse interests.

Because XIA and MF offer a limited and prespecified set of services, the designers of the architecture can work out the relationship between the router PHBs and the resulting end-to-end service. This allows the network to offer known service to the user. In Nebula and ChoiceNet, the end user composes the service out of a range of service elements that are not specified as part of the architecture. This approach allows the set of services to be open-ended (such as the example of the HIPAA-compliant path used by Nebula, or a path that avoids a particular region of the world), but it is the responsibility of the user rather than the network to determine how the service elements (the PHBs) combine to form an overall service.

The Role of the ISP

In general, these architectures give these ISPs a larger role in the operation of the network compared to the current Internet.

- NDN: ISPs are responsible for the dynamic caching of packets of data, optionally validating the legitimacy of the data, and other tasks.
- XIA: ISPs must implement the range of services tied to types of XIDs.
- Nebula: ISPs provide a validation that packets have followed the path that was generated by the data plane. They can offer a range of network-level services to the users, with different service qualities.
- MF: Like XIA, ISPs provide a range of services; they also host third-party computing services on their infrastructure and provide mobility-specific services such as short-term caching, redirection, and the like. ISPs are probably the actors that implement the GNS.
- ChoiceNet: The data plane is not specified in ChoiceNet, but it must provide a set of interfaces to the control plane, through which the data plane can be configured to deliver services. Delivery of enhanced services, and allowing the user to select them, is the central point of ChoiceNet.
- In contrast, network virtualization schemes give the infrastructure provider only the responsibility for installing routers and links and providing the means for ISPs to install code on those routers.

Functional Dependency

In chapter 3, I talked about the aspect of architecture that I called functional dependency. The architecture should identify and make clear which operations depend on others, and what services have to be up and running for the basic function of the network (packet forwarding) to operate successfully. I noted that the current Internet reflects a preference for a minimum set of functional dependencies. Several of the schemes I have described here have a richer set of functional dependencies.

- XIA: XIA is somewhat similar to the current Internet in its functional dependencies. Routers have a somewhat more complex task, since they have to compute more than one set of routes for different sorts of identifiers. XIA, like the Internet, presumes some sort of higher-level name service like the DNS.
- Nebula: Nebula depends on a presumably complex control plane (NVENT) that computes and authorizes routes. If the distributed control plane malfunctions, there is no way to send packets, since Nebula is "deny access by default."
- MF: In MF, forwarding depends on the GNS. The GNS is a complex, global system that must map from flat ID to network ID in real time. The robustness of this system is critical to the operation of MF.
- NDN: NDN depends on a routing scheme that provides routes for names, not numbers. However, it does not depend on anything else. It shares with the Internet the goal of minimizing functional dependencies. In order for a node in NDN to request information (send an interest packet), it is necessary for that node to know how to construct the name of the information (which may require a bit of context), but in principle the nodes can construct that context in real time if necessary, depending only on each other rather than any third-party service.

Dealing with Adverse Interests

Most architectures are described through the lens of aligned interests between sender and receiver, but most proposals also contemplate to some extent the problem of adverse interests—when the sender is not trustworthy or perhaps simply an attacker. An important question is how different schemes deal with adverse interests. One must look at both directions of the communication. Most of the architectures are described in terms of sender and receiver, client and server, or data requester and data provider. It is important to remember that no matter which actor initiates the connection, both ends will probably send packets to the other. Whatever the framework, either end can show malicious intent with respect to the other.

The architectures introduced in this chapter that configure services in the path from sender to receiver make their role explicit when interests are not aligned. All of the architectures use some sort of "firewall" device as an example of their utility. They deal to varying degrees with the problem of making sure that the attacker cannot bypass the protection device, but at least they acknowledge the issue. In contrast, the architectures that are focused on the delivery of data pay less attention to this problem—in particular, to the problem that the receiver of the data may need protection in case the data contains malware. The DONA proposal is an exception; it explicitly discusses inserting a middlebox into the path of the returning data, although the mechanism described seems clumsy, as the task of inserting the protection service is delegated to one of the nodes that does name-based routing. The TRIAD and NDN proposals do not dwell on the issue of protecting the requester from the data it requests; they describe a scheme in which the requester simply gets the data as sent. They focus on the (admittedly hard) problem of routing the data request to the closest copy of the data and do not discuss the problem of deploying services in the path from the data to the requester, either protection services or functional services such as format conversion, which is an example used in the DOA proposal (Walfish et al., 2004) of a useful service that might be delegated to an element in the network.

Counterpoint—The Minimality Principle

Some of the architectures I have described in this section offer a very rich delivery service. They support sending a packet in order to retrieve named data or to contact a service, or to send a packet to a destination (or set of destinations in the case of multicast.) A contrarian (and minimalist) view might be that no matter what the high-level objective of the sender is, in the end a packet is delivered to some location(s). These higher-level names (data or service) must somehow be translated into a destination so that the packet can actually be forwarded. The minimalist principle would suggest we ask why the network itself, in a manner specified by the architecture, should have the responsibility for doing those translations. Perhaps that function should be assigned to some higher-level service, as it is in the current Internet. The sender of the packet would query this high-level service, perhaps just before sending the packet, in order to get a low-level destination to which the packet is then directed. An example of such a design for multi-level naming is described in Balakrishnan et al. (2004).

Why is it useful to embed the translation service into the network itself? In some cases, the network may be in the best position to make the translation. If the goal is to reach a version of a service (or a copy of some data) that is the fewest network hops away, the closest in terms of around trip, or over a path that is not congested, the network might well have better access to this information. On the other hand, if the goal is to select a server that is not overloaded or not failing or compromised, the network does not know this. The application is in a better position to implement this sort of selection. By moving the translation service out of the network layer, it becomes easier to modify the criteria that control the selection, which could include criteria that reflect issues of policy and not just performance optimization. To make the problem harder, an application might want to make a selection based on both of these metrics, which implies some sort of cooperative decision-making—an idea that to my knowledge has never been embedded into an architectural proposal. Another reason to embed the translation service into the network is that there may be security benefits, but I need to defer that consideration.

The same minimality question could be asked as a challenge to architectures such as DOA, which provide architectural support for the placement of services into a path from a sender to the final destination. Perhaps this function should be implemented at a higher level. This is the way the Internet works; it provides no support for the configuration of services and still works. The placement of services into a flow of data is done either at the application layer (as with mail transfer agents in email) or by using devices such as "transparent caches," which depend on contingent (or topological) delivery of the data to perform their function. My suspicion is that the most compelling justification relates to protecting end points with adverse interests, which means the schemes have to display a level of resistance to manipulation.

A possible criticism of architectures that focus on delivery to higher-level entities is that they have pulled down into the network architecture part of what used to be an application-layer function, without pulling *all* of the service requirements into that layer. The design of email, with its application-specific forwarding architecture, allows certain sorts of protection services to be put in place in an application-specific way. In the Web today, providers of Web content use very complex schemes for data caching, finding the closest copy of some data, and other tasks. Pulling a data retrieval model that is Web-like into the network layer may actually make development of sophisticated application-specific forwarding architectures harder, not easier.

Expressive Power

It is interesting that most of the architectures I describe here do not include in their design any explicit arguments to service points (PHBs). The schemes with URLs as data identifiers can encode any sort of information in the variable-length ID, of course. Nebula has a very rich set of explicit arguments—the PoC and the PoP. In general, most of the architectures seem to assume that the service elements will operate on the data payload (doing format conversion, inspection for malicious data, and the like). Many of the architectures do distinguish the first packet in a connection, because that packet requires extra work to resolve IDs to more efficient network addresses. A careful review of all these architectures would include a catalog of any state in any service element, how that state is maintained, and other information.

Almost all of the architectures try to avoid contingent delivery except for the basic packet forwarding mechanisms. They use intentional delivery, with the sender and/or the receiver specifying the path of the packet across the service points. The use of intentional delivery probably aids debugging when things go wrong.

Where to Go from Here

Since this book is a somewhat opinionated walk through the space of architectural alternatives, an obvious question is which of the approaches I prefer, or what sort of architectural synthesis I might put forward as my approach. But there is a ways to go before I can answer that question, which I do contemplate in the final chapter. There are a number of requirements that I need to analyze in more detail, including many aspects of security, which are the topics of the next several chapters.

8 Naming and Addressing

Some aspects of network architecture take on a character that is almost philosophical, and one of these is how entities are named. What's in a name is a complex question.

Very early in the history of networking, John Shoch (1978) published a paper that provided a basic framework for naming that has stood the test of time:

- A name identifies an object,
- An address identifies where it is, and
- A route identifies a way to get there.

Jerome Saltzer (1982) elaborated on Shoch's work by emphasizing that to move from a name, to an address, to a route, some sort of *binding* is required, and the mechanism that provides the binding is key to the functioning of a naming system. Karen Sollins (2002) observed that names play a number of roles, which fit into three categories: location or access, identity, and meaning or mnemonics. All three of these categories will figure in the different examples of naming in this chapter.

While names serve a number of purposes, one important role is location or access: allowing the named object to be found. A name is often the starting point for a series of steps that end up with the location of the named object. *Name resolution* is the term given to this overall process, which may involve a number of bindings.

A real-world example may help. Imagine that a reader wants a book on the history of cooking. Initially, the reader may not know the name of a specific book, only the topic, but a *search* of some sort might yield the title of a suitable book. The next step is to find a copy. The reader might go to a library, where a search of the card catalog (or its online replacement) will provide a different sort of name for the book—probably its Library of

Congress Call (LC) number.[1] But the LC number does not yet indicate where the book is. What is needed are instructions of the form "second floor, room on the right, third set of shelves, fourth shelf from the bottom." A librarian might provide that information, or the library may provide maps that show where books with various LC numbers are shelved. Once the prospective reader gets to the correct shelf, some minor searching will be required to find the book on the shelf, which is generally a reasonable task, and, with luck, the books are in order by LC number. This example exploits several *bindings*: the card catalog binds the title to the LC number, and the map of the library binds the LC number to a location.

Alternatively, a prospective reader might go to a bookstore to purchase the book. In most bookstores, books are organized not by LC number but by general topic categories, and then perhaps by title or author within that category. It might not be clear, however, exactly where the store would shelve a book on the history of cooking, so the prospective reader might just ask if the store has the book in stock. Many bookstore owners know their inventory well enough that they can not only answer that question but also go directly to the shelf where the book is located. In this case, the binding between title and location is stored in the memory of the shop owner, which is an excellent scheme so long as the store is not too big. But nobody would expect a librarian at the Library of Congress to be able to remember all the titles and their locations. LC numbers were invented exactly to allow items to be found in such a large collection.

This example illustrates that different sorts of names (or *identifiers*) serve different purposes. The title of the book probably provides a better insight into the subject of the book than the LC number (it has more *meaning*), but the numerical ordering of the LC numbers makes it easier to find where the book is in the library.

The term *identifier* is a useful alternative to *name* because another role for names is to establish the identity of an object. This is where discussions of naming can get a bit philosophical. Are different press runs of a book the same book? If errata are fixed, is it now a new book? Are hardcover and softcover printings different books? And so on. A little thought will suggest that there is no consistent answer to these questions. Normally, the creator of an object (or an industry group that sets norms for an industry) will provide the answer in a given context, but the relation of name to identity implies that it is important that names be stable. The paper by

1. I do not mean to deprecate or dismiss the Dewey Decimal System, which many librarians and scholars still prefer.

Shoch I cited earlier has been published in a number of places, including as an Internet Engineering Note, but it has always kept the same title. If its title changed when it was republished, that would be quite confusing. It is the stability of the title that allows different readers to be confident they have read the same document, even though they found it in different locations. So another key role of certain kinds of names is the identity role identified by Sollins: to indicate when two data objects are different representations of the same entity or different entities.

What does this discussion mean for network architecture? The current Internet does not concern itself with names. It delivers packets to a *network address*, which is more like a shelf location than a book title. But some proposals for an alternative internet include the idea that the packet should carry something more abstract, such as a service name or the name of some data object, as the destination of the packet. These proposals have *names* rather than (or as well as) network-level addresses in their designs. If some sort of high-level name is part of a network architecture, this implies that (at least to some extent) the network must understand and perhaps control the binding from that name to a location. This raises the question of what sort of name it is and what its role is. Just as a book title and an LC number (both names or identifiers by my definition) play different roles, different sorts of names in an internet architecture may play different roles.

Service-Based Packet Delivery

The XIA proposal, as well as some of the other proposals from chapter 7, includes the concept of delivering a packet to a *service* without knowing where that service is. There may be many copies of the service (the binding from name to location is "one-to-many"), and the network will deliver the packet to the copy that is closest or best-suited by some measure, but the concept of service is abstract—calling something a service does not help a router forward a packet. What a router *can* implement is a concept such as "closest copy of something with the same identifier." In fact, this concept is the same as the routing mechanism I introduced in chapter 2 by using the term *anycast*. The anycast service delivers a packet to one copy of a destination, presumably one that the network finds preferable. While delivering a packet to a service is the same as anycast, the use of the term *service* hints at issues of scale and performance. The Internet today supports anycast, but only for a small number of end nodes. If an internet were to provide anycast to services, there might be hundreds of thousands of them, or perhaps even millions (but probably not billions, as there might be end nodes).

The current Internet uses the DNS in order to deliver a request to a particular version of a service. The DNS takes a name as input (such as www.mit.edu) and returns an IP address bound to that name. It can return a different address depending on where the requester is, so it can optimize the address returned depending on the location of the requester or any other factor that it is designed to take into account.

This DNS version of anycast differs in a number of important ways from an anycast function implemented in the routers. Perhaps most important is which actor has control or authority over how the name resolution is performed. With network-level anycast, the network operators control the resolution through the configuration of the routing protocols, whereas with the DNS, the owner of the name controls the resolution. The two options provide different benefits. The routing protocols can find a copy of a service that is close in terms of network hops or latency, but the network does not know about loads on different servers, or other considerations that the service provider may care about. The question of which actor (or set of actors) has control over a particular function (and name binding is a powerful function) is an important consideration in architecture design.

Routing to a service (or anycast) is more complex than simple unicast delivery to a specified end point. I described the Internet as "semantics-free" because what it does is so simple that there is essentially nothing to the specification of the service, but as delivery services get more complex, application designers will need to know something about how the service actually works and what its delivery model is. The specification will have to describe how service nodes join the anycast group (what sort of protection there is against unauthorized parties joining, for example), the basis for determining how the specific destination is picked from the anycast group, how consistent that binding is (is the anycast path at all "sticky"), and so on. The specification may be abstract: it might describe in general terms the result of the anycast routing scheme—lowest latency or shortest hop count—rather than specify the exact distributed routing algorithm. But a specification at some level will be required if application designers are to use it. There might turn out to be more than one variant of an anycast scheme, with different behaviors suited to different purposes. As network services get more complex, network designers need to understand application requirements in detail in order to offer services that are actually useful.

Data-Driven Packet Addressing

A digression on words for information The word *information* is used very generally as a high-level concept. In contrast, I will use the word *data* to describe a particular representation of an information object—a sequence of bits that capture the information in a specific representation. The term *content* is often used to describe commercial works—information produced to be sold or licensed for profit. The movie and music industries are content industries and produce commercial content. In this section, I will use the word *data* when I am talking about information reduced to a specific encoding as bytes and use content when I am talking about commercial content. A name for an information object can refer to that information independent of form, but a name for a data object normally refers to a specific encoding.

Several of the architectures I describe in chapter 7 support a form of name-based packet forwarding using the name of data to be retrieved as the destination in the packet. A scheme for naming data is much more complex than a scheme for naming services, in part because there are various sorts of data ("data" is not a homogeneous category) and in part because many naming schemes for information already exist. Some of those schemes have been invented by librarians and many by creators of commercial content—content created for the purpose of making money. The designer of a network architecture that includes names for objects must think about the relationship of network-level names to the many naming systems that already exist. I list here a few examples:

- Digital object identifier (DOI): Identifies any sort of digital object; for example, commercial content such as a book or video, a published paper, or one's personal files.
- International Standard Book Number (ISBN): Identifies a specific book, an edition of a book, or a booklike product (such as an audiobook).
- International Standard Serial Number (ISSN): Identifies serials, journals, magazines, and other periodicals irrespective of their medium.
- International Standard Audiovisual Number (ISAN): Identifies audiovisual works.
- International Standard Recording Code (ISRC): Identifies sound and music video recordings. It identifies the recording of a work, not the work itself.
- International Standard Text Code (ISTC): Identifies text-based works.

- International Standard Musical Work Code (ISWC): Identifies musical works.
- Library of Congress Control Number (not to be confused with the Library of Congress Call Number) (LCCN): Identifies the individual items in the library.

These naming systems have little to do with locating the named object. They capture the identities of objects at various levels (e.g., the ISWC identifies a musical work, while the ISRC identifies the recording of a work). Most of them play a role in tracking ownership, rights management, and the collection of royalties, for example.

As part of the overall scheme for managing identity, authorship, and royalties, multiple naming schemes also exist to name creators.

- International Standard Name Identifier (ISO 27729) (ISNI): Identifies people involved in the production of content. The ISNI International Authority searches many information sources to find individuals, and assigns them a number. An individual can also apply through a registration agent to be assigned a number.
- Interested Party Information (IPI): Identifies parties with rights to a particular work. The identifiers are maintained by the International Confederation of Societies of Authors and Composers.
- Library of Congress Name Authority File: Identifies creators of works in the Library of Congress.

A closer look at some naming schemes will yield further insights about the various ways names for data are used. The two subsections that follow give two examples.

Naming in the Web

Perhaps the most familiar data names on the Internet today are the URLs of the Web. URLs have two parts: the name of the machine where the data is located, and the name of the data (the "Web page") on that machine. The first part of the URL, the machine name, is a DNS name, so it names the default location of the item. The second part need have meaning only to that machine. The problem with URL names is that since a URL directly encodes the location, moving the data requires giving it a new name. URLs fail the requirement that names be stable, independent of where the object is located; the term *link rot* has been coined to describe URLs that name data that is no longer reachable using that URL.

I am using the term *URL* because it is commonly used and widely understood. However, in doing so I am being inconsistent with the current naming preferences of the IETF and the W3C (the organization that sets Web standards), both of which have declared that the term *URL* is deprecated, and the preferred term is *URI* (uniform resource identifier). URI is a more general term, which can apply to names that imply location, names that are more high level (and might be designed more for stability than location), or names with yet other attributes. There can be multiple *schemes* or classes of URIs, one of which looks like URLs. As part of the development of the URI concept, there was also a proposal for a specific form of URI called a uniform resource name, or URN (Masinter and Sollins, 1994). I invite the interested reader to look at the Wikipedia entry for *uniform resource identifier*, which will both provide much more information and illustrate why I sometimes described discussions about naming as philosophical.

In many cases, it may not matter if URLs are not stable. For commercial content in particular, going from name to location happens in many steps, and the URLs at each step have different functions. As a case study, consider the paper of mine that I critiqued in chapter 4, "The Design Philosophy of the DARPA Internet Protocols," which was published by the Association for Computing Machinery (ACM) in 1988 and is indexed in the ACM digital library. An online search reveals that the URL for that paper is http://dl.acm.org/citation.cfm?id=52336. However, that URL goes not directly to the paper but instead to information *about* the paper: an abstract, where it has been cited, and so on. On that page, there is no URL that points to the actual paper (the PDF representation) but rather a URL that lets a prospective reader sign in or pay to see it.

However, the overall system is much more complex than that. If I connect to that URL using my personal computer, there is an extension in my browser that does something so complicated I have no idea how it works—it knows that I work at MIT and that MIT has a site license for ACM content, so it rewrites that URL above as http://dl.acm.org.libproxy.mit.edu/citation.cfm?id=52336. If you (the reader) attempt to use that URL, the MIT library Web site will challenge you to log in using MIT two-factor authentication. Only if you pass that test will you get to the ACM page. But if you succeed, the page will look different. Instead of offering a URL where you can log in or pay, it now has a URL to the content itself. The URL is http://dl.acm.org.libproxy.mit.edu/ft_gateway.cfm?id=52336&ftid=59511&dwn=1&CFID=794436021&CFTOKEN=17429941. If you try to use this URL without first authenticating to MIT, you will get an error message.

That last URL is mostly meaningless to me, and to most people, I suspect. I can tell that there is a database query embedded in the name, but that is about all I can guess. More importantly, the URL has almost certainly been synthesized at the moment this page was sent to my browser. Cascades of URLs are being used to implement content management, access rights, and other tasks. This URL, which will actually retrieve the data (the PDF of the paper), is not intended to have any long-term stability. If the ACM wanted to move the location where the data was hosted, they would change the information that was used to generate this URL. The URL that needs to be stable and persistent is http://dl.acm.org/citation.cfm?id=52336. By using cascades of URLs (and explicit human involvement in following the links), the owner of content has achieved a number of things, including rights management and separating the URL that has to be stable from the URL that describes the final location of the data.

Digital Object Identifiers

One of the naming schemes I listed earlier was digital object identifiers, or DOIs. As the name implies, DOIs were created to name any sort of digital object. DOIs are assigned to an object by its creator or other interested party. The ACM assigned a DOI to my paper; it is 10.1145/52325.52336. The first part (10.1145) means that the ACM created the DOI; the second part is their ID for the paper. You will note that this second number also shows up in the URL. You (the reader) can also look up my paper by going to the Web site of the International DOI Foundation (IDF) and copying that DOI into the search box.[2] Try it and see where you end up. You will end up where you did before, at the ACM page that describes my paper, with the same invitation to sign in (or pay).

The DOI system is a very ambitious naming system for digital information. It was developed by the Corporation for National Research Initiatives and is now run by the International DOI Foundation, primarily for the use of the publishing industry. In contrast to the URL naming scheme, which does not have a lot of required mechanisms behind it (aside from the DNS, on which it depends), the DOI system is a more complete architecture, but it is a *naming architecture*, not a network architecture. The designers use a very straightforward method to bind DOIs to the next level of information, a conceptually centralized name resolution service run by the foundation. In contrast to the DNS, where names are resolved by a sequence of queries to different servers, or the PURSUIT system, where names are resolved by a sequence of queries to nested scopes, the DOI system just maintains a single mapping database. (I say it is *conceptually* centralized because it is under the control of a single actor, but it can be physically replicated for resilience and

2. https://www.doi.org/index.html.

performance.) At the moment, their Web site states that they keep track of about 150 million names.

DOIs are intended to be stable. Looking them up on the DOI Foundation Web site is a way to bind them to location. Its Web site states: "Clicking on a DOI link... takes you to one or more current URLs or other services related to a single resource. If the URLs or services change over time, e.g., the resource moves, this same DOI will continue to resolve to the correct resources or services at their new locations." Of course, it is up to the owner of the content to update the URL in the appropriate service to point to the new location.

The lessons from these examples are that naming schemes for objects are complex, mature, and designed to solve a number of problems, of which locating the actual data is only one. The designs would suggest that rights management is more important. Many of these naming schemes interact in complex ways (one can map from one kind of name to another in many cases). This space has to be seen as an ecosystem, with at least as much complexity (and range of actors) as the Internet ecosystem on which it sits.

Given this reality, how would a network-level data naming scheme fit into the larger naming ecosystem? If a network architecture proposes to incorporate some sort of data name into their scheme, the first question is why—which of the various purposes served by names will the network architecture address? For most of the schemes I have described, the answer is a narrow one—the role of the data name is only to facilitate locating and retrieving the data. This answer is plausible, but it raises a second question, which is where exactly to make the cut between what the network does and the larger ecosystem of content management. Various schemes that include object names as a mode of address make the cut at slightly different points, with significant implications. Two examples from chapter 7, XIA and NDN, will illustrate the range of options.

XIA, along with several other schemes, uses a name for data that is derived from the data itself—a hash of the bits if the data is immutable. These names have the following implications.

- *Self-certification*: The names can be self-certifying. Once a receiver has fetched some data, the software can recompute the hash and confirm that it received the correct data. Of course, if the wrong data was delivered, recovery is a different issue.
- *Name derivation*: The name (the hash) can only be created by someone who already has the object. In this scheme, there is no way to synthesize a name for an object in order to ask for it. Thus, this scheme requires a

higher-level naming ecosystem that can provide this name as a step in retrieving the object.

- *Immutability*: Because the names are derived from the data, a new name will be required if the data changes in any way, so these names are very low level, referring to a specific bit-level representation of the object.
- *Location*: Because the names are derived from the data, they cannot provide any guidance as to the location, so some other binding is required to map the name to the location. There are different ways that this can be done, as I will summarize.
- *Stability*: Names are stable. They will not change so long as the data does not change. (Dynamic data requires further discussion.)

NDN uses network-level names that resemble URLs. (The names identify packets, rather than complete objects, but that point does not matter for this high-level comparison.)

- *Self-certification*: The creator of a data packet is expected to compute a hash of that packet and sign it with its private key. Anyone with the public key can verify the signature and thus the authenticity and integrity of the data. However, there is no requirement that this hash be part of the name. It is associated with the data, not the name of the data.
- *Location*: NDN names provide a basis for locating the named object. In NDN, there is no equivalent of the DNS to map these names to a lower-level locator. The name itself is the address in the packet, and the router must make a binding from the name to a forwarding decision. A variety of schemes can be used, ranging from flooding the request to using a global routing protocol that precomputes routes for prefixes of NDN names. (In contrast to URLs, where the DNS name is distinct from the rest of the name, NDN allows the router to look at and route on as much of the name as appropriate.) Because a name indicates a location, if the location of the object changes, either the name changes or some sort of indirection is required to allow the name to continue to function.
- *Immutability*: Different sorts of information can have different conventions for naming the data, and those conventions will determine whether the data described by a given name can change.
- *Stability*: NDN does not require or prevent names from being stable. The owner (or hosting site) can decide how to manage the name of an object.
- *Name derivation*: Names for objects can be synthesized, if the receiver knows the rules (the naming conventions) for the class of object desired.

For an in-depth comparison of flat and hierarchical data names, see Adhatarao et al. (2016).

Digital objects can exist in many representations. For example, video can be coded in high definition (HD), standard definition (SD), optimized for a specific display, and other formats. At one level, those different representations are the same object, and might have the same name, just as hardcover and paperback version of the same book usually have the same title. But both XIA and NDN names must refer to a specific representation of the object (this is obvious in the case of a hash that is derived from the data itself, but it is also true of NDN names). To retrieve a specific representation of an object, there must be a binding that maps from a high-level name to a lower-level name that identifies a specific representation.

The designers of NDN posed a challenge that I call the "desert island" challenge. Imagine that there are two people stranded on a desert island but equipped with working wireless devices by some fluke of technology. One happens to have a copy of the *New York Times* on his device, and the other wants to read it. The two devices can communicate directly, but there is no connectivity to the outside world. One of the goals of the NDN scheme was to make this possible. In the NDN scheme, where request packets carry the name of the desired data, if one device requests it and the other is willing to provide it, the data can be fetched without the need for any system elements other than the two devices themselves. This approach is a beautiful example of what I called, in chapter 3, minimal functional dependency. However, two things must be true to achieve this objective.

First, the receiver must be able to construct the name of the desired object (the *New York Times* in this case). The receiver must know the naming scheme that the *Times* uses to name its objects.[3]

Second, there cannot be any other step required before the transfer happens. In particular, there cannot be any query to a rights management system, because there is no connection to the rest of the world. For NDN to solve the desert island challenge, its naming system must be complete and self-contained. It must include as a part of its design all the bindings

3. The alternative would be for the person holding the *Times* to give the name of the object to the person who wants it. There are a lot of data transfers that involve one person asking another: "What was the name of that file?" In this case, the most important requirement for the name is that it be easy to type. Or the scenario could get more complex, and one device could scan the screen of the other device. I dwell on this only to point out how quickly naming schemes can become complex.

necessary to go from a name that is meaningful to a user to the retrieval of the object.

A system that meets the desert island challenge would be a great tool for data sharing and would be very popular among those who share unlicensed content—otherwise called data piracy. The content owners would strongly resist such a scheme. To them, rights management is more important than sharing on a desert island. Of course, lots of content is free of rights management, rules that constrain sharing, but the requirements and expectations will differ. There is no single answer as to how data can or should be shared. A system that aspires to be complete or self-contained must solve all the problems of naming, including stability over time (capturing identity) and rights management, as well as location. A scheme like XIA, which has to sit inside some higher-level naming scheme, can leave some of these problems to that layer.

Scale The next challenge with packet routing based on names for data is scale. I estimated that there might be a million (to within an order of magnitude) service-based anycast addresses, but the number of data objects on the Internet is probably closer to a trillion (again as a very rough guess). It is daunting to imagine a practical routing scheme that can compute routes to a trillion items on the network, dealing with each of them individually, just as it is unrealistic to expect a librarian at the Library of Congress to remember the locations of all the items there. If delivery using names for data is to be implemented, some binding scheme must be devised that supports this sort of scale. Some sort of clustering or structure must be devised to help keep track of location.

One approach for dealing with scale, illustrated by NDN as well as TRIAD, is to embed some knowledge of the location into the name, but this approach creates problems for the stability of the names. The other approach is to associate some sort of location *hint* with the name. I find the idea of hint-based naming schemes appealing. When finding a book in a library, the LC number can be seen as a hint. A librarian queried about a particular book might say that the book is a current best-seller, and so it is shelved not in the normal location but on a "popular books" shelf. In that case, the librarian gives a better hint than the card catalog, but in all cases, the book still has the same title. When one uses a search tool on the network today, what is returned is a URL, but in a different naming architecture, a search tool could return a unique name as well as one or more hints as to where the search engine recently found it.

XIA, DONA, MF, and Netinf have data names that are not related to location—for data that does not change, the name is a hash of the data.[4] These names satisfy the requirement that names be stable but require either a massive database to provide bindings from name to location (like the DOI system or the GNS of MF) or some sort of hint scheme where the hint can be mapped more directly to the location. XIA takes the approach of using a network identifier (like an AS number) as a hint. The idea is that the network number is used to get the packet to a service or region of the network small enough that within that scope the system can keep track of the locations of all data stored in that scope. DONA uses the identity of the creator as a hint (names are of the form P:L, where P is the hash of the public key of the creator). This sort of hint will help if the data associated with a given creator is hosted at a consistent location.

Allocation of function In my view, this long and perhaps digressive analysis has not yet answered the central question: why should the network layer control the binding from data name to location? Given all the binding mechanisms that are provided in the complex naming ecosystems that already exist, there has to be some functional benefit from having this capability inside the network. NDN provides one clear answer. If the network knows about the names of objects (or packets, in the case of NDN), routers in the network can cache popular content and reply to a query with the object rather than forwarding the query all the way to the location where it is stored. So one way to decide whether network-level data names are valuable is to ask how often a caching scheme will prove beneficial. How often will data be sufficiently popular that short-term caching (in the routers of the network) will be effective? How often will providers of content insist that their content not be cached but instead fetched from an authorized origin in order to manage rights, track payments and royalties, and perform other tasks. Network architects need to understand application requirements in detail if they propose to offer more complex services.

Another benefit from having the network understand data names (if the names are self-certifying) is that the network itself can compute the hash of the data and detect whether it has been fooled into delivering the wrong object. If the network has control over how the data is obtained, it may be important for it to detect that it is being fooled. (The design of NDN

4. Other sorts of names would be derived based on different rules. For example, the name of an end point might be the hash of a public key associated with that end point.

makes this action much easier. Since NDN names refer to packets, not the complete data object, the validity of what is returned can be verified packet by packet, as opposed to having to reassemble the overall object from its component packets in order to compute the hash.)

Location-Based Addresses

Location or end-point addresses of the sort found in the Internet are much simpler than the higher-level names I have been discussing so far, but there are some important design considerations.

Names of this sort are usually organized in some sort of hierarchical or nested fashion. Postal mail has addresses of the general form <country, state, city, street, house number>. In addition, the postal system created ZIP codes, which again have a hierarchical structure: the first digit identifies a multistate region of the country, the next two digits identify the sectional center facility (SCF) where mail is sorted, and the last two digits identify a village, town, or district of a city. They are structured to make the binding from ZIP code to the routing and delivery sequence as efficient as possible.

Phone numbers were similarly designed to make routing efficient: the area code and the next three digits identified a switching center, and the last four digits identified the specific circuit within that center. But the telephone numbering system broke down when users demanded that they be able to take their number with them (or *port* them) when they switched carriers. Users wanted to port their numbers because other people knew those numbers—they were serving as a form of identifier (they had *meaning* to people) as well as a locator. The telephone numbering scheme now includes a massive local number portability database that records whether a phone number has been ported. If so, the database stores a second number (that looks just like a phone number), called the location routing number (LRN), which actually provides the information necessary to route the call to the correct switching center. The lesson here is that since almost all naming systems have cascades of identifiers, location names should *never* be used for any purpose other than location. For any aspect of identity, use different sorts of names that are designed to be used as identifiers and to allow efficient binding to a location. Pragmatically, design them so that they appear to have as little meaning as possible.

There is a nice example in the data-naming space of how names can avoid suggesting any meaning. The elements of DNS names identify the service where the next-level name can be resolved, but they are text strings (like www.mit.edu), and people associate meaning with text strings. MIT means something to many people, and the consequence of this is trademark fights over who owns which name. The early designers of the DNS believed that the names had to have meaning, because they expected people to remember them type them in, for example. I thought it would be better if they were meaningless, but I lost that argument. In contrast, DOI names convey no meaning. The DOI of my architecture paper is 10.1145/52325.52336. The first part means that the ACM assigned this DOI to my paper. If the first part of the string had been "ACM" instead of "10.1145," then a tussle might have broken out over which organization owns the name "ACM."[5] No one is going to fight over the string "10.1145," but no one is going to remember it either. These names *only* serve as the input to some further step in the name resolution process.

Internet addresses are not too memorable in most cases, and with the conversion to IPv6, where the addresses are 16 bytes long, there is little risk that anyone will attempt to remember them. But there will always be a temptation to use them as some form of identifier, and this temptation must be resisted.

The complexity with Internet addresses today is that while they initially had a two-level hierarchical or nested structure that supported an efficient forwarding operation at each router, this structure eroded as the Internet started to run out of addresses. Actors that had control of a block of addresses started to give a subset of those addresses to another actor that was reachable by a different path through the Internet. It is as if in the postal system, some part of an area served by a single ZIP code was somehow transported to a different city. Running out of addresses is a disaster in any addressing scheme, which is why the Internet is slowly converting to the larger addresses of IPv6.

Many of the proposals in chapter 7 had a clean, two-level scheme, with regions, and end points within the regions. In principle, a new architecture could have a multilevel scheme, with global regions, subregions within those regions, and so on. The challenge is that Internet routing is not just geographical. Because different ISPs operate in the same region, packets need to be routed to a specific region belonging to a specific ISP. The telephone system was tripped up by a similar problem. When the system was operated by a single major provider in the United States (the Bell Telephone System or AT&T), routing based on geographic region made sense. But when competing telephone companies entered the market, calls had to be routed

5. This is just hypothetical. I do not know of any trademark disputes over the name "ACM."

to a region inside a specific company. Today, the telephone system routes calls by using the first six digits to identify the call center, independent of what provider owns the call center (except for numbers that have been ported, as discussed earlier). The Internet works the same way. The first part of an Internet address maps to a region of a provider, and the routing system deals with them as flat identifiers. There are about 59,000 ASs in the Internet, but these identify ISPs, not regions. Across the Internet, there are about 680,000 provider-specific regional blocks, and the routing computation deals with them as flat identifiers.

If I were designing a new network-level addressing scheme, I would first think very hard about whether there is some way these ISP-specific regions could be structured to facilitate delivery, and how this number will grow in the future. I have not seen any good answers to these questions.

Conclusion

In the final chapter of this book, I will return to the comparison of different delivery modes when I discuss the future of networks, and my own views on architectural options, but for now I must put off further discussion, because I have not yet discussed all of the issues that will influence the design of a naming scheme. In particular, chapter 10, on security, will make it clear that the purpose of a naming scheme is not just to let good things happen but also to prevent bad things from happening, so it is time to change topics and move on.

9 Longevity

Introduction—The Goal of Longevity

Compared to many artifacts of computing, the Internet has lived to an old age—it is over 35 years old. Opinions differ as to the extent that it is showing its age, and among some researchers there is a hypothesis that the Internet of 25 years from now might be built on different principles. Whether the network of 25 years from now is a minor evolution from today's network or a more radical alternative, it should be a first-order requirement that this future internet be designed to survive the test of time. I have used the terms longevity and long-lived to describe this objective. The objective is easy to understand, but the principles that one would use to achieve it are less well understood. There are a number of theories about how to design a network (or other system) that survives for a long time. In this chapter, I argue that many of these theories are relevant and that one can achieve a long-lived network in different ways by exploiting various combinations of these theories to different degrees. While some theories are incompatible, many are consistent with one another.

The approach I take in this chapter is inspired by the book *Theories of Communication Networks* (Monge and Contractor, 2003). The topic of that book is not networks made of routers but rather social networks made out of people and their relationships. The authors identify many theories proposed to explain the formation and durability of social networks, and their thesis is that many of these theories are valid to different degrees in different social networks, so it is necessary to have a multitheory, multilevel framework in order to explain the character of any given social network. Since there are many examples of social networks in the real world, one can do empirical research to try to determine how (for example) theories of self-interest, collective action, knowledge exchange, homophily, and proximity shape a given network. While there are fewer data networks to study compared to social networks, I will attempt to catalog various theories that might explain why a network might or might not be long-lived. I will also discuss how the designers of the various proposed architectures I have described have made arguments about longevity. Some of these theories have been articulated by system designers, and some I have named in order to give structure to my discussion.

Classes of Theories

With some oversimplification, I will classify these theories of longevity into three subclasses, as follows.

Theories of change These theories presume that, over time, requirements will change, so a long-lived network must of necessity change. Theories of this sort sometimes use the word *evolvability* rather than longevity to describe the desired objective, since they assume that a network that cannot change to meet changing requirements will soon cease to be useful. The word *change* as used here usually has the implication of *uncertain* change; if the future trajectory of the requirements on a system were understood, one could presumably fold these requirements into the initial design process if the cost were not prohibitive. The XIA and FII proposals are examples of architectures based on a theory of change.[1]

Theories of stability In contrast to theories of change, theories of stability presume that a system remains useful over time by providing a stable platform on which other services can depend. The NDN proposal falls in this category.

Theories of innovation These theories assume that change is not just inevitable but beneficial. These theories stress the importance of change and innovation as economic drivers. The FII proposal is specifically an example of this category.

These classes of theories are not incompatible. Theories of innovation are often theories of stability, in that the stability of the network as a platform allows innovation on top of that platform by what innovation theory would call complementors. Taking an example from operating systems, the stability of the interfaces to the operating system invites application

1. The various architectural proposals I use as examples here were introduced and discussed in chapter 7.

designers to take the risk of developing and marketing new applications for that system.

Architecture and Longevity

I have defined the term *architecture* to describe the basic design concepts that underlie a system like a network, such as the top-level modularity, interfaces and dependencies, and the assumptions that all parties must take as globally consistent. Within a theory of stability, architecture plays a natural role: it is part of what defines stability. With respect to theories of change, however, the relationship is more complex. If architecture defines those things that we want to have longevity, how does architecture encompass change?

Stable architecture that supports change In this view, the architecture embodies those aspects of the system that do not change. It is the stability of the architecture that permits the overall evolution of the system. The XIA proposal, with its flexible address header, is an example of this category.

Evolving architecture In this view, the architecture itself can (and does) evolve to address changing needs. If the architecture cannot adequately evolve, this leads to violations of the architecture, which (according to these theories) leads to a gradual loss of function and makes further change difficult, an ossification of the system that gradually erodes its utility. The FII proposal is an example of this category, where the higher-level architectural framework allows the introduction of new embodiments over time.

The Theory of Ossification

The theory of ossification was first proposed with respect to operating systems by Belady and Lehman (1976). They posed their first law of program evolution dynamics, the law of continuing change, which states that a system that is used undergoes continuous change until it is judged more cost-effective to freeze and re-create it. According to this point of view, eventually systems lose the ability to evolve over time and have to be redone from scratch in order to allow continued change. This theory of change is therefore an episodic theory, which predicts that systems (or architectures) have a natural lifetime and need to be renewed from time to time by a more revolutionary phase.

The idea that systems are subject to episodic replacement is an example of a more general theory of episodic revolution. Thomas Kuhn, in his book *The Structure of Scientific Revolutions* (1962), argues that the progress of science is defined by periods in which a theory is incrementally improved and embellished (what he calls *normal science*), punctuated by moments in which a theory is overthrown by an alternative that has better explanatory value, simpler structure, or more power to resolve anomalies. One could see the NSF FIA program as an invitation to the research community to contemplate such a revolution.

New theories of design suggest that it may be possible to derive an architecture from a set of requirements by a rigorous and formal process (see, for example, Chiang et al., 2007; Matni et al., 2015). It is an open question how such an architecture will deal with change. If one changes the requirements and then derives a new architecture, the differences may be pervasive: essentially a new design rather than a modification of the old one. But if one takes an architecture derived in this way and modifies it after the fact, all of the theory that applied to the original design process no longer applies. This sort of action is like taking the output of a compiler and patching the machine code. It is thus possible that architectures that have been algorithmically derived from requirements will be brittle with respect to change or (in terms of these theories) easily ossify.

The Theory of Utility

All discussion of longevity must occur in the context of a network that is used. A network that is long-lived is a network that continues to be useful over time, so it is a precondition of a long-lived network that it be useful in the first place. (Chapter 4 lays out my framework for considering the extent to which an architectural proposal is fit for purpose.) So any theory of longevity must have inside it some theory of utility, which explains why the network is useful. The first theory of longevity I identify is based on a specific theory of utility.

The Theory of the General Network

According to this theory, a fully general system, which could meet all needs, would not need to evolve and would thus be long-lived. The theory of the general network is thus a theory of stability. The challenge is to define exactly what a general network might be.

The theory of the ideal network and impairments According to this theory, networks provide a very simple service that can be described in its ideal (if unrealizable) form. One candidate formulation is as follows:

> An ideal data network will reliably deliver any amount of data to (and only to) any set of intended and willing recipients in zero time for zero cost and zero consumption of energy.

Of course, such a network cannot be realized. Some limits, such as the speed of light, are physical limits that cannot be violated. Others, such as cost, seem to improve over time as a consequence of innovation. Taken together, these limits, or impairments, define how far any practical network diverges from the ideal. In this theory, a maximally general network minimizes the various impairments and, to the extent possible, allows each set of users to trade off among the impairments. Thus, queuing theory captures a fundamental set of trade-offs among speed, cost (utilization), and delay. A network that (for a given class of traffic) does as well as queuing theory would predict, and allows the users to move along the performance frontier defined by queuing theory, is a maximally general network by this definition.

According to this theory, a network that is maximally general with respect to the fundamental impairments (a theory of stability) and can evolve in response to impairments that change over time (a theory of innovation) will be long-lived.

I believe that the Internet is a good example of a general network by this definition and see its longevity as a consequence of that fact. The use of packets as a multiplexing mechanism has proved to be a very general and flexible mechanism. Packets support a wide range of applications and allow for the introduction of new technology as it evolves. Mechanisms that allow control over QoS allow (in principle—though they are not deployed on the public Internet) an application to control the trade-off among such parameters as cost, speed, and delay.

The Theory of Real Options

Real option theory captures the idea that one can attempt to quantify the cost versus benefit of investing now to keep options open, or, in other words, to deal with uncertainty. It is thus a theory of change, to the extent that change equates to uncertainty. It is also a theory of the general network, but in economic terms, in that it suggests that one can spend money now to purchase flexibility later to respond to uncertain change. It does not describe what the resulting general network is (in contrast to the definition offered earlier) but just postulates what generally is often to be had but at a price; price is one of the impairments to the definition of the idealized general network I proposed earlier.

Real option theory is perhaps more often applied to the construction of a network (e.g., how much spare capacity to purchase now) than to the architecting of a network. However, Gaynor and Bradner (2001) discuss how real option theory can shape the process of designing standards, arguing (among other things) for modular systems (as I discuss in chapter 3).

The Theory of Tussle and Points of Control

The previous discussion of the ideal network does not fully capture what happens inside networks, because the ideal is stated from the perspective of one class of actors—the parties desiring to communicate. The statement of the ideal does not afford any attention to other actors, such as governments that want to carry out lawful interception of traffic, to employers and others who want to limit what can be carried over their networks, and others. The variety of stakeholders in the current Internet is substantial, and each tries to put forward its interests, perhaps at the expense of other stakeholders.

In Clark et al. (2005), we called this ongoing process *tussle*. We argued that tussle is a fundamental aspect of any system (such as the Internet) that is deeply embedded in the larger social, economic, and regulatory context. According to the theory of tussle, systems that are designed to minimize the disruptive consequence of tussles will be long-lived; the design goal should be that tussle does not lead different actors to violate the architecture of the network in pursuit of their interests. Various aphorisms describe how a system should be designed to tolerate tussle:

> The tree that bends in the wind does not break.
>
> You are designing not the outcome of the game but the playing field.

The idea behind these aphorisms is to design your systems so that they do not attempt to resist tussle and impose a fixed outcome but instead are flexible in the face of the inevitable. However, they give little practical guidance as to how one might do it, other than to hint that one can tilt the playing field to bias the resulting tussle consistent with the values of the designer.

Tussle isolation One design principle that emerges from the consideration of tussle is a new modularity principle called tussle isolation. Computer science has a number of theories of modularity, such as layering (e.g., the avoidance of mutual dependency). The idea behind tussle isolation is that if the designer can identify in advance an area where there is likely to be persistent tussle, then the design should isolate that area so that the resulting tussle does not spill over into other aspects of the network. Here are some examples of areas in which this might have been done.

- DNS: If the early designers had understood that the DNS would include names over which there would be trademark disputes, they could have designed a separate service to deal with those names in order to minimize the scope of trademark disputes. Naming systems where the names are designed to have no meaning, such as DOI names, will not trigger trademark disputes.
- Secure Web access (TLS): If the designers of the certificate authority system to secure access to Web sites had understood that the real tussle would be over which actors would be trusted to vouch for different Web sites, they could have designed a different trust framework to limit or prevent these tussles.

Placement of interfaces In addition to isolating tussle, one can move it around by the placement of critical interfaces within a system. Another way of saying this is that functions can be moved from one module to another, shifted across the interface. Modularization of tussle is an example of a nontechnical principle for modularizing a system. The relationship between how modules are defined and the resulting shifts in power among different classes of actors can be rather subtle, but here are a few examples.

- In the current Internet, there has been no interface specified between two key router functions—the actual forwarding of the packets and the routing computation that populates the forwarding table. The developers of software-defined networking (SDN) defined such an interface, which allows the routing computation to be performed by a separate device that is not under the control of the router vendor. This option shifts the landscape of competition among providers of routers by allowing the ISP to purchase a forwarding element from one vendor and the route computation device from a separate vendor.
- When a packet arrives at a destination end point, it has to be dispatched to the correct application on that host. The dispatching is controlled by a field in the packet header called the port. Early on, there was a debate as to whether the port field should be part of the TCP header or the IP header. The decision was to put the port field in the TCP header. If the port field had been in the IP header, the implication would have been that the routers would look at it, which would have had the effect of revealing to the ISPs more information about what the end points were

doing. In fact, routers were programmed to peek at the TCP header to extract the port number and modify the packet treatment accordingly as early as the 1980s. When the specifications for encrypting packets were debated, the ISPs argued that the port field in the TCP header should not be encrypted, even though TCP is nominally end-to-end, because they needed to see it. Advocates for privacy wanted the port field encrypted; network operators (and probably those interested in metadata surveillance, although I cannot document that) wanted the port field revealed.

Removal of interfaces Another point of view about interfaces is that by intentionally removing them and making the system less modular and more integrated, one can increase the power of the firm that owns the system and limit competitive entry as well as other forms of tussle. This theory is an example of a theory of stability through market power and hegemony, which I discuss later in this chapter and on page 241. This method of reducing tussle may not be in the best interest of the ecosystem.

Asymmetric struggle Many tussles are shaped by the fact that different stakeholders have access to different tools and methods. Network architects define module interfaces and shift functions around, governments pass laws and regulations, and network operators make investments and configure the physical network. Each of these actions can advantage particular stakeholders and disadvantage adverse interests. Given this fact, it is worth studying how these various methods interact with each other, but a full exploration is beyond the scope of this chapter. Like a game of rock, paper, scissors, they sometimes seem to circle around each other in endless cycles. I call one theory that characterizes some of the interactions the *theory of the blunt instrument*, the idea that while each stakeholder has distinct powers, the design of one part of the system can blunt the tools of control others have, rendering them less effective. Thus, for example, the use of encryption as part of the network design greatly limits the ability of other actors to observe (and thus to impose limits on) what users are doing. In the extreme case, the network operator is reduced to carrying all traffic, blocking all encrypted traffic, or refusing to serve the relevant customer—an example of blunting the network operator's instrument of control.

Tussle and Longevity

The theory of tussle might be seen as a theory of change, but in fact it may be closer to a theory of dynamic stability. Stability need not imply a fixed system; it can also imply a system that has checks and balances, or feedback, to home in on a stable point. Tussle is such a mechanism—a set of forces that tend to bring a system back to a stable compromise point if some new input (e.g., a technical innovation) shifts it away from that point. Tussle can be seen as a dynamic and ongoing mechanism that contributes to system longevity: over time, the compromise point may shift (as social norms shift over time), and the stable point may be different in different cultures. Tolerance of tussle is the alternative to a rigid system that attempts to impose global agreement in contexts where such agreement is not feasible. Tolerating variation in outcome is a contributor to longevity.

The Theory of Building Blocks and Composable Elements

The theory of the general network assumed that one could describe what an ideal, or fully general, network would do, but this theory conceived the network as having a very simple core function. Another view is that a network should be capable of offering a much richer set of services (perhaps not all at the same layer). By this theory, the success of a network would not be measured by how well it limits the impact of impairments but rather by how easy it is to incorporate new sorts of services between communicating parties. In this view, a network built of fixed-function routers is limiting rather than stabilizing. My discussion about expressive power in chapter 6 gets at this tension: should a network strive for minimal expressive power or a rich set of tools to add new PHBs as needed. The i3, DOA, and Nebula proposals allow the composition of arbitrary services, which can potentially enhance longevity by allowing new services to be invented as needed.

The concept of building blocks becomes much richer if one looks not just at the simple, packet forwarding layer but also services above that layer, which might do things such as convert information formats, validate identity, provide security services, and the like. In this layered view, one would then ask whether the packet layer was optimally suited to support the deployment and configuration of these higher-level services. For example, to ensure the proper operation of security services, it would be important to make sure that packets cannot bypass the services. So the desire to deploy these higher-layer services may change and expand the requirements at the packet level, even if these services are higher-layer services.

There seem to be two, perhaps contradictory, theories of building blocks and composable elements—the maximal and the minimal. In the maximal theory, a network will be long-lived if it has rich expressive power, so that new service elements can be added. An architecture can increase expressive power by adding more powerful addressing modes (such as source routing, which could route a packet through a series of service elements) and additional fields in the packet header to convey additional information to the service elements. The minimal theory about service elements and expressive power arises within the theory of tussle. In this point of view, any service element will be a point of contention and tussle, as different stakeholders try to control the service. Thus, ISPs sometimes block access to third-party mail transfer agents in an attempt to force a customer to use their mail service; by doing so, the ISP may be able to impose limitations on what the customer can do. The minimal theory guides the network design to limit the expressive power of the design deliberately (perhaps at certain layers) in order to limit the points of tussle in the network, leading to longevity through stability.

Theory of Programmable Elements (Active Networks)

The theory that building blocks bring beneficial flexibility has an aggressive version in which elements within the network are programmed dynamically, perhaps by means of programs carried within the data packets themselves. I discussed this approach to design, active networks, in chapter 7. Active networks can be seen as reducing tussle rather than facilitating it, since it tilts the playing field toward the end user and blunts the instruments of control that belong to the stakeholders that control the network. The programs come from the edge; the stakeholders operating the network only provide the platform for these programs. They cannot regulate what those programs do except by imposing limits on the expressive power of these programs. The various Active Network schemes I discussed in chapter 7—ANTS, the Smart Packet scheme, PLANet, and PLAN—illustrate different approaches to limiting the expressive power they allow in programs. With only a "blunt instrument" capability to limit how programs are composed, the result (according to this point of view) is a stable platform on which innovation can be driven from the edge.

The Theory of the Stable Platform

The theory of the stable platform has to do with innovation. According to this theory, innovation at one layer (which represents a valuable form of change) is facilitated by a stable platform (the layer below) with a stable interface and service definition. In the language of this theory, those who innovate on top of the platform are called complementors. If the platform itself is unstable and subject to change and innovation, this change increases the cost of building complementary systems (e.g., applications) that exploit the platform, as the application must be upgraded to keep pace with the platform changes, and increases the risk because of uncertainty about changes in the platform that might reduce the functionality of the application. Thus, the stability of IP encourages application developers to innovate on top of that layer. The stable platform is thus both a theory of change (at the level of the complementors) and a theory of stability. For an extended discussion of platform theory as it relates to the Internet, see Claffy and Clark (2014). I hypothesize that platforms can be stable for some period but will, over time, be subject to increasing pressure to evolve, which may lead to an episode of replacement. Alternatively, a platform that starts out with the goal of being stable may be augmented in ways that do not disrupt the old functionality but offer the option of enhanced functionality.

The theory of the stable platform can be stated in dynamic form: to the extent that complementors use their power to argue for the stability of the platform, more complementors may be induced to join, a positive feedback situation. The reverse of this dynamic is also a part of the theory: if a platform is not useful, it makes no difference whether it is stable. I see the packet forwarding service of the Internet as a good illustration of a stable platform that permits innovation on top of that platform; in my view, the stability of the IP platform (and TCP running on top of that platform) is an important part of the longevity of the Internet.

The Theory of Semantics-Free Service

The theory of the stable platform does not say anything about what function the platform should implement in order to be useful and general. The theory of the general network provides one answer to that question: the platform should provide a general service that is as close to the ideal (the minimum set of impairments) as the designers can fashion.

The minimalist version of the theory of the general network is that the network should just deliver bytes. In contrast to the theory of composable building blocks, in the minimalist version, the network should not have any model of what those bytes mean or what the high-level objective of the application is; the network should be semantics-free. This version of the theory has also been called the transparent network, or (in more colloquial terms), "what comes out is what goes in." If the network begins to treat packets based on a model of what an application is doing, this can be seen as a highly beneficial optimization but also a loss of generality and an erosion of network neutrality.

I argue that the longevity of the Internet results from its semantic-free design and the refusal of its designers to allow the protocols to be optimized to the popular application of the day. In this respect, semantics-free service is an example of the theory of utility, but it is not clear what line of reasoning would be used to make this point in advance. However, the theory of the general network may imply the theory of semantics-free service, since (as stated earlier) the general network was defined as delivering data, which seems to imply a semantics-free service.

The semantics-free service is a close relative to the end-to-end argument (see page 11), but in the beginning the end-to-end argument was about correct operation, not about generality. The argument was that packet networks (especially those that do not have per-flow or per-packet state in the routers) will always have some residual unreliability because of the statistical nature of traffic arrival. Since different applications will have different requirements for reliability (some applications do not require perfect reliability), network designers should not attempt to build additional mechanisms to make the network perfectly reliable by some definition but instead let the end nodes correct any network errors in ways that meet the need of the particular application—the application, not the network, should have control over how the trade-offs among various impairments are managed. The interpretation of the end-to-end argument as an argument for generality can be found implicitly in the original paper (Saltzer et al., 1984) but has become more explicit in some of the subsequent writings about the argument.[2]

The Theories of Global Agreement

One aspect of network architecture that I proposed in chapter 3 is that architecture defines those aspects of the system about which there must be global agreement: architecture defines those parts of the system that work the same way everywhere. In this context, there are again two variants of global agreement and longevity: the minimal theory and the maximal theory.

The theory of maximal global agreement This theory postulates that the more aspects of the system are well-defined, the more stable the platform. By providing a well-specified functional specification for the platform, the

2. The literature elaborating, misreading, and attacking the end-to-end argument is almost boundless. I apologize for not offering a few citations, but a critical review of the literature would take a whole chapter, if not a book.

implementation challenges and risk to the complementor is minimized. The word *maximal* is probably an overstatement of this theory—the more careful statement would be that, up to a point, increased specification and careful definition are good.

The theory of minimal global agreement This is a theory of change. It states that the fewer things we all have to agree to in common, the more we will be able to accommodate a range of uses with different needs. As long as the platform remains useful, having fewer points of global agreement is beneficial and will allow the network to evolve over time without disrupting the utility of the platform. So in contrast to the maximal or "up to a point" theory, this is a "down to a point" theory, or perhaps (to paraphrase, as I did before, Einstein's quotation): architecture should be made as minimal as possible, but no less. The FII proposal is an explicit example of this theory.

False agreement In either version of the theory, one question concerns when a global agreement is really an agreement and when it is the illusion of agreement. An example from the Internet is the initial assumption that the Internet was based on the global agreement that there was a single, global address space. We thought that this agreement was important, and one of the basic tenets of the stable IP platform, but then NAT devices were introduced, and the Internet survived. Some would say that because NAT devices impair certain classes of applications (in particular, servers located behind NAT devices), we should view NATs (and the loss of global addresses) as a significant violation of the stable architecture. The IETF has now developed protocols, discussed in chapter 6, that allow state to be installed dynamically in NAT devices (perhaps an example of the theory of the building block), with the potential to support essentially all the applications from the era of global addresses.

Whatever lesson there is in this example, the more general question about agreement is how one would test a proposed point of global agreement to see whether it is actually required in order to have a stable platform. Clever reconceptualization may allow what was seen as a global agreement to be set aside with no loss of power.

One might pose an informal test-of-time approach: designers should judge a presumed point of global agreement only in hindsight, based on whether people actually depend on it. But this approach seems like a poor design principle. On the other hand, it seems difficult to take the position that we can force dependency to force stability. The theory of utility suggests that if a feature is not useful, it does not matter whether it is stable or whether it is a point of nominal global agreement.

The Theory of Technology Independence

The theory of technology independence is another theory of stability in the face of change. This theory states that a system will be long-lived if new generations of technology can be incorporated into the system without disrupting the stable platform for complementors. Since technology evolves rapidly in the computer world, a long-lived system must not be rendered obsolete by new technology.

Again, I will use this theory to explain the longevity of the Internet. The simple, packet-based platform of the Internet can be implemented on top of all sorts of communications technologies. The Internet has accommodated circuits that have increased in speed by at least six orders of magnitude during its lifetime. It has accommodated multiaccess local area networks, wireless networks, and the like. The applications running on top of the IP interface are largely unaffected by these innovations.

The Theory of the Hourglass

The combination of the theory of the stable platform and the theory of technology independence leads to a conception of an architecture structured as an hourglass: a narrow waist representing the common point of agreement on the IP layer, with great diversity in technology below and applications above, as illustrated in figure 2.1. Once the structure implied by the hourglass was seen as a theory of longevity, further study revealed that the Internet had many hourglasses in it: the reliable byte stream on which email sits (the Internet standards for email work quite well on transport protocols other than TCP), HTTP, and others.

The Theory of Cross-Layer Optimization

The theory of cross-layer optimization is contrary to the theory of the hourglass. This theory, which I discussed in the context of wireless networks in chapter 7, states that, over the long run, the evolution of technology will be so substantial that a stable, technology-independent platform will become limiting, and eventually uncompetitive, compared to an approach that allows the application and the technology to adapt to each other. The application designer will have a harder task than with a stable platform, but in exchange for doing the additional design work so that the application can adapt to different technologies, the designer will achieve greatly improved performance and function.

In the past, a number of technologies have caused their inventors to argue for cross-layer optimization, perhaps starting with multiaccess local area networks. In the past, the theory of the stable platform dominated. Today, cross-layer optimization is proposed for wireless networks, especially those for very challenging circumstances, such as battlefield networks. It is not clear whether longevity is a primary requirement for tactical battlefield networks.

The Theory of Downloadable Code

The theory of downloadable code is a theory of the stable platform, or perhaps of innovation. It states that the ability to download code into communicating elements can minimize the need for global agreement. Agreement is achieved not by the mandate of global standards but by a local agreement to run compatible software.

If code is downloaded into routers, this is what I described as active networks. Active networks have not achieved much traction in the real world. However, code that is downloaded into the end node (most commonly at the application layer or as a supporting service to applications) has been a powerful tool to support innovation. New formats for audio and images (still, animated, and video) can be easily introduced if end nodes can download new rendering code. Creators of standards such as PDF, Flash, various representations of audio and video, and the like made free software available for download as a way to enter the market. Pragmatically, once a format can be implemented in downloadable software (as opposed to hardware, for example), proliferation of competing standards does not seem to be an impediment to progress and longevity.

The theory of downloadable code is an example of the stable platform. In this case, the platform is a software platform, not a network service platform (such as the IP layer). The browser of today, with its "plug-in" architecture, becomes a stable platform on which innovation (e.g., new downloadable modules) can be built.

This observation suggests the question of what parts of the network could be based on downloadable code rather than on global agreement. Today, for example, transport protocols such as TCP require more or less global agreement. For performance reasons, TCP has usually been implemented in the kernel of the operating system rather than with the application, which makes it much harder to download an alternative. However, is this a fundamental consequence of some aspect of transport protocols or just a historical accident? It might be possible to design a platform onto which different transport protocols could be downloaded, just as the Web browser provides a platform for downloadable code at a higher level. Were this framework demonstrated, the theory of downloadable code might be a better path to longevity than the theory of global agreement, even at the transport layer of the protocol stack.

Change: Hard or Easy?

More abstractly, the theory of downloadable code challenges us to take a rigorous look at what makes change hard or easy. The need for global agreement seems to make change hard, especially if everyone has to change at once.

Version numbers are sometimes used as a technique to manage change. Version numbers in protocols can allow two incompatible designs to coexist, either transiently during a time of change or (more realistically) forever. Version numbers work so long as it is possible to verify that all the components that will be involved in some operation support at least one version in common. Proposals such as XIA and FII try to facilitate change (in different ways) by making it easier to make changes gradually, across different parts of the network at different times.

Changes to production code are sometimes hard to make, or at least not quick to make. Vendors need to be convinced of the need for change and then schedule the change into the development cycle. Such changes can take years, especially if the change is based on a standard that requires broad agreement. However, one should not mistake the time it takes to make a change with fundamental difficulty. What makes a change easy or hard to implement is its interaction with other parts of the system. As interactions increase over time, consistent with the theory of ossification, the complexity of change will increase over time. On the other hand, when the change is more a bug fix and the need is urgent (as with the discovery of a security vulnerability), changes are often made in a matter of days or weeks, and the current trend to automate the downloading of new versions (e.g., of operating system and major software packages such as Microsoft Office) can allow substantial deployment of updates in days.

Overall, there is a trend in the current Internet (and the systems attached to it, such as operating systems) to make change (updates, patches, releases of new versions) easier to accomplish. This trend raises the question of which changes are hard to make and why. The theory of minimal global agreement would suggest that with the right tools to allow software to be replaced, there is little that cannot be changed in principle, and more and more that can be changed in practice. With the trend of moving functions from hardware to software (e.g., software-defined radios), functions that had traditionally been viewed as fixed and static have turned out to be amenable to change.

The FII proposal and the DTN work bring out an aspect of the current Internet that, while not a formal part of the architecture, seems to have frozen in a way that resists change. Today, most applications get access to the Internet via a software interface to the TCP service (the so-called socket interface) that presumes a two-way interactive reliable flow among the end points. In contrast, in a DTN, many nodes may only be connected intermittently, and applications must tolerate a more store-and-forward mode of transport between the end points, so a more general network API may be an important part of building a more general version of the stable platform. FII includes in its required points of agreement a set of tools to allow the network API to evolve.

The Theory of Hegemony

The theory of hegemony is a theory of stability. It postulates that a system will be long-lived if a single actor is in charge of the system that can (if benevolent) balance change against stability and balance the needs of the various stakeholders in an orderly way. By taking tussle out of the technical domain and into the planning or administrative (regulatory) context of the controlling actor, the platform becomes more predictable and thus more appealing. So the theory of hegemony is a theory of innovation based on stability. However, a hegemon is not always benign and may exploit its position to extract high profits while not particularly stimulating innovation in other parts of the ecosystem.

The telephone system, for most of its life, was an example of a system managed according to the theory of hegemony, with single providers, in most countries (but not the United States) a part of the government, and standards set through a deliberative body, the ITU (or earlier the CCITT). One interpretation of history is that this approach led to a very stable system that was easy to use but inhibited innovation. The movement in the 1970s and 1980s to privatize the telephone system in many countries was in part an explicit attempt to stimulate innovation. However, the innovation spurred by these policy decisions was nothing compared to the innovation triggered by the Internet. The low rate of innovation can be explained by the theory of utility: the platform for innovation provided by the telephone system, the 3 kHz audio channel, was not very general, so the failure of innovation resulted as much from the limited utility of the platform as from the presence of the controlling interest.

The Present Internet

I have identified a number of theories as contributors to the observed longevity of the Internet: the theory of the general network, the theory of the stable platform, the theory of semantics-free service, the theory of technology independence, the resulting theory of the hourglass, perhaps the theory of minimal global agreement, and (to some extent increasing over time) the theory of downloadable code in the end nodes. The design of the Internet seems to reject the theory of hegemony, and the theories of composable services and downloadable code in the network.

Global Agreement

The early designers of the Internet assumed that substantial global agreement would be a necessary step toward an interoperable network. (In those days, downloadable code was not a practical concept.) Over time, in an application of what I called the test-of-time approach, the real need for global agreement has emerged.

Addressing The original addressing model has mutated into a more complex structure, in which there are lots of private address spaces, with translation among some of them by using NAT devices. However, there is still a single common addressing region in the core of the network where a pool of addresses are given a consistent common meaning. Services that want to make themselves widely available obtain an address in the common addressing region so that other end points can find them.

TCP The original designers were careful not to make TCP mandatory—they believed that alternatives to TCP would be needed. The socket interface to TCP is not specified as an Internet standard. Over time, however, in an example of the emergent form of the stable platform, enough applications have used TCP that it is mandatory in practice, which means that other applications take on little risk in depending on it, and TCP has emerged as a required global agreement.

TCP-friendly congestion control This idea was not part of the original design—in the beginning, the designers did not have a clear idea about how to deal with congestion. However, in the 1990s (more or less), as congestion control based on the slow-start algorithm and its enhancements matured, there was a sense that every application, and every transport protocol, needed to behave in the same general way, so there was a call for a global agreement on the congestion behavior called TCP-friendly. To a considerable extent, this norm was accepted, but it seems today as if there is a drift away from this approach (based on economic issues and the theory of tussle). It remains to be seen whether the network will take on a more active role in controlling congestion in the future.

DNS The architects of the Internet have been ambivalent about whether the DNS is a core part of the architecture. It is not strictly necessary; one can use other tools to translate names into addresses (as some applications do) or just type IP addresses where one would normally type a DNS name (e.g., in a URL). However, as a practical matter, the DNS is a necessary component of the Internet for any real use, and the amount of tussle surrounding the DNS (trademark, diverse alphabets, governance, creation of top-level domains (TLDs) etc.) makes it a prime illustration of tussle. It is actually not clear to me exactly what aspects of the DNS require global agreement. The simple DNS interface (send a name to the DNS, get an Internet address back) is a stable interface under which all sorts of changes have happened.

Web The Web standards are a platform critical to the growth of the Internet. While the Web is just one application among many, it is clearly (as of now) a dominant application and, as such, embodies many attributes that can be explored using these various theories—tussle, platform, downloadable code, and so on. Without global (if rough) agreement on many aspects of the Web, the Internet experience would not be what it is today. On the other hand, the use of downloadable code has enabled rapid innovation and growth in complexity.

The packet header Participation in the Internet requires agreement on how a packet is formatted and what (at least some of) the fields mean. The address field may be rewritten as the packet traverses NAT boxes, but the specification does impose some constraints to which all players must conform (e.g., the length, the TCP pseudo-header, and the like). The IP header seems to be a manifestation of the stable platform rather than something that is shaped by a theory of change, which is in part why the current effort to convert to a new header format, IPv6, is proving so difficult.

The Future

I have discussed a number of theories about how to design a future network so that it is long-lived. Various architectural proposals I have discussed take different approaches toward longevity (e.g., stability versus change, minimality versus rich services), but these choices become interesting only if the architecture passes the basic test: the theory of utility. If the network is not useful, it will not be given a chance to demonstrate its ability to be long-lived.

In the chapters that follow, I turn to a detailed look at some of the critical design requirements I identified in chapter 4, starting with security. In the final chapter of this book, I will look across all of these requirements and offer a few opinions about designs for a future Internet, taking into account these requirements, including longevity.

10 Security

Introduction

In this chapter, I discuss the mix of historical, architectural, and pragmatic reasons why the Internet of today is generally considered to provide a poor level of security.[1]

This chapter will not resemble most papers on security, which typically identify a particular vulnerability and pose a solution. It addresses the more general challenge of how to identify and classify the range of security problems that arise in the context of a global internet. It is concerned with *security architecture*, and a more traditional security paper might take up where this chapter leaves off. I will draw two general conclusions from this process of classification. First, different security problems arise in different parts of the network ecosystem and must be dealt with by different actors. Second, many of these issues cannot be addressed by a change in the architecture itself, but changes in the architecture may raise new security issues.

Defining Security

The first issue is to consider what the word *security* actually means. Without a clear definition of what is meant by the word, it is not very meaningful to discuss whether we have enough of it. The concept of security captures a range of issues that may not actually have that much to do with each other—security is a "basket word" for many different issues, like the word *management*, which I will unpack in chapter 13.

Computer science tends to define security in terms of the correct operation of a system: a secure system is one that does what it is supposed to do

1. This chapter has benefited greatly from discussions with Josephine Wolff, Shirley Hung, John Wroclawski, and Nazli Choucri at MIT.

and does not do unacceptable or unintended things, even when it is under attack. This conception, of course, requires that the functions of a system be well-specified. There is an old saying among security experts: "A system without a specification cannot fail; it can only present surprises."[2]

A user of the Internet might not think about security in the same way. What a user cares about is whether the overall probability of bad events is low enough to tolerate. Users care about outcomes; technologists tend to address inputs. An analogy from the physical world may help. A home security expert might say that a home has a secure door if it has a good lock and is strong enough to resist being kicked in, but what the homeowner cares about is whether, all things considered, the probability of being burglarized is low enough.

Alternatively, a political scientist of the realist school might define security by saying that a nation is secure if it can sustain peace at an acceptable cost or, alternatively, if it can prevail in war. Security is not automatically equated with peace; unconditional surrender will create a state of peace but not one of security, since the price of unconditional surrender is presumably high. In this framing of security, there is no attempt to define what "correct operation of the system" would mean; that would be nonsense with respect to a nation taken as a whole.

While users may care about outcomes—keeping the risk of harms to a realistic level—network designers are forced to work in the space of inputs. Exactly because the Internet is a general system, designers must address security by making the components strong (correct). Just as we designed the Internet without knowing what it is for, we have to design its security components without knowing exactly what security problems will arise as the network is used. Most people understand that it would be nonsense to ask for a door to be designed without knowing whether it was for a house or for a prison cell, but dealing with that sort of uncertainty is the price of building a general-purpose network. Perhaps it will turn out to be fundamentally easier to deal with functional generality than security generality; perhaps we just have not yet figured out how to think about security in this general, abstract way. But that is the challenge I must address in this chapter.

The rest of the chapter proceeds as follows. First, I offer a way of sorting out the landscape of security to provide some structure to the discussion that follows. I will look specifically at the range of issues that make up *network security*. Building on that, I focus on the issues of trust and trust

2. I cannot determine who first said this. I have questioned a number of elders in the field, all of whom agree that they said it but believe they got it from someone else.

management as a key to better overall security. I then proceed to a narrower topic that brings me back to the theme of the book, the relation of architecture to these various aspects of security; I consider how architecture, within the minimalist framework, can contribute to better security.

Defining Network Security

Security experts often frame security by breaking the problem into three subgoals: confidentiality, integrity, and availability (the CIA triad),[3] and I will refer to that structure when it is relevant, but in fact, for many security problems, this framework is not very helpful. To begin, I structure my discussion of network security by looking at the structure of the system, a taxonomy that derives loosely from the layered structure of the Internet, and asking where a malicious action manifests. Later in the chapter, I will return to a harms-based taxonomy of security and ask what this view can teach us about how to think about the inputs—the correct operation of the system elements under attack.

Here is my taxonomy based on where the attack manifests:

- **Attacks on communication** This problem, sometimes classified as information security, arises when parties attempting to accomplish mutual communication are thwarted by an attack, perhaps launched by the network or by some party that has gained control of some critical control point.[4] In the discussion of expressive power in chapter 6, this class of attack maps onto the case where the communicating actors have aligned interests but some element in the network is hostile to those interests. This is a space where the traditional triad of confidentiality, integrity, and availability (CIA) has some validity, as I will discuss. Another issue in this category is *traffic analysis*, the form of surveillance where the observer is looking not at what is being sent but who the sender and receiver are. Knowledge about who is talking to whom can be as revealing as what exactly is being said.

3. Not to be confused with the Central Intelligence Agency.

4. A few years ago, there was a furor in the United States because Comcast blocked a peer-to-peer music-sharing application (BitTorrent) by injecting forged packets into the data stream. This blocking was characterized at the time not as a security event but as a violation of the norms of service, but in the language of security, blocking was without a doubt an attack on a communication by the network. End-to-end encryption would have detected this particular attack, but since this was intended to be an attack on availability of service (discussed later), Comcast could have used other approaches.

- **Attacks on the attached hosts** Attacks on attached hosts can occur as a result of communication with a malicious party (who uses the capabilities of some layer to deliver an attack) or as a result of an unsolicited incoming packet that somehow exploits a vulnerability to launch a successful attack. In the discussion of expressive power in chapter 6, this class of attack maps onto the case where the interests of the various end points to a communication are not aligned. The receiver may choose to draw on resources in the network (PHBs) as a means of protection. Both the attacker and the defender will draw on the expressive power of the network to achieve their respective goals.
- **Attacks on the network itself** These include attacks on network elements, the routing protocols, critical supporting services such as the DNS, and the like. In many cases, one part of the network attacks another—in a global network, not all parts are equally trustworthy. Since the core functions of the current Internet are actually rather simple, there are only a few critical services; the interesting question is why they remain insecure. I return to this later. To the extent that the network layer cannot detect and remedy the consequences of failures and attacks internally, higher layers will have to take actions to mitigate the consequences of such attacks.
- **Denial of service attacks** Denial of service attacks (usually called distributed denial of service, or DDoS, attacks, because attackers usually use many machines to launch an attack) do not fit cleanly into any of the prior categories. They can be classified as attacks against the network if they exhaust the capacity of a link or switch or as attacks against a host if they exhaust the capacity of that host. I consider this class of problem separately.

A Historical Perspective

Critics have complained that the early Internet architects, including me, did not think about security from the start. This criticism is to some extent valid, but in fact we did consider security; we just did not know at that time how to think about it. We made some simplifying assumptions that turned out to be false, which I will describe. Interestingly, much of our early advice about security came from the intelligence community (IC), and their particular view biased our thinking.

The IC had a very simple model of protecting the host from attack: the host protects the host, and the network protects the network. They were not prepared to delegate the protection of the host to the network, because they did not trust the network. So our job was to deliver everything, including attacks, and then the host would sort it out. We now see that this view is overly simple and not totally realistic.

Within the CIA framework, the IC gave the highest priority to confidentiality—the prevention of declassification and theft of secrets. Their view was that once secrets are stolen, the damage is done. What we now see is that users care about availability in the CIA framework—their ability to get the job done.

Because the IC assumed an attacker with a very high skill level and motivation, they argued only for mechanisms that were perfect. To them, a mechanism that only provided a partial degree of protection just defined how much effort the adversary would have to expend, and they assumed the adversary would expend it. Today, we see that many attackers are economically motivated and are not prepared to expend effort that does not give a good return on investment. In this case, mechanisms that do not strive for absolute protection but just make the job of the attacker harder are justified. The typical homeowner I described does not expect his house to be resistant to every form of attack but just secure enough to deter the typical criminal.

The CIA framework separates the world into two sets of actors—those who are authorized and those who are not. Confidentiality is violated if information is disclosed to an unauthorized actor. If an actor is authorized, they are allowed to see the information and (if suitably authorized) allowed to modify it. If an actor is not authorized, then the goal of the system is to deny them access.

This framework is deceptive, but it shaped our early thinking. We knew that some routers might be suspect, so there was no way we could ensure that a router did not make a copy of a packet—the packet forwarding layer could not itself provide confidentiality. Also, a malicious router might modify a packet—the packet forwarding layer could not itself provide integrity for data in transit. We took a very simple view, which is associated with the "end-to-end" mode of thinking: only the end points could undertake to mitigate these vulnerabilities and achieve these objectives, because only they could know what the objective was, and only they were (presumably) trusted and authorized to exchange this data. End-to-end encryption is the obvious approach: if the data is encrypted, it is useless for an attacker to make a copy, and the recipient can detect any modification.

When the Internet was initially developed, encryption algorithms were too complex to implement in software. Only implementation in specialized hardware had adequate performance. This reality was a barrier to deployment; not only would every machine need such hardware, but there would have to be broad agreement as to which algorithm to use, which was hard to negotiate. However, the early designers expected that the Internet could move to the use of end-to-end encryption at some point in the future.

This approach theoretically resolved confidentiality and integrity, and left only availability as a challenge for the network designers. Of course, all the network does is deliver packets, so it would seem that availability is a core requirement. In this context, it is interesting that the network design community has no theory of availability, the subject of chapter 11.

Why was this conception of security deceptive? It implied a simple world model—mutually trusting parties communicate, and parties that do not trust each other do not. This worldview, with its emphasis only on information security among mutually trusting actors, distracted us from the insight that most of the communication on the Internet would turn out to be between parties that were prepared to communicate but did not know whether to trust each other. We agree to receive email knowing that it might be spam or have attachments that contain malware. We go to Web sites even though we know (or should know) that they can download malware onto our computers.

Taking into account both challenges, communication among trusting parties and one user attacking another, using end-to-end encryption is not a complete solution. Encryption addresses the problem of protecting communication between trusting users from disclosure or corruption but fails to address the mirror problem of adversarial end points using network protocols to attack each other. An analogy may help. If trusting parties want to send a private physical letter, they want assurances that the letter is not opened in transit, but if recipients suddenly realize that they may get a letter full of anthrax, then their security objective reverses—they want that letter opened and inspected by a trained, trustworthy (and well-protected) intermediary. End-to-end encryption between an attacker and target may be the last thing the target wants—it means that the target can get no help from trusted third parties. An encrypted exchange with an untrustworthy party is like meeting them in a dark alley—there are no witnesses and no protections. The problem of operation in an untrustworthy world has to be handled by involving the higher layers in the system, specifically the application layer, and the early designers neither clearly articulated this design problem nor explored how to remedy it. In this respect, the end-to-end principle is not wrong, just incomplete and in need of reinterpretation (for a reconception of the end-to-end principle in the context of trust, see Clark and Blumenthal [2011]).

After a short digression, I will look in more detail at the four security challenges I listed earlier.

A Brief Tutorial on Encryption

Even though I have said that the persistent security problems on the Internet are not technical, it is necessary to understand how encryption works. Encryption is a building block of many mechanisms to enhance security. Conceptually, encryption is a simple idea. The heart of an encryption scheme is a procedure that transforms some material (the *cleartext,* or *plaintext*) into a different representation that obscures the meaning of the original material. This procedure can be reversed to yield the original material. Normally, cryptographers do not want to keep the transformation procedure (the *encryption algorithm*) secret—they want others to inspect it for flaws, implement it, and so on. So the encryption algorithm is designed to take the cleartext and a *key* and produce the encrypted version (the *cypertext*). The key must be protected as a secret among those authorized to encrypt and decrypt the material. The success of an encryption scheme depends both on the strength of the encryption algorithm—a strong scheme makes it essentially impossible or very difficult to guess the plaintext from the cypertext without the key—and the ability of the communicants to keep the key secret. I will give several examples in this chapter that suggest that keeping the keys secret (*key management*) is the weakest aspect of most deployed encryption schemes. Here is another saying from a wise security expert: "Amateurs think they have to break the crypto scheme, professionals just steal the keys."[5]

Historically, encryption schemes used the same key for encryption and decryption. Decryption was the reverse of encryption. This class of scheme worked well when the number of people communicating was small, at best only two. If two people share a secret key, it may remain a secret in practice, but this class of scheme, called a symmetric key scheme, becomes impractical as the number of potential communicants grows. Consider the challenge of implementing encrypted communication in a popular Web site, which might have tens of millions of subscribers. If the Web site had one key that it shared with any party that wanted to communicate, that key would no longer be secret and there would be no protection. On the other hand, if the Web site tried to issue a separate key to each of the tens

5. I do know the attribution for this statement; it came to me from Earl Boebert, who is a rich source of short, insightful statements about security.

of millions of subscribers, the problems would include how these keys are given out, how the Web site keeps track of which key goes with which subscriber, and what to do when a subscriber loses its key.

The problem of scale was solved by a breakthrough in cryptography called asymmetric or public-private keys. In this class of encryption algorithm, different keys are used for encryption and decryption, and knowing one key does not make it possible to derive the other. The scheme most commonly used on the Internet today is called RSA, after the names of its inventors, Ron Rivest, Adi Shamir, and Leonard Adleman.[6] Public-private key systems can be used to let a huge number of people send encrypted messages to a Web site as follows. To oversimplify, the Web site makes a pair of keys and gives out one key of the pair, the public key, to everyone, but keeps the private key secret. Anyone can use the public key to encrypt a message, but only the Web site can decrypt it. This is the scheme used today in many Internet applications. The standard that defines encrypted communication with a Web site is called transport layer security, or TLS, and one way to tell if it is being used is that the URL begins with "https" as opposed to just "http." Another use of asymmetric encryption is in the mechanism called IPSec, which encrypts packets (not complete messages) and is used to create encrypted paths across the Internet, for example in support of virtual private networks (VPNs).

Asymmetric key systems can also be used in reverse to solve another problem, which is assuring the authenticity of an object, in other words *signing* it. To sign an object, the creator of the object computes a hash[7] of the object and then uses its private key to encrypt the hash, which then serves as the signature. Only the person with the private key can sign the hash, but anyone who has the object can compute the hash (the hash algorithm is not secret) and then decrypt the signature to see if it yields the same hash.

There is one remaining problem in the practical use of asymmetric key systems: making sure that the mechanism by which a Web site gives out its public key is not corrupted itself. If an attacker can replace that key with a different public key and then somehow deflect packets going to the legitimate Web site to a malicious copy that holds the false private key associated with that public key, the unsuspecting client will have an

6. I do not claim to be a historian of cryptography, and the invention of asymmetric encryption involves parallel invention in the classified and unclassified communities. At the risk of being un-scholarly, I point the reader to the Wikipedia site on public key cryptography for those interested in more, rather than listing a bunch of citations here.

7. See the discussion of a hash function on page 114.

encrypted communication with the wrong (malicious) Web site. In the sections that follow, I will point out that the problem of public key distribution is a crippling challenge in many encryption schemes.

With this as background, I will now discuss the four security problems I cataloged earlier.

Attack and Defense of the Network Itself

The physical layer of the Internet is made up of links, routers, servers, and the like. Routers and servers are computers and thus potentially susceptible to remote attack across the Internet. Links themselves seem more immune to this sort of attack and are mostly susceptible to physical attack based on close access—cutters and explosives. There are physical responses to close-access physical attacks: links can be hardened against attack (against both destruction and tapping), and operators can place routers in physically secure facilities.

The Internet is the global collection of links and routers that forwards packets from an entry point to an exit point. The functional specification of this service, as I have said, is rather weak: these components are expected to do what they are designed to do except when they don't. The best-effort service model means that the network is expected to do its best, but failure is accepted. We know that links can fail, routers can crash, and so on, and it would be foolish to pretend that these components can be completely dependable.

In a system where failure is accepted, each layer must be designed to take into account the anticipated failures of the layer below. The Internet layer deals with link and router failures—it includes a dynamic routing scheme that finds new paths if a path fails. The end-to-end transmission control protocol (TCP) copes with the loss of transmitted packets. TCP numbers the packets, keeps track of which of them are received and which are lost, resends the lost ones, gets them in correct order at the receiver, and then passes the data up to the next layer.

The overall resilience and function of the system is based not on precise specification of the service each layer provides but on a pragmatic balance of effort at each layer. The better each layer works, the less the layer above it has to do, and (probably) the better the resulting behavior. Operators can engineer different parts of the Internet to different levels of performance and reliability (driven in many cases by pragmatic considerations of cost), and each layer above must be designed to cope with this variation. Investment at the lower layers benefits the next layer's functions, but over investment at the lower layer may add unnecessary cost to the service. The working out of this balance is not a part of the Internet's specification; the interplay between performance and reliability at different layers is a point of adaptation as the Internet evolves.

But given this weak specification, how would we think about the security specification? How would we characterize the security of the packet forwarding service of the Internet? A formally correct but useless response would be that since the network is allowed to fail, it need not concern itself with security. Pragmatically this is nonsense. There are well-understood expectations of Internet service today, and an attack that materially degrades that service is a successful attack. But degradation is a matter of degree, and degraded service may still be useful.[8] With a loose specification of the Internet's function, the determination of how to make the system resistant to attack is potentially ad hoc. One must look at specific mechanisms, not the specification, to see where attacks might come from. The critical services that the Internet provides are the forwarding function itself, the routing protocols, and the DNS. Thus, one could look to the routing protocols and ask whether they are robust to attack (they are not, as I will discuss). But the core function of the Internet is simple. If there are links connecting routers, the routers are working, and the routing protocols are computing routes, packet transport is mostly working.

The Internet provides a general service useful for many applications in many circumstances. This generality raises a security conundrum: different contexts will face different security threats. There is no uniform threat model against which to design network defenses. Nonetheless, designers must face the security challenge and make pragmatic decisions about how robust the forwarding service must be to different sorts of attacks. But any security analysis must begin with an assessment of the range of motivations behind an attack on the network. What we see today is that in most cases an attack on the network is not the end objective but a means to carry out a subsequent attack on either an attached host or (more likely) on communication. So in order to address the challenges of preventing attacks on communication, it would be good to have a sense of how robust and reliable the packet forwarding service of the Internet is, but there is no consistent answer to this question.

8. Security experts understand that the most dangerous attacks are those that might cause massive, correlated failure of components, for example attacks on routers that exploit a common failure mode and take out so many routers that the dynamic routing algorithms of the network are overwhelmed and the network essentially ceases to function.

A Case Study: Why Is Securing the Network Hard? Securing Interdomain Routing in the Internet

The challenge of securing interdomain routing in the Internet is a good case study of the barriers to better security; it illustrates the challenges caused by lack of trust and difficulties of coordination. Each of the autonomous systems (ASs) that make up the Internet must tell the others which addresses belong to it and how that AS is connected to others in order to create the interconnected Internet. The way this exchange of information works is that each AS announces to its neighbors the addresses it contains, which in turn passes this on to its neighbors, and so on, until this information reaches all of the Internet. Each such message, as it flows across the global network, accumulates the list of ASs through which a packet would go in order to reach that original AS and its addresses. Of course, there may be many such paths—a particular AS may be reachable via many neighbors, so a sender must pick the path it prefers or, more precisely, each AS computing a route back to a particular set of addresses must pick among the options it receives and then offer that option to its neighbors in turn.

Originally, there were no technical security controls on this mechanism. A rogue AS could announce that it was a route (indeed, a good route) to any other AS in the Internet.[9] If other ASs believe this announcement, they will forward traffic to that false destination, where it can be dropped, examined, or otherwise used. False routing assertions are not uncommon in the Internet today, resulting in failures along all dimensions of CIA.

Today, these attacks are managed operationally. Network operators monitor the system, problems of reachability are reported from the edge by end users, and over some period, perhaps a few hours, the offending AS is identified and the false assertions are stopped or blocked. However, there are few tools in the current Internet ecosystem to discipline this bad behavior.

Ignoring details, a technical fix would be to sign these assertions originating at each AS by using a public key cryptography scheme, so that they cannot be forged. Indeed, this was the path down which the designers of a secure BGP started, but there are two barriers to this approach, one having to do with migration to the new scheme and the other having to do with trust.

9. Designers understood as early as 1982 that an AS could disrupt routing by making a false statement. RFC 827 (Rosen, 1982, section 9) says: "If any gateway sends an NR [neighbor reachability] message with false information, claiming to be an appropriate first hop to a network which it in fact cannot even reach, traffic destined to that network may never be delivered. Implementers must bear this in mind." This situation was identified as a vulnerability but not a risk. The advice to "bear this in mind" could have multiple interpretations.

The migration problem is easy to understand. In the global Internet, there is no way that everyone is going to switch to the more secure scheme at once. Unless operators of ASs impose some draconian discipline such as disconnecting nonconforming ASs from the network, some ASs may refuse to undertake the effort of upgrading and continue to originate unsigned route assertions. An AS that receives an unsigned routing message can either reject it (which is the draconian outcome of disconnecting the sending AS) or accept it (in which case a malicious AS and a lazy AS will look the same); until the last AS starts signing its assertions, the inter-AS routing system will contain vulnerabilities.

The issue of trust is more complex. When an AS signs an assertion (for example, when MIT signs the assertion that it is AS 3 and that it owns a particular set of addresses), it has to use some encryption key to sign that assertion. The designers proposed a scheme based on public-private or asymmetric keys, where MIT has a private (secret) key used to sign the assertion and a public key it gives to everyone so they can decrypt the signature and confirm that MIT signed it. So far so good, but why is that public-private key pair trustworthy? If MIT can just issue itself a set of keys and start signing assertions, it might seem that we are no better off, because a malicious actor could do the same thing—make up a public-private key pair and start signing assertions that it owns AS 3, controls those addresses, and so on. To prevent fraudulent assertions from being effective, the designers proposed creating a trusted third party that could confirm, based on its own due diligence, which public key was actually associated with the real MIT. But why in turn would anyone trust that third party? A scheme like this ends up in a hierarchy of trust, which requires a *root of trust*, a single node that all parts of the Internet trust to tell them which second-level parties to trust and so on until we get to the party that asserts that it knows who the real MIT is.

An engineer might think this was a simple, elegant scheme, but it runs aground in the larger world. First, what single entity in the world would all the regions of the world agree to trust? The United Nations? This issue is serious, not just hypothetically but concretely. Several countries (including Russia) asserted that they would not assent to a common root of trust with the United States. The third party who has the power to validate these assertions will almost certainly have the power to revoke them. Imagine a world in which the United Nations, by some sort of vote, revokes the trust assertion about some nation and essentially ejects it from the Internet. What about entities that are within some legal jurisdiction and thus subject to the legal regime of that region? This fear is not hypothetical. The institutions that allocate Internet addresses are the Regional Internet

Registries (RIRs). The RIR for the EU is Réseaux IP Européens (RIPE) and is located in Holland. The Dutch authorities brought a police order for them to revoke the addresses of a specific AS. RIPE correctly said that it did not have the technical means to revoke an allocation. However, if they were issuing certificates of authenticity for AS allocations, then they would no longer be able to make that claim. The hierarchy of trusted certifiers and a single root of trust is technically robust to the extent that it will give the right answer *if the trust relations are valid and accepted by all the parties*. This approach may be technically robust but is not socially robust.

Given these considerations, would signed but revokable BGP assertions make the Internet more stable and secure or less? Once people understood the social consequences of this scheme, there was substantial resistance to its deployment. The problem with adding a "kill switch," such as revocation, to the Internet is controlling who has access to it. Once we grasp the complexity of functioning in a space where not all the actors share the same incentives and not all are equally trustworthy by different measures, and that these actors of necessity are in the system, the design problem becomes complex. Management of the trust relationship, the expression and manifestation of that relationship, and the balance between centralized and localized decision-making about trust, not the exact manner in which cryptography is used, becomes the defining feature of a successful scheme.

What happens today is that the Internet does not try to solve these problems by using technology. Operators fix some of these problems by using management—oversight of the system by trained network managers. We just tolerate some of the residual consequences.

I will now sketch an alternative to the "root of trust" scheme described earlier that illustrates a different approach to social robustness.[10] Earlier, I dismissed the idea that MIT just make up a public-private key pair and start signing its assertion. What would happen if interdomain routing was based on this scheme? At first, various regions of the Internet might get conflicting assertions if there were a malicious AS in the system at the same time that MIT started to send its valid assertions. That situation, while not desirable, is exactly what we have today—conflicting assertions. But over time—hours or days—it would become clear what key went with the real MIT. Each AS in the network could learn this for itself, or groups of mutually trusting ASs could cooperate to learn it. If necessary, actors could exchange valid public keys through side channels. Once the other ASs in the Internet have

10. To my knowledge, this idea has not been seriously evaluated. I am offering it here as an example of an alternative way to think about engineering a social system.

decided which key to trust, they have independent control of that decision and there is no authority that can compel a third party to invalidate their trust assumptions. The scheme decentralizes control: any AS can decide on its own to stop forwarding traffic to MIT, just as they can today. This scheme is less technically robust (one cannot prove that it will give the correct answer under certain assumptions) but is more socially robust.

What this scheme exploits is not a technical scheme for propagating trust but instead a social protocol called "getting to know you," which humans have been running, probably for millions of years. We can be fooled, but in fact we are pretty good at it. It is also simple. It requires no trusted third parties, little administration (except that each AS should try very hard not to lose its own private key), and great adaptability to changes in the landscape of trust. Through this lens, the landscape of security becomes a landscape of trust. Regions of mutual trust will be more connected, more functional, and more effective, and regions that don't trust each other will still try to communicate, but with more constraints, more limitations, and perhaps more failures, especially with respect to availability. This pattern will be found within any application that tries to tailor its behavior to the degree of trust among the communicating parties, whether the function is exchange of routing information or email.

Attacks on Communicating Parties

This category of attack relates to parties that are attempting to communicate and are being attacked by some malicious actor. The three traditional CIA subobjectives (confidentiality, integrity, and availability) make sense here. Information should not be disclosed except to parties authorized to see it, it should not be corrupted, and it should be available. With respect to network communication, these goals take a rather simple form, particularly with respect to integrity. Since the current Internet does not perform computations on data, the simple form of integrity is that data is transmitted without modification.[11]

As I discussed, cryptographic algorithms fit into the CIA triad by giving strong assurance that data is not disclosed and a strong indication if data

11. If PHBs that transform the data in transit are added to the network, a more complex theory of integrity will be needed, such as the one offered by Clark and Wilson (1987). That framework takes the approach that PHBs (or transactions generally) that transform data must be audited by a trustworthy party to ensure that the transformation is valid. The scheme my coauthor and I devised is based on a foundation of trust, not a technical approach to prevent incorrect transforms.

is modified. Cryptography is a powerful tool to improve security. However, it is important to see how cryptographic methods tend to function in the larger context. They protect the user from failures of integrity by halting the communication. They map a wide range of attacks into a common outcome—cessation of communication. But this outcome, while potentially better than a failure of confidentiality or integrity, is just a failure along the third CIA dimension—availability. Essentially what these schemes do is turn a wide range of attacks into attacks on availability, which is not the desired outcome—we want to offer assurances about all dimensions of CIA.[12]

If the best we can do using cryptography is to turn a range of attacks by untrustworthy actors into attacks on availability, what system designs can improve availability? Availability is such an important issue, involving factors other than security, that I defer my major discussion of availability to the next chapter. However, in brief, there are two ways to try to deal with untrustworthy actors: constrain or discipline them or reduce our dependency on or even avoid them. Imposing constraints on untrustworthy or malicious actors that both compel them not to misbehave and compel them to perform at all are hard to devise. The only way to compel correct operation is to design the larger ecosystem so that the cost to the actor from expulsion from the system outweighs the cost from forgoing malicious behavior. This might work for an ISP that is hosting both legitimate customers and spammers (and ISPs have been driven out of business after being expelled from the Internet for hosting spammers), but malicious individuals show great resilience to constraint and discipline, especially across jurisdictional boundaries. This leaves the other option as a path to availability, accept that untrustworthy actors are in the system but avoid them, an approach I explore in the next chapter.

The real problem with the cryptographic schemes used in the Internet, such as the TLS protocols used to secure Web transfers, is not the cryptography itself but public key distribution. As a concrete example of the problem, imagine a user that wants to connect securely to Google, at www.google.com. That user needs the public key of Google to encrypt his messages. TLS solves this problem as follows. When the user first connects to Google, the server sends the public key to the user. But for this to be trustworthy, there needs to be some validation of that public key. Without

12. This observation provides one explanation as to why many users deal with dialog boxes warning about potential security hazards by clicking the "proceed anyway" option—what they want is to make progress. Another reason, of course, is that the content of those warnings is often inexplicable.

validation, a false Google Web site could just send a false public key to the user, who would then initiate an encrypted connection to the false Web site. To prevent this sort of mischief, what the Web server actually sends is a *certificate* that contains the URL of the Web site and its public key and is signed by a trusted third party, a *certificate authority*, or CA, that has verified that the public key in the certificate is associated with the real Google, not a malicious clone. But this approach only pushes the problem up a layer: why should the user trust that this CA is actually valid and not a malicious clone? In the TLS scheme, that CA must present a certificate of its own validity that is signed by a higher-level CA and so on. Again, as with the trusted third-party hierarchy that I described for validating BGP routing assertions, there has to be some root of trust. However, the designers of TLS wanted to avoid the problem of a single root of trust, so the scheme allows multiple roots of trust. Our hypothetical user attempting to connect to Google must be pre-equipped with a list of these trusted root servers. Today, that list is embedded in the browser—it is included in what is fetched when a user downloads a browser. So the providers of the major browsers—Google, Apple, Microsoft, and the Mozilla Foundation (developers of Firefox)—are the ultimate arbiters of what roots of trust are actually trustworthy. There is an industry group called the Certificate Authority/Browser Forum that meets to deliberate about which root CAs to include in the list.

The problem with this scheme is that some of the root CAs (or a second-level CA beneath them) have proven to be untrustworthy. The problem can be corruption, penetration by malicious parties, or just adverse interests. A Dutch CA, DigiNotar, was penetrated, apparently by the Iranians, who used their access to issue false certificates for sites, including Google and Facebook. Eventually, the Dutch government forced DigiNotar to shut down, which had the unfortunate side effect of invalidating the certificates for many of the Dutch government Web sites. I do not know what sort of harm resulted to Iranian citizens who thought they had a secure, protected path to Google, but the outcome may have been dire in human terms. As another example, a subsidiary of the state-run Chinese root CA issued false certificates for Google, presumably to allow China to intercept communications from its citizens that were ostensibly protected by encryption. This incident led Google to declare that the Chinese root CA was untrustworthy and remove it from the list included in its browser. Events of significant diplomatic consequence are occurring not just between states but between states and powerful private actors.

A possible remedy to this situation, called certificate transparency, was developed by two Google employees and is being pushed by Google. The idea of certificate transparency is that any CA can issue any certificate that it wants (including ones that the owner of the name considers false) but browsers will not accept a certificate sent to it by a Web server unless that certificate can also be found in a public log. This scheme does not prevent the creation of false certificates, but since the creator must post the certificate to the log, the existence of the certificate is public. Anyone (most obviously the owner of a Web site) can scan the log to look for false certificates. A CA can lie, but not in secret. Google is in a position to push this change to the CA ecosystem because it has the capacity to run a secure public log (but other actors must run logs as well in order for the scheme to work) and because it controls an important browser (Chrome) and can unilaterally make the change to require a lookup of a certificate in a public log.

This scheme illustrates what I have called socially robust design, as opposed to technical correctness. It does not prevent a CA from issuing false certificates; it just shines a light on their behavior so that various pressures can be brought to bear on them (including, in the worst case, ejection from the list of trusted root CAs). The scheme also illustrates the same lesson that I took from the attempts to secure BGP. The central challenge in the design of these schemes is understanding the landscape of trust (and the presence of untrustworthy actors in the ecosystem) and designing a system that can reflect degrees of trust. Trust or its lack, not some conception of technical correctness, is the dominant factor that determines how different actors interact.

Traffic Analysis

The term *traffic analysis* describes a form of surveillance in which the observer looks not at what is being sent but at the source and destination of the communication. In the Internet context, the observer can capture information such as the IP addresses and port numbers in the packets. This sort of logging, sometimes (for historical reasons) called pen register/trap and trace logging, has its roots in the logging of telephone numbers for phone calls. From a legal perspective, in the United States (and many countries) it is easier to get a court order allowing pen/trap logging than for data capture, which has led to a set of legal debates about what sorts of data can be gathered using pen/trap. Such data is often called metadata because it is data about other data. The complexity of the Internet makes the distinction between data and metadata contentious: since a packet is a sequence of headers, each with information about the next header, one layer's metadata is another layer's data.

From a technical perspective, encryption limits what observers in the network can see, but the headers that routers (and other PHBs) process must be visible (barring very complex uses of encryption, such as TOR), so there seem to be limits on the extent to which a network design can substantially shift the balance of power with respect to traffic analysis. One exception is the NDN proposal, which through the use of the per-packet state in each router removes any source address from the packet. In NDN, an observer can tell that a piece of information has been requested but cannot easily tell which source requested it. In chapter 15, I describe a scheme that is even more robust to traffic analysis.

It turns out that an observer can deduce a great deal of information by observing an encrypted stream of packets (see, for example, Chen et al., 2010; Wright et al., 2008). It is possible to deduce a great deal about what is being communicated, about the communicants, and other information. This kind of leakage is a serious problem in high-security contexts and may become a serious problem for typical users as tools for analysis improve, so while encryption may protect the data in transit, the idea that encryption protects the communicating users from harm related to confidentiality should be viewed with some skepticism.

One way to limit the harm from traffic analysis is to avoid routing packets through regions of the network that are more likely to practice this form of surveillance. Attacks on the routing protocols of the Internet often have the effect of misrouting packets through additional regions, perhaps so that those regions can analyze the packets. There is no obvious way to detect in real time that packets are being subjected to traffic analysis, but if a group of users can make a judgment about which regions of the network are less trustworthy, and have some control over routing, they may be able to mitigate the peril somewhat. Nebula provides this sort of control. The Nebula packet header contains the sequence of regions through which the packet is to pass, so the packet rather than the routing protocols controls its path (assuming that the sequence in the packet header was correctly constructed in the first place).

Attacks on the Attached Hosts

Today, we see a wide range of attacks in this category, ranging from attacks that involve a malicious sequence of packets sent to a machine that was not a willing participant in the communication (an attack that exploits an unintentionally open port, a flaw in the network software, and the like) to attacks that use an intentional act of communication (receiving email or going to a Web site) to download malicious code.

Again, it may be helpful to return to a historical perspective to understand the current situation with respect to these classes of attacks. As I said earlier, the security experts that we consulted in the early days of the Internet were mostly from the intelligence community, and their primary concern was confidentiality—preventing disclosure of classified information. This framing of security tends to ignore the issue of communication among parties that do not necessarily trust each other. This framing also tends to divide the world cleanly into trusted and untrusted regions of the network. In the context of classified work, it made sense to accept that there were trusted regions of the network, typically inside facilities where users had clearances and computers could be trusted. These regions might be connected together over a public, untrusted Internet, but in this case the packets across the public internet would be encrypted and wrapped in outer IP headers that only delivered the packet to the distant trusted region. This concept, called encrypted tunnels, made sense from a technical perspective, since only one encryption device would be needed at the interconnection point between the trusted region and the public Internet. At the time, encryption boxes were expensive, and even a point-to-multipoint device was pushing the state of the art. Having such a device for each host was not practical. The concept also made sense in the security calculus of the day. There was no way an untrusted computer on the public Internet could make a connection to a trusted computer in a trusted region, because the encryption device would not accept a packet that was not encrypted at another region. End nodes did not need to worry about being attacked, because within the trusted region the prospect of attack was discounted, and from outside the region packets were totally blocked.

The security analysis of this sort of architecture became quite sophisticated. There were concerns about the possibility that corrupt insiders could leak information by hiding it in *covert channels*, low-bandwidth communication channels exploiting features such as the timing of packets in the channel. The *confinement problem* was understood early on (Lampson, 1973). These concerns did not end up being the real threats, and a focus on this framing may have distracted the early thinkers from a broader consideration of the security landscape, such as the need for users with clearances to talk to people without such clearances.

This simple division of responsibility between network and host has proved flawed, for several reasons. First, of course, the operating systems of today are flawed. Second, application designers have favored functionality over security and designed applications with rich features (e.g., the ability to download and execute programs), so the applications have become the vector of attack. Early in the design of the Internet, Daniel Edwards of the NSA identified the problem of a "trojan horse" program. Experts from the intelligence community made it clear that, in their opinion, if executable code was transferred across the network, the only practical protection would be to transfer it from trustworthy sources—trying to vet code for malicious content was a losing game.

So here we are today, with a need to reconsider all of these assumptions more or less from scratch. First, we have started to depend on (in other words, trust) at least some elements in the network. Firewalls provide crude protection from attacks that involve packets sent to a machine that did not want to participate in the communication. Firewalls block traffic to applications that should not be reached and (if combined with NAT) hide the IP addresses of machines. For this protection to work, the topology and routing protocols of the network must prevent traffic from bypassing the firewall. This sort of trust is both simple and local, but it reflects a recognition that the hosts being protected and at least the local region of the network to which they are attached must share responsibility for protection.

The next question is what services the packet forwarding layer might provide to make the security job of the host and the higher layers easier. In chapter 6, I asked how the expressive power of the network could be designed to help the defender in when the interests of the end points are not aligned. The network cannot make the end points secure, but perhaps it can be a part of the solution rather than delivering the attacks with best effort. Perhaps there are new sorts of elements or new actors that could provide protection services. In the language of chapter 6, what PHBs can we devise to protect an end point, what actor would be best trusted to deploy and operate them, and, finally, what architectural support, if any, is needed to utilize them. If we allow ourselves to rethink this framework for security from scratch, new design approaches might emerge that have additional actors and services beyond the host and the network.

There has also been considerable progress in devising ways in which the operating system of the end node, in addition to being more robust itself to attack, can help protect the application running on the end node from attack. The concept of sandboxing describes an approach where the code of the application is placed in a confining environment before it interacts with the network, and this environment is discarded at the end of the interaction, thus discarding any malware or other modifications that may have resulted from the interaction.

The Role of Applications

The Internet, as I have repeatedly stressed, is a general network that moves packets. But packets only flow because some higher-layer software chooses to send and receive them. It is applications that define what actually happens on the network. It would be nice if the packet carriage layer of the Internet, perhaps properly augmented by innovative PHBs, could protect one host from attack by another, independent of the application being used, but this hope is not realistic. The simple semantics of the Internet—best-effort delivery of packets—is (for the moment) about all the network can do. Higher-layer software, the application, translates information sent over the Internet into actions on the end nodes, and it is actions that can cause harm. Many of the security problems we deal with today arise because of design decisions at the application layer, and it is to that layer that we must turn for an overall improvement in the landscape of security. Applications can, by their design, either create security vulnerabilities or limit them.

The lesson we learn by looking at the design of applications is in some respects a bleak one. Applications today, in the pursuit of more powerful functionality and appealing features, have incorporated functions that are known to be risky, and were known to be risky at the time they were designed. The ability to download active code (e.g., Javascript) from a Web site and execute it on a client machine was understood as risky from the beginning and was decried by the security community at the time. The designers of Web protocols implemented this capability anyway. We must accept that applications today are *insecure by design*, and we must figure out how to deal with this, since this preference for features over security is not likely to reverse.

One answer lies in the operating system, where features such as sandboxing can potentially prevent malicious code from having any persistent consequences. Another answer may lie in designing applications so that they only enable risky modes of operation when there is good reason to trust the communicating parties. Actors that choose to trust each other may want to exploit applications in a more potentially risky mode that imposes fewer constraints and allows more flexible communication, while actors with less mutual trust may want a mode that provides more protection. As another answer, since applications define and control the patterns of communication among the entities, applications can be designed to invoke PHBs as part of their security architecture.

Applications can play another key role in an overall framework for security. My analysis to this point has swept a serious problem under the rug. What if the attempt to protect the host from attack fails, and the host falls under the control of a malicious actor? At this point, that malicious actor may undertake activities (e.g., data transfers) that seem entirely legitimate with respect to the network (they seem like transfers between mutually trusting parties), but the security goal is to block them. In other words, in the case where a machine has been compromised, the security goal reverses. The goal is to "attack" (block) what otherwise would be legitimate communication.

Perhaps some cases of this sort can be classified as malicious by behavioral monitoring—a user who suddenly transfers gigabytes of data out of a secure area might attract attention in any case. But a general way to think about this situation is to distinguish between penetration of a host and a harm to that host. The harm arises when applications are used in ways that lead to unacceptable outcomes. Applications can be designed so that they reduce the risk of harms even when a machine has been penetrated. For example, a design might require that multiple machines concur before allowing potentially dangerous actions. A firewall might block all outgoing data flows above a certain size unless a second machine has first authorized the transfer. The second machine should be implemented in such a way that penetration of the first machine by a malicious actor does not provide a means to penetrate or subvert the function of the second machine. To minimize disruption of normal work flow, applications could be designed to notify the second machine when validation is required, which could then carry out some independent check of identity, authorization, and the like.

I have described here a sophisticated design challenge for the application developer, but suggesting that potentially dangerous tasks should require dual authentication is not a novel idea. My point is that this sort of constraint must be built into, or at least controlled by, the application, not the network. I discussed earlier the basic design approach of the Internet, which is that layers should be designed to deal with failures in the layers below them. TCP deals with lost packets and related problems. What I propose here is just the application of this approach at a higher layer—the design of the overall system must take into account the possibility of failure in the layers below, in this case corruption of machines on which (part of) the application is running. The network can have a role, which is to ensure that only authorized flows take place. One could imagine that software-defined networking (SDN) technology could be used to allow only flows that are consistent with the application-defined security policies. I discuss SDN in chapter 13.

The Role of Identity

My repeated reference to trust raises a more basic concern: it is nonsense to talk about whether actors trust each other unless they have sufficient information about each other's identity, so identity management must be part of any framework that depends on trust management. This fact, in turn, raises the question of which entities or layers within the system should implement the mechanisms of identity management.

One view is that the network itself (the layer that forwards packets) should specify how identity is managed. There have been calls for an "accountable Internet" (see, for example, Landwehr, 2009), which might imply that the architecture assures the identity of the participants in all interactions. I think this is a bad design approach, as Susan Landau and I have argued (Clark and Landau, 2011). Society uses identity in a very nuanced way—sometimes we need strong, mutual confirmation of identity, and sometimes we function well with total strangers. We do not walk around with ID numbers written on our foreheads. Context determines the need for identity, and on the network, applications define the context. So it is the application that must establish the correct level of mutual identification and use this information to deploy the correct level of protection.

Application designers should not have to solve problems of identity management from scratch with each new application; they should be provided advice and guidance, perhaps applicable to a class of applications, that suggests how to approach these problems. What is needed is a collection of application design patterns for developers to use. Thinking about design patterns in an organized way should yield another benefit in that by looking across applications to see common needs, new ideas may emerge for common services that the lower layers can offer to help improve the security of applications. It is highly unlikely that some new service at the packet forwarding layer will suddenly make applications secure, but there may be supporting services that can make the task easier. The way to find these services is to look at application requirements, generalize from them, and see what concepts emerge.

Denial of Service Attacks

Denial of service attacks flood a part of the network or an end node with so much malicious traffic that normal operation is degraded or halted. Reasoning from first principles, the success of DDoS attacks could represent a fundamental indictment of the Internet architecture as a layered system—the essence of a proper layered design is that the (mis)behavior of a higher layer should not disrupt a lower layer. Since DDoS attacks can disrupt the Internet layer just by sending packets, the design of the Internet layer is definitionally flawed. However, one should not be too harsh in judging the Internet design. Simple approaches to protecting the forwarding layer, such as separating flows by using fair queuing and limiting the flow rate for large flows, can only do so much if the attacker can assemble a million attack machines.

Another point of view is that the problem arises because the sender can send at will, without the permission of the receiver. If the Internet required the permission of the receiver before delivering a packet, as in NDN, certain sorts of DDoS attacks would not be possible. However, many of the machines that are attacked on the Internet offer services such as providing Web data. These machines need to accept traffic from anyone if they are to fulfill their intended purpose.

Yet another point of view is that DDoS attacks are possible because, in many cases, users pay a flat rate for access rather than a usage-based charge. This pricing model reduces the incentive of the user to remove malware—if the user suddenly got a large and unexpected monthly bill because his computer had been corrupted and had participated in a DDoS attack, he might have a larger incentive to remedy the situation.

Perhaps over time there will be sufficient mechanisms in place to prevent successful attacks on computers so an attacker can no longer assemble the necessary set of infected computers to launch the attack. (See the previous section, discussing attack on the end node.)

One approach to defense is to replicate the service often enough that the attacker cannot marshal the resources to flood all the copies. Either the attacker concentrates on one machine and leaves the others functional or attacks all of them, which diffuses the attack to the point where it is ineffective. Network architectures that include the concept of indirection (i3 or DOA) may be able to diffuse an attack launched against the indirection nodes. Some commercial DDoS mitigation services on the current Internet work this way: the DNS is used to disperse an attack to a large set of machines that then attempt to filter out the attack traffic by using various heuristics.

In my view, once we take into account the need for services (server machines) to be open to connections from anywhere and the potential scale of DDoS attacks, realistic mitigation techniques will have to identify the attack traffic as such and drop it to a level that renders the attack ineffective. However, once we contemplate the idea that certain traffic will be classified as malicious, we must then think through the potential of this mechanism itself as a vector of attack. We must ask which entity would have the authority (or be trusted) to declare traffic as malicious and which actors would be expected to honor this declaration. Again, this is an exercise in crafting the expressive power of the architecture so that the "right" actors can preferentially exploit that power. I return to this topic when I discuss architecture and security later in this chapter.

Balancing the Aspects of Security

The preceding discussions suggest that there are four general security problems to address: protecting regions of the network from being attacked, protecting communication among parties with aligned interests, protecting parties with adverse interests from harming each other, and mitigating DDoS attacks. It would be very nice if these could be resolved independently, and to some extent they can, but I will argue that there are tensions between protecting the host and protecting communication, and part of the overall security design of an architecture will be to balance these objectives.

One could imagine the design as proceeding as follows:

- First, make sure that algorithms that support key PHBs are secure. Interdomain routing is the obvious example—since routing as currently conceived is a distributed algorithm in which all regions of the network participate, it creates opportunities for one region to attack another. There are other PHBs, such as anycast and multicast, that may need better security.
- Second, put in place schemes to protect communication. Assume that applications (which define the patterns of communication) will deal with issues of confidentiality and integrity by using encryption, and assume that the application will intentionally route communication to any service elements they need. To deal with the goal of availability, the application must design its communications and take advantage of any expressive power provided by the system to detect and localize if and where a PHB or service component is malfunctioning, and reconfigure itself to avoid it.
- Third, put in place PHBs that can prevent or constrain communication among untrusting or hostile end points. Assume that the application can modulate its behavior based on sufficient identity information and can add or remove protective PHBs as necessary.
- Fourth, put in place suitable mechanisms to diffuse or disable DDoS attacks.

This assessment is both glib and incomplete. It is glib overall in that it trivializes very hard tasks, even if they are well-defined. In more detail, it is glib first with respect to the issue of localization of malfunctioning and availability. However, since the current Internet does nothing in this respect, any new capability would be better than what we have today. Second, it is glib with respect to the degree to which it depends on the developer of the application to get the design right. For this approach to work, the application designer will need help and design guidance, even with respect to issues that are not embedded in the architecture.

Nonetheless, if this breakdown of the security challenge is well-structured, it can outline a research approach, even if it trivializes the challenges. However, I also described it as incomplete. The list begs the question of whether these tasks are independent—whether we can proceed with each separately, doing the best we can at a given time. In fact, I believe that the tasks are not independent; it is possible that the design space of secure operation implies a trade-off between two perils: attacks on communication and attacks by one party on another. In brief, the more protections that are put in place to protect one end point from the other (in the language of chapter 6, the more PHBs), the more points of attack are created that attackers can use to disrupt the communication. A clean, encrypted channel between two end points is a simple concept, with few modes of failure and few points where an adversary can exploit a PHB to disrupt the communication.

The entanglement of these various goals represents a challenge to policymaking in this area. Better encryption enhances privacy but impairs interception of traffic, even if lawful. PHBs to protect one node from attack by others increase the chance that these elements can themselves become subverted and turned against the users of the network. To repeat what I said earlier, security is not a single-dimensional objective where more is better but rather a balance among objectives of actors that may not have aligned interests. Finding the balance among those objectives is not a technical challenge but a policy challenge. Simple assertions of policy objectives such as a call for total privacy or giving the state access to all encrypted communication do not help in finding that balance.

The Role of Architecture

The preceding sections propose a way to view the landscape of network security. The focus of this book is architecture; the final question for this chapter is what architecture has to do with security. My minimality argument states that architecture should say as little as possible, but no less. Architecture does not, by itself, determine how a system meets its requirements (such as a requirement for secure operation) but rather provides the framework and sufficient starting points for the subsequent design to meet its requirements. The previous discussion about expressive power and its potential dangers suggests a starting point of view, but there are some more specific concepts. I look again at my breakdown of security into four objectives, this time through the lens of architecture.

Attacks on the Network

The network can be attacked in two ways: by an attack on a device in the network (a router or a device implementing some other PHB) or an attack on the control protocols. Devices in the network are themselves computers, and in principle can be attacked, as can end nodes. However, there are practical reasons why they may be less susceptible to attack. They are not general-purpose platforms that run independently developed and potentially risky applications. It may be practical to engineer them to a higher standard of resistance to attacks. I am not aware of successful attacks that involve penetration of a router (but successful attacks often are not publicized).

However, once we recognize the existence of intermediate elements (and their PHBs) as a part of the system, a part of the design should include an analysis of how to deal with attacks on these devices themselves. To the extent that PHBs are protective—first-line elements exposed to the full open Internet to shield resources behind them—they will be an inviting target for attack. More generally, we have to ask about DDoS attacks against these devices. Intermediate elements with their PHBs will be attacked if this provides a way to disrupt the service provided by the PHBs, so the ability to protect first-line elements from DDoS attacks is a general problem the architecture should solve.

The network-level security problem the Internet faces today is attack on the control protocols by some of the participants in those protocols. I discussed on page 199 why securing the routing protocols of the current Internet is not a simple technical problem solved by the use of encryption but instead a complex problem embedded in a space of trust management and tussle. To the extent that designers add complexity to the network (for example, additional PHBs with distributed control algorithms), the failure modes and attack modes may become more complex—an argument for simplicity.

Designing a distributed control protocol that is resistant to attack by one of its participants is a problem that must be solved by the creator of the protocol, not by the network. The problem is known in computer science as the Byzantine generals problem (Lamport et al., 1982)[13] and takes its name from the problem of how a set of generals can coordinate a decision to attack or retreat when some of the generals may be deceitful and some of the messengers may modify the messages. A Byzantine failure is one where the failing (or malicious) device can behave in an arbitrary way, as opposed to a simple failure that just causes the device to stop. Distributed algorithms that can survive a certain number of components with Byzantine failures have Byzantine robustness or Byzantine fault-tolerance. Radia Perlman, in her PhD thesis (Perlman, 1988), demonstrated a distributed network control scheme with Byzantine robustness. The scheme is quite complex, with all messages being signed using asymmetric keys, a global trustworthy server to distribute the correct public keys of all the devices, and complex capacity allocation in all the devices.

One key design choice with important security implications is the expressive power in the packet header to represent interdomain routes. In the current Internet, where BGP computes the interdomain paths (the path vectors) incrementally for each AS along the path, any corrupt AS can disrupt the proper delivery of packets to the destination, leaving the source with no option for routing around that AS. Schemes like those in Nebula and SCION in XIA allow the sender to put a cryptographically signed interdomain source route in the packet. The *pathlet* proposal in (Godfrey et al., 2009) similarly requires that the packet header have sufficient expressive power to describe a sequence of pathlets. In schemes like these, the sender or its agent composes the path from routing assertions made by the different regions of the network. The resulting delivery sequence can still be corrupted if these basic routing assertions are not trustworthy, but the sender has more control over the final route, since the route is computed by the source or an agent the source has reason to trust.

Considering the preceding discussion and all the architectural alternatives that I reviewed in chapter 7, I do not find any architectural innovations that can help solve the problem of malicious nodes in distributed control algorithms. But any architecture that includes a distributed control algorithm to bind IDs to locations, such as the distributed hash table used

13. In an abstract to this paper later published by Lamport (https://www.microsoft.com/en-us/research/publication/byzantine-generals-problem/), he said that he picked the term *Byzantine* after rejecting Chinese and Albanian in an attempt to find a title that would not offend any current ethnic groups. He then said that the major purpose of the paper was to assign a new name to the problem. Whimsey is alive and well in the CS community.

by i3 and DOA or the global name resolution system in MF, must include an analysis of how actors with adverse interests could usefully manipulate those schemes, and how to deal with this sort of attack.

Attacks on Communication

To the extent that confidentiality and integrity are managed using encryption, the remaining problem is availability. Availability is the topic of the next chapter, but I include some thoughts about architecture and adversarial interests here.

Attacks on communication mostly arise as a result of the packets flowing through some hostile node in the network.[14] The node may be hostile because it has been penetrated by some attacker or because the owner of the node has interests adverse to the communicating end nodes. If an architecture gives end nodes control over which PHBs in the network it uses, and can detect and eject untrustworthy PHBs from the path of communication (the second challenge), perhaps the risk of attack could be minimized. But we see today in parts of the Internet a situation that brings this issue into sharp focus—more repressive or restrictive governments that require their ISPs to act as agents of the state to regulate communication, censoring and blocking what the state considers unacceptable. In this case, there are PHBs in the network that are, from the perspective of the users, untrustworthy, and the users cannot avoid using them. In this case, the analysis of the threat implies that those PHBs may take arbitrary actions. It is not realistic to base any proposal to mitigate this situation on the assumption that the attack PHBs are limited in what they can do.

For the users in that country, there is no way to avoid using that network (it may be the only network available), so communication becomes a cat and mouse game in which the expressive power of the network is used by both sides to achieve their goals. A sender can encrypt as much as possible to limit what the adversary PHBs can see. The PHBs of the state may try to force more revelation by blocking encrypted packets. The sender may tunnel to an exit node, and the PHBs may respond by blocking those destination addresses. By revealing less, the sender tries to prevent the PHBs from doing fine-grained discrimination, forcing on the PHBs a "blunt instrument" response such as blocking all encrypted flows, which may cause more

14. I am ignoring a few sorts of attacks here, such as jamming a wireless network with a strong radio signal. Those attacks are specific to a particular technology, and I am not aware of an architectural response, except to emphasize the importance of resilience and diversity to improve availability.

collateral damage than the censor can tolerate. So part of what an architecture (through the lens of expressive power) can do is provide tools to shape this game, and perhaps bias the outcome. This is a classic example of tussle, carried out using the tools of the architecture.

In this context, rich expressive power may be intrinsically dangerous. It may reveal to the adversarial PHBs more that allows them to exercise more fine-grained control. An application that allows the user to pick among multiple modes of operation is also potentially dangerous. Making an encrypted connection to Web pages is a simple example of how choice can be dangerous. Today, a Web server makes the decision as to whether to use TLS. The client has no control. If a censorship PHB blocks encryption, it blocks access to all TLS Web sites, but if the use of TLS were a choice under the control of the user, blocking all encryption would only force the user to choose to communicate in the clear. Choice can be a bad option if the end node can be coerced into making bad choices.

The less explicit information in the packet (the less the end nodes try to exploit the expressive power of the architecture), the less opportunity is available for adverse elements to disrupt communication. As another example, in the discussion on applications, I proposed that applications might want to adapt their modes of operation based on the degree to which the end points were prepared to trust each other. To the extent that one of those modes reveals more about what the user is doing, the network might do selective blocking to force the end node to use that mode. When the goal is to protect communication from attack by the network, the best architectural design point may be minimal expressive power with no choice given to the end nodes over how that expressive power is exploited. This approach, almost of necessity, will make the target of availability much harder to achieve.

Another aspect of threat analysis is whether explicit parameters in the packet need special protection. The gross corruption of a header by some hostile PHB is perhaps not worth considering—if an element is that malicious, the problem can only be dealt with by avoiding that element (if possible). The more interesting question is spying on the information, or more purposeful modification of information in the packet to somehow break a PHB further along the path. As an extreme remedy, each explicit parameter in the packet could be encrypted or signed somehow. This implies considerable processing overhead and requires some reliable way to get the right public keys. The Nebula proposal has this sort of complexity, similar to the TOR system, so we have evidence that users are willing to tolerate this overhead when it brings value. However, this high-level speculation is probably not informative; specific designs will require a specific threat analysis and plan for mitigation.

Concerns about corruption of a packet header are only an example of the more general problem I discussed earlier—if an intermediate element is untrustworthy, in general the only option is to reduce one's dependency on it to a sufficient degree, perhaps avoiding it altogether. This approach depends as much on the ability to detect and localize a problem as to prevent it, so the design approach that the architecture takes around explicit parameters in the packet header should focus on fault localization as well as avoiding elements with adverse interests. Again, the context of the repressive government must be a cautionary thought at this point.

Attacks on Hosts

To a large degree, the application layer creates the opportunities for attack, and the application (supported by mechanisms in the end nodes such as sandboxing) will have to mitigate these risks. What can the network, and in particular the network architecture, do to help mitigate these problems? One answer is that the network can provide means to prevent data flows that are not authorized by trusted elements in the network. If applications are designed so that authorization from trusted elements is obtained before dangerous data flows are permitted (e.g., exfiltration of data from a trusted region), then the network should prevent rogue applications (perhaps based on malware) from initiating these flows. Trusted elements might be able to use mechanisms such as SDN, which provides a means to download forwarding policies into routers, to prevent patterns of data flow that are not authorized at the application level.

Another potential role for architecture is to add some way to convey identity information to the expressive power of the packet header so that hosts and applications can discriminate between trusted and untrusted actors earlier in the initiation of communication. I discuss both benefits and risks of this idea, and a designer should think carefully about whether there is benefit in having any greater indication of identity visible in the packet or whether this information should be conveyed at a higher level (end-to-end, perhaps encrypted) so that issues of identity become private matters between the communicating end points.

Architecture and Identity

I argued that it would be a bad idea for a future internet architecture to include a fixed method to manage identity as part of its specification. Doing so would (in abstract terms) be embedding too much semantics into the network, but perhaps as part of the expressive power of the header there should be a field into which the sender could put any sort of identity information needed by the receiver. Different applications in different contexts could require specific information in this field, so that the credential could be checked upon receipt of the first packet, perhaps by an element in the network that had credential checking as its PHB. The NewArch proposal included a *rendezvous field* in a session initiation packet, whose meaning was private to the communicating nodes.

As with any proposal to add some form of expressive power to the architecture, this must be examined from all perspectives, both protecting a receiver from a sender and protecting communication from attack by the network. For example, a conservative government might demand that a sender add some explicit identifying information to the packet as a condition for making a connection out of the country. Today, there is no practical way to demand that, because the packet header is not expressive enough. As we make the header more expressive, we have to consider how the design shifts the balance of power among the various actors.

DDoS Attacks

DDoS attacks must be mitigated (at least to some degree) at the network layer. A network must have a way to manage and protect its resources—this problem cannot be kicked up the layers to the application. But again the question is what sort of architectural support would be useful in mitigating DDoS attacks.

Since I consider the mitigation of DDoS attacks to be the responsibility of an internet architecture, I am going to explore in some detail the prior work on DDoS mitigation. This segment will be rather technical. Feel free to skip over it if it gets too dense.

There are several ways to think about dealing with DDoS attacks. DDoS attacks are usually launched from a collection of end nodes that have been penetrated by a malicious actor, and which are then used as a source of traffic. End nodes used in this way are called a *botnet*. One way to control DDoS attacks is to increase the barriers to the construction of botnets to the point where they become impractical. Perhaps with careful attention to all the issues discussed so far in this chapter that might be possible, but I set that approach aside for the moment. A second approach is to make it easier to disrupt the control of a botnet once it has been created. Again, a different architecture might make that goal easier, but since there are many ways to communicate with an infiltrated machine in order to control it, blocking the control path would require a rethinking of the basic communication paradigms of the architecture.

Proposals such as NDN do not allow the sending of unsolicited data packets, so all attackers can send are interest packets. This limitation certainly changes the landscape of attack.

Assuming it is possible for an attacker to assemble and control a substantial set of infiltrated machines, which can then send traffic to overload a target resource, the mitigation of this form of attack seems to have two components: first to determine which machines are sending the traffic and second to block the malicious flows. In the context of the current Internet, much of the work on DDoS has focused on determining which machines are originating the attack. Determining the source of an attack is an issue in the current Internet because it is possible for a sender to put a false source address in a packet. Putting a source address in a packet that is different from the address of the sender is not always a malicious action; there are some legitimate reasons to do this: mobile IP (RFC 5944) requires that the source address in a packet be that of the "home agent" (a persistent address) rather than the current address assigned to the mobile device. The IETF tried to address false source addresses by proposing a "Best Current Practice" (BCP 38, RFC 2827) that recommends that ISPs check and validate the source addresses of packets their customers originate. However, there is no requirement that an ISP conform to BCP 38 (other than some mild and probably ineffective peer pressure and shaming), and complying with BCP 38 imposes additional costs and complexity on ISPs. BCP 38 was promulgated in 2000 and has achieved some, but by no means complete, compliance.[15]

Perhaps some alternative design could prevent an attacker from forging a source address. A designer could take the naive approach of making a test similar to the one proposed in BCP 38 mandatory for the ISP, but how would the design enforce such a mandate? As I noted in chapter 6, features described as part of an architecture that are not actually necessary for the operation of the network have a tendency to atrophy over time. An addressing scheme could make checking the source address an intrinsic part of packet forwarding; for example, a routing scheme that determines how to forward a packet based on where it came from as well as where it is going. The NIRA addressing scheme (Yang, 2003) has this feature.

Some architectural proposals explicitly allow the source address (the address to which a return packet should go) to differ from the address (location) of the sender. DOA, which concerns itself with delegation of services to other points in the network, makes it clear that the sender of a packet can

15. See https://spoofer.caida.org/, which reports on their attempts to measure compliance. Their report as of August 2017 is that about 50 percent of the ASs they tested detect false source addresses.

indicate that the return packet should pass through a sequence of services. The address of the sender is not in the packet being sent but instead the packet carries the address of the service to which the return packet should be sent. This design can give a sender some protection from attack by the other end of the connection but seems to open up many opportunities for DDoS attacks. The DOA paper (Walfish et al., 2004, see chapter 7) discusses the use of an intermediary to protect a server from a DoS attack but does not discuss the problem of malicious source (return) addresses. At the other extreme, schemes like Nebula, which require a sender to obtain a PoC from the control plane before sending a packet, would seem to preclude the forgery of source addresses.

Tracing traffic back to its source A number of *traceback* schemes have been proposed to allow a victim to identify the sender of malicious traffic (to within some range of accuracy) even if the source address has been forged. In general, these schemes exploit some mix of two mechanisms–*packet logging,* where the routers are augmented to keep track of packets passing through them, and *packet marking,* where routers add to packets they forward some indication of that router's identity. It would seem that, independent of architecture, the first sort of scheme imposes significant processing costs on every router, and the second must deal with how to write this information into the packet in a practical way. An example of packet logging, the source path isolation engine (SPIE), is described in (Snoeren et al., 2002), where routers compute and record a hash of every packet they forward.[16] A victim can compute the hash of a single attack packet and send a query into the network that follows paths where the hash has been recorded in successive routers in order to determine the source of that packet. While the details of how the hash is computed clearly depend on the specifics of the IP header, this scheme would seem to be generalizable to different architectures.

Most proposals for *packet marking* make the assumption that it is impractical to record the complete sequence of routers forwarding a packet into that packet. The IP record route option did provide this capability, up to a fixed maximum number of hops.[17] A simple packet marking scheme requires each

16. The scheme proposes the use of a bloom filter to allow efficient recording. A bloom filter is a data structure that efficiently records whether an element is a member of a set.

17. This option itself is not useful for tracking attack packets. The function is not widely implemented today and further depends on the sender inserting the option

forwarding router to record its identity into a single field in the packet with some probability. A victim receiving enough packets, will get the identity of each router along the path and (armed with some topology information) can reconstruct the path back to the sender. Perhaps a better scheme is for the packet to provide enough space to record the addresses of *two* routers. A router receiving a packet would record its address in the first field of the packet with some probability if the field has not already been filled in, which would then trigger the next router along the path to put its address into the second field. The marked packet thus records a link or segment of the path between two routers. Again, with enough packets, a victim can reconstruct a path to the attacker by concatenating these records. See (Savage et al., 2000) for a description of this *edge marking* scheme. Song and Perrig (2001) describe different encoding schemes for the link, including more security for the marking. Wang and Xiao (2009) describe a hybrid scheme in which the packet records the exit router from the source AS and the entry router into the AS of the victim, and these routers log information about the packet.

Many of these packet marking schemes are shaped by the fact that they are designed to work in the current Internet, and thus spend much of their effort fitting the marking into the existing IP header. The only real option available is to repurpose the usually unused *fragment offset* fields. In fact, the papers so focus on the necessity of conforming to the extreme constraints of the IP header that they do not give much insight as to how best to do packet marking in a different architecture where the header could include expressive power designed for this purpose.

Blocking the attack Assuming that the victim knows the actual address of the malicious traffic, how can the victim exploit this information? The key to DDoS mitigation is blocking the traffic, not just figuring out which machines are sending it. Most of the traceback papers cited here do not address the question of how to do blocking in a secure fashion, which may be one of the reasons why none of these schemes has caught on in practice.

A number of schemes have been proposed to block unwelcome traffic, including the indirection schemes I mentioned in chapter 7, such as i3. SOS (Keromytis et al., 2002) protects an end point from DDoS attacks by putting in place a set of filters in the region of the network that hosts that end point, such that all traffic must flow through those filters. (In the language of chapter 6, SOS depends on topological delivery to force the traffic to flow

into the packet, which an attacker is not likely to do. A scheme that can track attackers must be mandatory, and not subject to defeat by the attacker.

through the filters.) They try to prevent a further class of attack by keeping the addresses of the filters secret. Mayday (Andersen, 2003) elaborates on the design approach of SOS. A number of papers propose the idea of putting a *capabilty* into packets sent from valid senders so that filters can distinguish valid traffic from unauthenticated or malicious traffic (Anderson et al., 2004). These include TVA (Yang et al., 2005), Pi (Yaar et al., 2003), SIFF (Yaar et al., 2004), and Portcullis (Parno et al., 2007).[18]

These papers do not describe their proposals as a new internet architecture, but they all require new fields in the packet header (new expressive power), new functions in routers, and new protocols and mechanisms for connection setup. They should thus qualify as new architectural proposals and should be evaluated using all the criteria I have laid out in this book. The concept of putting a packet filter into routers raises many questions. One has to do with incentive—why should a router that may be distant from the victim agree to provide this service? Another critical aspect of blocking is to ensure that any blocking mechanism cannot itself be used as an attack vector. If a malicious machine can forge a request to block data from a source to a destination, it can shut down valid communication—yet another form of availability attack.

The framework for internet innovation (FII) takes a different approach to limiting DoS attacks, the shut up message, or SUM. The FII proposal is overall an exercise in just how minimal an architecture can be, and much of its complexity relates to blocking DoS attacks, which its authors (Koponen et al., 2011) argue is the only aspect of security that requires architectural attention. The SUM scheme requires that each sender be associated with a trustworthy component that can verify the source address of that sender, and block traffic from that source to a given destination when the destination sends a SUM. The authors define two fields in the packet header that must have global meaning: a valid address for this trusted agent and an identifier that this agent can use to map to the actual sender. (In this design, the actual identity of the sender is not revealed to observers in the network. Only this trusted agent can bind the identifier in the packet to the actual source.) Koponen et al. discuss the design of the SUM mechanism in detail, with the goal of preventing attackers from abusing the mechanism. The resulting scheme is complex and requires substantial cryptographic processing by the trusted agent.

One aspect of blocking an attack has to do with the design of source addresses. Setting aside for a moment the issue of forged source addresses

18. For a more complete discussion of these schemes, see the Appendix.

and forged blocking requests, what information in a source address would be useful (or in fact necessary) to implement a blocking scheme? Many of the architectures I have described use some form of separation between identity and location. The identity is intended to be robust, but the location information may be transient. A scheme like NewArch, in which the only mandatory information in the packet is the locator, may not provide a useful framework for blocking unwelcome traffic. A number of addressing proposals (including IPv6) allow a sender to use a variety of locators, which partially masks the ability of an observer to map from locator to actual machine or identity information. Obviously, the AS within which the locator is meaningful must be able to resolve the binding from locator to specific machine, but the scheme prevents this binding from being done in general, for privacy reasons. In such a scheme, the only region of the network in which effective blocking can be done is the AS hosting the source. Closer to the victim, there is, by intention, no robust mapping from source address to specific machine.

Assumptions of trust Any scheme to mitigate a DDoS attack will end up depending on the trustworthy operation of some set of components in the network (e.g., a trustworthy network interface card on a corrupted sender, or a trustworthy router along the path). Mitigating DDoS is another example of a situation where the design of a scheme depends not just on a good technical design but also on the design of a system that makes the correct assumptions about trust and incentive. In general, the architectures I have discussed do not devote full attention to the issue of DDoS, which is perhaps a missed opportunity, since the range of options for DDoS attacks may depend very much on the specifics of the architecture.

Conclusions

Security problems in the current Internet are in part a result of technical design decisions, but the flawed technology is not the result of error or lack of attention. Applications that are insecure by deliberate design, driven by larger economic considerations of appeal and usability, are perhaps the most difficult challenge. Barriers to the security of distributed systems such as the inter-domain routing system, email, or the Web are problems of coordinating and incentivizing collective action, dealing with negative externalities and costs imposed on first movers, understanding how to cope with a lack of uniform trust across the system, and the like. Overcoming these barriers will require good system design, but that design is not exclusively technical; it must include complementary aspects of technology, operational requirements, governance, and the like.

In the beginning of this chapter, I offered three definitions of security, from the perspectives of computer science, the user, and political science. The computer science definition of security—that a system will only do what it is specified to do, even under attack—prevents unexpected outcomes or behavior but does not necessarily prevent specific harms. It is an appealing definition to a system engineer, because it bounds the security problem to the system in question. Framing security in terms of preventing harms (consider preventing credit card fraud) brings many more elements into its scope. For example, the best way to prevent or mitigate some forms of credit card fraud may be to modify the credit card clearing system. This definition of security frustrates the designer of a component, because the solution is no longer within the scope of what the designer can specify. Of course, if the system in question has multiple components with multiple actors responsible for parts, even the computer science definition of security may be hard to apply, but only by contemplating potential harms can one begin to determine the level of effort to put into defending the system elements. As I said in the introduction to this chapter, the design of the Internet presumes that the lower layers are not perfect, so the degree of effort to put into hardening them must be a matter of judgment, not the pursuit of perfection.

What has repeatedly emerged in this analysis is that technical mechanisms intended to improve security are embedded in a larger context of trust management. Trust management is the ugly duckling of security, compared to cryptography. Cryptography is lovely and complex mathematics and has provable bounds and work factors; it is an amazing capability. Tools to manage trust are messy, socially embedded, and not amenable to proofs of correctness. Sadly, cryptography is almost always wrapped inside one of the larger, messy contexts. At a minimum, the problem of key management is always present. The challenges of securing the routing protocols of the Internet (or the DNS, for another example), improving availability, or allowing applications to adapt their behavior based on an apparent threat all depend on trust as a central issue.

The actor that gets to pick which element to trust has the power to shape the security landscape. Mechanisms that create points of control may initially be easier to reason about, but given the potential tussle that arises around any centralized point of control (e.g., certificates), a better real solution may be to prefer more socially stable solutions such as highly decentralized control and decision-making.

11 Availability

Since the most basic task of a network is to deliver data, its ability to carry out this function, even under adverse conditions, is a primary consideration. The term used to describe this general character is *availability*. The term *resilience* is also used in this context, and captures the idea that the challenge of availability is to function when things are going wrong, not when everything is working as it should. An available network is a network that is resilient in the face of failures. Failures of availability are *outages*.

Characterizing Availability

A definition of availability only makes sense within the scope of the particular functional specification of a network. A delay-tolerant network that promises to deliver email within a day under normal operation (the utility of such a network is a separate question) would presumably define a failure of availability differently than a real-time delivery service that promised delivery within a factor of 2 of the latency of light.

There seem to be at least two dimensions to availability (or its lack): time and scope. For how long was the service not available, and over what portion of the network did the failure apply? The dimension of time is essential here. In the current Internet, we do not consider the loss of a packet as a failure of availability. TCP retransmits the packet, and application designers anticipate and deal with this sort of fluctuation of delivery time as normal. Failures of links or routers can cause a loss of connectivity that lasts long enough to disrupt some applications, such as a voice call (Kushman et al., 2007), so it might be reasonable to describe these events as a transient loss of availability. However, short disruptions like this would not rise to the level at which a regulator tracking outages would be interested. An outage that lasts for hours or days is much more disruptive.

The dimension of scope is similarly essential. A user of the Internet in the United States might not even notice the loss of connectivity to a small country in Africa. For the citizens of that country, the disruption would be severe. The measure of availability (and the assessment of the importance of a failure) is a matter of the observer's point of view.

These examples also suggest that availability is a concept that requires a layered analysis, just as with security. Higher layers can (in many cases) compensate for failures at a lower layer. For example, caching data in many locations can improve the availability of the data even in the presence of failures of the communications infrastructure, but cached data may not reflect the latest version if the data is changing.

A Theory of Availability

I start with the assumption that a system in which its components are working according to specification is available. While networks may have different service commitments, it makes little sense to talk about a network that fails its definition of availability under normal operation. This framing ties a potential loss of availability to a failure of part of the system. When something fails, two sorts of corrections can occur. First, the layer in which the failure occurred can undertake to correct the failure. Second, a higher layer can undertake corrective or compensatory action. Ideally, actions at different layers will not be in conflict, which implies that a valuable capability of a layer would be some way to signal to the layer above it the nature and duration of a failure, or perhaps specify the normal duration of a failure.[1] I return to this issue of interlayer interfaces for management in chapter 13.

In order for a layer to recover from a failure, either the failed component itself must recover or there must be enough usable redundant elements to restore service. There is a division of responsibility in exploiting redundancy to achieve high availability. The system as deployed must include enough redundancy to cope with failures, and the redundancy must be usable—the design of the system must allow for full exploitation of whatever redundancy is there. In order for a system to recover from failures and restore availability, redundancy must be both present and usable. There is thus, at

1. For example, the ring topology of the SONET (synchronous optical networking) technology was designed to recover from a single fiber cut within a bounded time. The design of the higher layer should be to wait this amount of time before undertaking any adaptive steps to deal with a failure in order to see if the SONET recovery was successful. If connectivity did not return within that time, the problem was more severe, and higher-layer action was justified.

an abstract level, a series of steps that must be part of a scheme to cope with failures:

- It must be possible to detect the failures.
- It must be possible to localize the failed parts of the system.
- It must be possible to reconfigure the system to avoid depending on these parts.
- It must be possible to signal to some responsible party that a failure has occurred.

Each of these may seem obvious, but all can be tricky. The preceding list is deceptive. It begs the question of *which actor* has the responsibility for each of those steps.

Detecting failures With respect to detecting a failure, simple fail-stop failures are the easiest to detect. The hardest are failures where a partially operational element responds (for example, to management probes) but does not fully perform. A mail forwarding agent that has totally failed is easy for a sender to detect (and the sender can use information in the DNS to switch to a backup forwarding agent). A mail forwarding agent that accepts the mail but does not forward it onward is harder to detect. Some future design for an internet might argue that its architecture allows the network layer to detect *all* failures in its operation, but I would find this a bold assertion. It is probably possible to enumerate all the elements in the network (although even this task gets more difficult as more PHBs creep into the network, especially if the execution of the PHB is based on contingent delivery). However, as network functions get more complex, to enumerate all the failure modes (or to create a robust taxonomy that covers all classes of errors) seems a rather daunting challenge, especially in the context of security-related failures. In general, only an end-to-end check can confirm whether something is failing (e.g., the mail is not getting through), but the end-to-end check does not help with the second step—localizing the problem. So the resolution of which actor should detect failures is that while a layer should do all it can, the end nodes must play an essential role as a last resort.

This line of reasoning about detection of errors applies specifically to availability issues that arise in the context of attacks on communication (see page 202 in chapter 10). Given that faults that arise from malice may be crafty and Byzantine, both detection of faults and their localization may be difficult.

Consider a simple example—a router that drops or adds packets to a packet flow. This sort of action does not break the forwarding layer, just the end-to-end communication. Should the packet forwarding layer keep count of packets and exchange these counts to see what is lost or gained? Or consider the more subtle attack of changing a bit in an encrypted packet. This attack disrupts the higher-level flow. Should the network recompute the encryption function at each node to detect that (and where) the packet was corrupted? That kind of validation may be a useful mechanism in specific cases, but the complexity and performance cost seem daunting.

Localizing the fault For simple faults, where the layer itself can detect the failure, localization is often a direct consequence of discovery. Routers send recurring messages to each other, which serve to construct routes and also to confirm that the remote router is still functioning. Dynamic routing protocols are designed to recover when they detect a failed node.

A more complicated situation arises when end nodes detect a problem. In this case, there does not seem to be a single general approach to localizing the fault. One approach would be monitors or "validation" units at interconnection points within the network that could record what passes—by comparing records from the various monitors, it might be possible to localize where the packets had been manipulated. ChoiceNet (page 138) includes this mechanism, but the question remains as to what sort of architectural support would facilitate this sort of scheme—perhaps some sort of control flag in the packet that would trigger logging and debugging. Another approach is route diversity: trying selective reconfiguration of the system, avoiding different parts of the system in turn, to see whether the problem persists.

As I discussed on page 100 in chapter 6, it is not always desirable to assure availability. When an end node is being attacked and prevents the attack from reaching it (perhaps by using some PHB in the network), it has intentionally caused a loss of availability (as seen by the attacker). In this case, where the interests of the sender and receiver are not aligned, the sender should not only be deprived of tools to remedy this impairment but also should not have tools that can localize the source of the impairment. The current Internet has no obvious way to resolve this dilemma if the resolution depends on the network being able to tell whether the sender and receiver have aligned interests.

Reconfiguring to avoid failed elements Designers fully understand the idea of reconfiguration to remedy failures in specific contexts. For example, the email system uses the DNS to allow a sender to try a backup forwarding agent, and dynamic routing uses probing to try to detect failed elements and route around them.

However, with respect to enhancing the availability of packet forwarding (the essential element of network-layer availability), the current Internet faces a serious conundrum. Routers can detect some simple failures, but in general, failures, especially those resulting from a malicious attack, can only be detected at the end points. Assuming that the end point detects a failure, what can it do? In today's Internet, the end points have very little or no control over network functions like routing. If communication between end nodes is disrupted, whether by failure or malice, the end nodes have no general way to localize the problem and no way to route around it.

There are (or were) some means in the Internet to give the user control over which entities to trust. The mechanism of source routing, which would have supported this sort of control, was specified in the original standard from 1981 but has mostly vanished over time. There are several reasons why source routing was deprecated. One is economics: giving the user choice over routes might increase costs for the provider, so should choice be linked to a way of charging the user for the resources he chooses to use? Another reason is security. If the network design gave the end nodes a choice among routes, that control might be exploited as an attack vector. Mechanisms that give the user more choice have to be designed with great care.[2]

Today, what happens is that we accept that a range of attacks will result in loss of availability. If a network must function at high levels of availability, we use nontechnical means to make sure that only trustworthy components and actors are in the system. Therefore, to achieve the full complement of CIA, both technical means and operational and management means are required.

At the higher layers of the system, the current Internet gives the end node a degree of choice among versions of a service. A sophisticated user may know to select a different DNS server manually if the default server is not acceptable. As I mentioned, the email system uses the DNS to provide the sender with alternative forwarding agents if one is not responding. A very sophisticated user may know that it is possible to edit the list of CAs that he chooses to trust. I would claim that while these features do exist, they are not part of an overall conception of how to improve availability.

Reporting a failure I defer the problem of reporting a failure or loss of availability to chapter 13, on management.

2. For one discussion of potential security risks associated with source routing, see https://www.juniper.net/documentation/en_US/junos12.1/topics/concept/reconnaissance-deterrence-attack-evasion-ip-source-route-understanding.html.

Availability and Security

The discussion to this point suggests that the objective of availability is enhanced by allowing both the system and the end users to have enough control to select elements for use that are functioning correctly. A more nuanced version of that objective is required in the context of an attack on communications. What the end user should be doing is selecting elements that *do not attack him or otherwise disrupt him.* A crude approach might be to select alternatives at random until one is found that avoids the problem. A more constructive approach is to allow the end nodes to structure their interactions so that they only depend on elements they consider trustworthy. If there is a malicious ISP, do not route through it. If there is an email sender that seems to send only spam, block receipt from it (this sort of treatment is what antiabuse organizations such as Spamhous try to coordinate). Essentially, if we want all of the CIA triad for communication security, we must organize the system so that even if untrustworthy actors are in it, we do not depend on them. We tolerate them if we must, but we do not make any interactions among mutually trusting actors depend on untrustworthy elements unless they are so constrained that we can rely on them.

This point of view has been slow to come into focus for some designers of security mechanisms, because it is a shift in mind-set. For some in the security community, with its history of focusing on confidentiality and integrity, the idea that availability must depend on assessment of trust rather than a technical mechanism is perhaps disconcerting, and perhaps disappointing.

Routing and Availability

While attacks by third parties or the network itself can cause loss of availability for a set of communicants, a more basic aspect of security from the perspective of availability is an attack on the network itself that disrupts availability, most obviously by disrupting the routing protocols. Clearly, the stability and proper operation of routing is essential for network availability.

In addition to making the routing mechanisms more resistant to attack, having multiple routing schemes running in parallel might be a way to improve the resilience of the network when it is under attack. XIA implements this idea, with its different sorts of end-point identifiers (content, service, network, host, etc.), different routing schemes for these different classes of entities, and the ability to fall back to a different scheme if the preferred one is not available. The emergence of new routing schemes operating at the same time raises the question of how the users signal what service they want. A field in the packet header (somewhat like what XIA has done) would allow the sender to indicate which routing schemes should be used. Using different destination address ranges to trigger different routings (as the current Internet does for multicast) gives the receiver control over which schemes to use to reach it (by controlling which sorts of addresses it gives out for itself) and allows the sender to pick among them (by picking which destination address to use). By tying the choice of the routing protocol to the address range, third parties in the network cannot override the end-node choices by rewriting a field in the router. The NIRA scheme (Yang, 2003) uses addresses to control routing in this way. This might allow senders and receivers to select between more availability and more protection, based on their specific needs.

Architecture and Availability

A key challenge for a future architecture is to resolve the conundrum I identified earlier: if only end nodes can detect failures of availability caused by attacks, but end nodes cannot be trusted to reconfigure the network lest this be another attack vector, how can failures be resolved? Working around this conundrum is a challenging design problem. One approach to it is the creation of new components trusted by the network that intermediate between the user and the network in order to provide a measure of choice and control to the end node that has detected a failure or attack.

Another potential role of architecture is to facilitate fault localization. As I noted, fault localization by the sender is not always in the interest of the receiver, if the receiver is being attacked. Perhaps it is worth exploring a shift in the basic architecture of the Internet. The Internet of today is "deliver by default": a sender can send to any receiver at will. Perhaps there is merit in an approach that is closer to "deny by default," so that the receiver has to take some action to indicate its willingness to receive all traffic or just receive traffic from a particular set of senders. Several of the architectures I discuss in this book are closer to deny by default.

The balance of interests with respect to fault localization may be clarified by referring to my discussion of invoking PHBs on page 86. There I proposed a rule stating that (with the exception of routing itself) any invocation of a PHB that facilitates communication between willing parties should be intentional—the address in the packet should be the location of the desired PHB. A more abstract way of saying this is that when senders and receivers have aligned interests, the end points should be explicit about what PHBs are being invoked. (This action may be done by an application on behalf of the end node, in which case it will be the application that has to attempt to localize the point of failure when something goes wrong.)

This rule implies that third parties should not claim to be helping out an application by inserting PHBs into the path that neither the sender nor the receiver know about. Once this rule is in place, use of encryption can limit the failure modes that manifest from unknown PHBs that show up in the path. It may be cleaner (and more secure in practice) to use an encryption scheme that lets selected PHBs decrypt a transfer (or parts of it) rather than have an ambiguous relationship between sender, receiver, and arbitrary PHBs in the path.

The different proposals in chapter 7 provide a range of mechanisms for dealing with aspects of availability. To some degree, all those schemes depend on the end node as the ultimate detector of loss of availability. With respect to localization, Nebula and XIA (both with its basic addressing scheme and in particular the forwarding scheme called SCION) provide a way for the user to pick different routes through the network, potentially making choices to avoid untrustworthy regions. ChoiceNet provides monitoring elements at region boundaries that are intended to confirm that the user is getting the promised service, but it is unclear what range of problems the monitors will be able to detect. ICNs raise a slightly different version of the availability challenge. ICNs attempt to exploit all the redundancy in the network, often through some sort of anycast search for a nearby copy of the data. The anycast search may avoid some failures. The attack that can disrupt availability in ICN is a malicious provider that provides a malformed version of the data. Even if the false copy can be detected, the anycast mechanism cannot easily avoid it and find another copy that is valid. DONA provides an enhancement to the FIND operation that allows the user to ask for the *n*th closest copy rather than the closest copy. NDN allows the receiver to include the public key of the sender in the data request packet, so that nodes along the path can check for themselves the validity of the data and reject malformed copies.

Conclusion

Availability is an understudied aspect of network design. While mechanisms exist to deal with specific failures, we do not have an answer to the general question about the balance of responsibility between the end node and the network to reconfigure operation in the presence of faulty or malicious elements. Giving the end node more control over which resources to use can allow it to avoid failed elements (assuming that it can localize the failure) but may at the same time provide new opportunities for a malicious end node to attack the network. The challenge of fault localization is itself an unsolved problem in general.

If the attacks that are potentially created by giving the end node more choice regarding which resources to use have the character of putting an extra load on selected elements, one mitigation may be to charge the user for usage. ISPs in general have been resistant to giving the user more control (e.g., over routing, as with source routes) because the result could be higher costs to the ISP. Translating those costs into revenues may be an effective discipline for a variety of user (mis)behaviors.

12 Economics

Introduction

The viability and success of an internet architecture in the real world is strongly intertwined with the economics of its deployment and operation. However, there has been little attention in the literature to the relationship between architectural alternatives and economic viability.

As a way to understand the issues, I will look at the current Internet as a case study and then generalize from what we learn there. The current Internet is composed of regions (ASs), which are deployed and operated by different actors. There were about 59,000 of them active in the Internet as of 2017. Of these, about 5,000 are ISPs; they offer transit service (Internet connectivity) to other parties. The rest are customers of these providers. Many of these service providers are private-sector, profit-seeking actors. The Internet comprises the interconnected mesh of these ASs.

This Internet has taken on the status of essential or critical infrastructure. The Internet may not be as important as water, sewers, or roads, but it is now infrastructure on which society depends. In contrast to these other infrastructures (roads or water systems), the current Internet is provided largely by these private-sector actors. Roads and water systems are normally built by governments, and historically, in many countries the government also provided the telephone system.[1] There has been a movement in many countries to privatize the telephone system, in part to stimulate investment

1. The United States was distinctive in that its telephone system (its previous communications infrastructure) was built by the private sector. There were initially many small telephone companies in the United States, most of which were merged into the Bell Telephone System. The Bell System was a highly regulated firm that operated as a government-sanctioned monopoly in its service area, not at all like the ISPs of today.

and innovation. The Internet shifted from government investment to private-sector investment in the mid-1990s and is now largely shaped by that private sector.

If today the private sector (in large part) brings us the Internet, what the future? The current Internet exists because ISPs chose to invest. Nothing forced them to do so. Looking backward, why did they do so? Looking forward, will this situation continue? Why should we assume that the private sector will continue to invest at a level suitable for meeting the needs of society? Is it even doing so today? Perhaps investment will stagnate. More troubling, if the Internet is societal infrastructure, why should we assume that the private sector will build and operate the Internet that society needs? Private-sector ISPs might mutate their offering into something different, perhaps more closed or dedicated to specific purposes such as delivery of commercial video.

The highest-level question in this chapter (and in chapter 14) is who will shape the future of the Internet. Should the future of the Internet be whatever the private sector chooses to deliver, or does society as a whole want a say in its future? If so, by what means can society shape what the private sector does? How can *society*, whatever that term means, even discuss and decide what its internet should be? Should governments step in and define the future of the Internet? The link from this question to economics is straightforward—attempts to shape the future (as opposed to letting it be whatever the private sector chooses to build) may reduce the incentive of the private sector to invest.

In fact, governments are starting to shape the future of the Internet, for better or worse. The most obvious evidence in the United States is the sequence of efforts by the FCC to impose network neutrality rules on the Internet.[2] One response (or threat?) by the ISPs covered by these regulations is that regulation will stifle investment.

This FCC activity seems to reflect a regulatory assumption that industry is doing more or less the right thing and only needs to be nudged. Many states and municipalities have made the more radical decision to use public-sector funds to build out parts of the Internet, either because they

2. The initial action by the FCC was a policy statement of principles (Federal Communications Commission, 2005). Following that, the FCC made three attempts to translate these principles into enforceable rules, the first two of which were overturned by the courts. The third (Federal Communications Commission, 2015) has not been challenged in court, but the Trump administration (as of 2017) has put in place a rule-making process to undo this regulation. Who knows what the future may bring?

believe that there is inadequate competition or because in some parts of the country (in particular, rural areas) the private sector is *not* investing.

Looking to the future, an optimist can see the Internet as so compelling that it will obviously continue to be there, and a pessimist may see several paths to the future, all fraught with perils. To the extent that society continues to depend on private-sector investment to build the Internet, it must pass between Scylla and Charybdis, those perils between which society always sails, avoiding on the one hand stifling investment and on the other hand getting an Internet that is not suited for society's needs. If that navigation fails, there may be another major shift for the Internet, with public-sector funding replacing private-sector funding in more places. The question of whether ISPs will continue to invest and whether they will build the internet society wants is not abstract. The answer is being worked out today as we watch. I personally worry that we may not be able to count on the private sector to build our future Internet and that the public sector will have to invest as they do in roads or water systems. Today, we see public-sector municipal networks either enhancing competition or bringing service to the unserved.

Technologists (e.g., internet architects) may try to distance themselves from high-level questions like this about the future of the Internet, saying things like "It's above my pay grade." However, network designers cannot ignore this question, for two reasons. First, different architectural designs might change the incentive of the private sector to invest. Second, an architecture well-suited to the private sector may not be equally suited to an internet funded by the public sector. The technology can drive the outcome, and the outcome may need to drive the technology.

For much of this chapter, I will look at the current economic landscape, which has been shaped largely by the private sector. At the end, I will return to the question of how architecture relates to the different outcomes for the future.

A Look Back

How is it that the Internet (and the investment that brought us to this point) actually happened? I reviewed the history of the Internet in chapter 2 but here are some highlights that directly relate to economics and incentives. In the early days, the government-funded Internet operated over circuits owned by the telephone company—the long-distance circuits in the ARPAnet were 50 kbps telephone circuits. The option of constructing new long-distance capacity dedicated to the Internet was not practical for many reasons (economic, political, and timing), so the early Internet engineers worked with what ARPA could purchase. ARPA did invest in alternative technologies such as packet radio, but one of the reasons the Internet protocols were designed to "work over anything" is that using what was at hand was the only way to move forward.

In the mid-1980s, the NSF took over the operation of the national backbone from ARPA and built (again using public-sector funds) the NSFnet to replace the ARPAnet as the wide-area backbone of the Internet, upgrading its capacity when it was practical.[3] This public-sector leadership and investment demonstrated the viability of the Internet as an information technology and provided the incentive for profit-seeking private-sector actors to enter the market ARPA and the NSF had created. It is not clear whether the private sector would have chosen to enter the market if the NSF had not taken this high-risk step (from a private-sector perspective) of building it to see what the demand was.

But even in the mid-1990s, when NSFnet was being decommissioned in favor of private-sector alternatives, the appeal of the Internet as a product was not clear to many private-sector actors. In a conversation about broadband to the home around 1995, a telephone company executive said the following to me:

> If we don't come to your party, you don't have a party. And we don't like your party very much. The only way you will get broadband to the home is if the FCC forces us to provide it.

Of course, he thought the copper telephone wires into the house were the only option for broadband access. Perhaps this executive could have been more forward-looking, but given this attitude, why did the telephone companies start to invest in residential broadband? To a considerable extent, it was because of the emergence of the cable industry as a competitor in the residential market, using a separate physical infrastructure to deliver broadband access. Competition can be a powerful driver.

I identify the following issues concerning the economic viability of an internet:

- What are the incentives of the private sector to invest in infrastructure? Can investment be sustained?

3. As I discussed in chapter 7, the ARPAnet was a distinct technology that predated the Internet, with a different architecture, which I describe in the Appendix. The Internet operated as a spanning architecture over the ARPAnet. In contrast, NSFnet was itself based on the Internet architecture. This approach marked a transition in thinking, where the Internet technology was used both to build an Internet by interconnecting networks, and to build those networks themselves.

- To the extent that society as a whole wants to have a say in the future of the Internet, what methods can shape the behavior of the private sector to get the outcomes that society desires?
- To the extent that there is competition in the access market, will competition continue or fade? Will the future be defined more by direct regulation than by competition? Can regulation stimulate or protect competition? Is there evidence that competition will drive the private sector to build an Internet that best meets the needs of society?

For each of these issues, there is the related question about architecture:

- Can architectural decisions shape (or reshape) the Internet ecosystem to better provide incentives for investment? Should we make decisions about architecture based on the assumption that the private sector will continue to build an internet? Would different decisions be preferable if a future internet were a public-sector infrastructure?
- Can architectural decisions serve to nudge the outcome of private-sector investment in directions that meet the needs of society?
- To the extent that one accepts competition as a desirable discipline in shaping the internet of the future, can architectural decisions improve the potential of competition?

Earlier, I used the term *tussle* to describe contention among actors with different and perhaps misaligned interests who seek to shape the Internet. I believe the *fundamental tussle* is between ISPs who assert that they should be able to use the infrastructure they paid for in any way they see fit and people (and governments) who wish to constrain how that infrastructure is deployed and operated in pursuit of an Internet suited to the needs of society.

What Shapes Industry Structure?

Can economic theory tell us anything about the relationship among system design, the resulting industry structure, and the incentives of the actors in the system to contribute their part to a healthy economic ecosystem? In fact, there is a wonderful framework that helps to explain this space, based on the work of economist Ronald Coase.

Coase received a Nobel Prize for his theory of the firm (1937), which builds on his concept of *transaction cost*. When firms engage to buy and sell, there are costs aside from the actual cost of the product or service itself: the costs of searching for the providers, the costs of bargaining, and the costs that arise from a lack of accurate information about other firms. Collectively, these are transaction costs. When transaction costs are low, efficient interfirm competition can occur, but if transaction costs are high, a firm may incur a lower total cost by realizing the service or function internally. Competition drives down costs in principle, but not in practice if transaction costs are high. One conception of this situation is that interfirm competition and intrafirm planning and coordination themselves compete to deliver the lowest cost of production. Large firms exist when and if the cost savings from internalizing the function exceed the cost savings from competition.

My link between Coase's theory and network architecture is the role of well-defined interfaces between modules. If an interface between modules is well-defined and easy to understand, then exploiting this interface as a basis for competitive interaction among firms may have a transaction cost low enough to be viable. If, on the other hand, there is no clear interface at a particular point, it is hard to open up that point to interfirm action, and it is likely to remain internal to the firm. So the modularity of a system like the Internet, which a technologist might think of in functional terms, is also likely to shape the industry structure of the system.

Many years ago, I was visiting one of the organizations that standardized the technology of the telephone system. I inquired about what different groups did, and when I asked about one particular group, I was told that their job was removing interfaces and making the system less modular. I asked why this was a good idea, and I was told that if the system had clean interfaces, the FCC might identify them as a point where the system could be opened up to competition. Their goal was to make that sort of regulatory intervention technically infeasible. They could have been joking, but I don't think so.

Defining the Internet Service Provider

The Internet protocol (IP) defines three interfaces that have shaped the industry structure of the Internet, in particular the ISP.

The Internet service IP defines the packet carriage service of the Internet—the service the network provides to the higher layers. The service specified by the IP is the basis for the service provided by an ISP. If IP had been specified differently, the business of an ISP would be different. For example, if IP had specified reliable delivery, ISPs would have the responsibility for reliability. The service specified by the IP spec is minimal. RFC 791 (Postel, 1981b) says:

> The internet protocol is specifically limited in scope to provide the functions necessary to deliver a package of bits (an internet datagram) from a source to a destination over an interconnected system of networks. There are no mechanisms to augment end-to-end data reliability, flow control, sequencing, or other services commonly found in host-to-host protocols. (1)

Perhaps a different specification of the service, with more attention to the features of the service, would have created different revenue-generating opportunities for the ISPs. (On the other hand, later attempts to augment the IP service specification with new features such as QoS have failed in the marketplace, even though they might have generated additional revenues. I return to this later in the chapter.)

The interface to the communications technology A second interface created by the IP specification is that between the service itself and the lower-layer network technology used to deliver it, often called layers 1 and 2. The IP specification says nothing about the technology other than that it be able to forward sequences of bits. This decoupling, while perhaps implicit, means that the specification allows (and thus encourages) innovation in network transmission technology. Over the decades since IP was specified, any number of network technologies have been invented, including local area neworks (LANs), WiFi, and cellular networks; the limited requirements that IP places on these technologies facilitate these innovations.

The interface between the ISPs The third interface that is relevant to the IP layer is the interface between ISPs. This interface was not specified in the original design of the Internet. In the original Internet, the designers downplayed the importance of this interface (sometimes called the network–network interface, or NNI). That decision was perhaps shortsighted. The original designers were not thinking about industry structure—they were government-funded researchers designing a system of networks connected by routers. Getting that function right was the initial focus of the designers.

When the architects started to focus on the fact that different entities owned and operated different parts of the network, there was some confusion as to how to implement the interconnection. One view was that there would be two routers, one belonging to each of the providers, connected to each other. Technically, this seemed inefficient—why have two routers next to each other? The alternative view was that there might be one router, jointly operated by the two providers. This idea reduces the number of routers in the system (a serious consideration at the time) but would have required some sort of division of responsibility within this one element. Which provider would be responsible if the device was failing? What if the router starts to disrupt the correct operation of one of the providers? Only after it became clear that a shared router was totally unworkable for operational and management reasons did the question arise as to what information two routers belonging to different providers would need to exchange with each other. (Today, we do see the other configuration, with a common switch connecting the routers of a number of ISPs. This configuration is called an Internet Exchange, and a neutral third party usually operates that shared switch.)

RFC 827 (Rosen, 1982), published in 1982, provides some insight into the level of understanding about the inter-AS interface at that time.

> In the future, the internet is expected to evolve into a set of separate domains or "autonomous systems," each of which consists of a set of one or more relatively homogeneous gateways [now called routers]. The protocols, and in particular the routing algorithm which these gateways use among themselves, will be a private matter, and need never be implemented in gateways outside the particular domain or system. (1)
>
> The Exterior Gateway Protocol enables this information to be passed between exterior neighbors.... It also enables each system to have an independent routing algorithm whose operation cannot be disrupted by failures of other systems. (3)

The Expressive Power of the Control Plane

Chapter 6 distinguished different architectures by their expressive power: some packet headers have richer expressive power than others, which can enable a richer set of services in the network (as well as introduce new potential security issues). Another dimension of expressive power is the set of protocols defined for the *control plane*. Control plane protocols define the messages that can be exchanged among network elements. The messages that shape the interactions among ISPs are those exchanged between routers at interconnection points. The interdomain routing protocol (border gateway protocol, or BGP) is the only significant control plane protocol specified for that interface in the current Internet.

BGP is perhaps the first example in the Internet (or at least the first substantially worked-out example) of a protocol designed to shape industry structure. The predecessor of BGP was designed in the era when the ARPAnet was the backbone of the Internet. Exterior gateway protocol, or EGP, assumed a hierarchical pattern of interconnection among the regions (the ASs), with ARPAnet and then NSFnet at the top of the hierarchy. If EGP had become the routing protocol for the commercial Internet, a single commercial provider would have taken the place of NSFnet, in a position that seems close to an architecturally created monopoly. A specific goal for BGP was to allow multiple competing wide-area ISPs.

At the same time, BGP has limited expressive power, and these limitations have arguably limited the business relationships among interconnecting ISPs. In particular, the expressive power of BGP limits the choices that individual ISPs can make when selecting among alternative routes to a specific destination ISP. If different ISPs make arbitrary choices among competing routes, it is possible that traffic to that destination will travel in a loop among ISPs rather than reaching the destination, so there are certain potential business relationships that cannot be expressed in BGP. See Feamster (2006) for a discussion of the limited expressive power of BGP.[4]

At the time of the transition to the commercial Internet, industry had a clear understanding of the importance of critical interfaces. As I discussed in chapter 7, the Cross-Industry Working Team laid out their vision of an architecture for an NII (Cross-Industry Working Team, 1994), and a central component of their conception was a set of critical interfaces, including the interface between ISPs. They understood that careful specification of this interface was critical to the health of the emerging private-sector industry. The designers of Internet protocols may not have fully appreciated the importance of this interface.

Alternative Architectures

To see that architecture can influence industry structure, consider the implications of information-centric networks such as in NDN. Compared to IP, the NDN forwarding layer has a much richer set of responsibilities, and access to a much richer set of explicit information in the packet headers. ISPs in an NDN internet will be able to see the names of what is being sought rather than end-point addresses, which opens up different opportunities for traffic discrimination. NDN routers may include data caches to support efficient operation of the protocols. ISPs would have control over what is cached and on what basis. To better understand the economic implications of the architecture, one must look at a scheme such as NDN not just through the lens of a mesh of interconnected routers but also as a mesh

4. For those interested in a deep dive into technology, there is a feature in BGP that does improve its ability to express preferences among routes: the *communities* attribute.

of interconnected ASs with independent profit-seeking motivations. The designers of the original Internet may not have evaluated their design in this way, but in today's world this analysis is mandatory.

As another example, Nebula and ChoiceNet make negotiation over how and whether traffic will be carried an explicit part of the design. The designs include a control layer (in ChoiceNet an *economy* layer) in which negotiations over service and payment can occur. The designers attempted to bring out the economics of the architecture and make it an explicit part of the scheme. The scheme includes new components, such as packet monitors at AS boundaries to verify the service being provided; the designers stress that their control plane creates new opportunities for competition and user choice.

Money Flows

If architecture defines the industry structure and at least some of the features of the relationship among the actors, the next question is what defines how money flows among these actors. A while back I had a conversation with a well-known economist who studied the Internet. It went like this:

Economist The Internet is about routing money. Routing packets is a side effect. You screwed up the money routing protocols.

Me I did not design any money-routing protocols!

Economist That's what I said.

So should we, as the economist said, have designed money-routing protocols? There have actually been attempts to add money routing to the Internet architecture.

One of the earliest proposals was by MacKie-Mason and Varian (1996). Their paper proposed that in addition to a possible fixed cost for sending data, there should be a *congestion price,* which reflects the cost one user imposes on another by sending when the network is fully loaded. Their approach was a *smart market,* where users specify their willingness to pay, but the price charged to the users that are served is the price specified by the marginal user—the user with the lowest willingness to pay that can be accommodated. This idea, which is a form of a Vickery auction,[5] provides

5. In a Vickery auction, bidders submit sealed bids and the highest bidder wins. However, the price the winner pays is the second highest bid. Since the winner pays no more than is necessary to win, the Vickery auction encourages bidders to bid their actual value.

an incentive for users to disclose their true willingness to pay, since they will not be charged that price unless it is the minimum price that gains them admission.

Mackie-Mason and Varian described this scheme as preliminary, and there were indeed issues—for example, does the user want to pay for individual packets or for an overall transfer? The authors were clear that there were many details and modifications that would arise if this scheme were implemented. The implication for architecture was that the price was in the packet header and that money would flow to ISPs along the path according to the congestion there. They were exploiting expressive power to implement a pricing scheme—money routing.

Other pricing schemes were proposed in the 1990s. I dabbled in this area. In 1989, I wrote RFC 1102 (Clark, 1989) on policy routing, which included a flag in the control plane routing messages to indicate whether a particular ISP in the middle of the path should be paid by the originating ISP, the terminating ISP, or separately by the sending party (as was then happening with long-distance telephone service). In 1995, I proposed a more complex scheme with marking in the packet header to indicate the boundary between the region where the sender and the receiver should pay for forwarding the packet (Clark, 1997).[6] Needless to say, none of these proposals went anywhere. There was no commercial interest in making these sorts of complex payment models part of the ecosystem.

Money Routing in the Current Internet

There have been a number of changes in the pattern of money flows since the entrance of commercial ISPs into the ecosystem. The early pattern of packet (and money) routing among ASs was *transit*: small ISPs paid large, wide-area ISPs to carry their traffic to the rest of the Internet. Payment went from the customer to the provider, independent of the direction of the packet flow. Packet counts were sufficient as input to the billing system. The idea of "sender pays" versus "receiver pays" (by analogy to the 800 numbers in the telephone system) never emerged and did not seem of any interest, although it might have emerged if the necessary indicators in the packets had been there from the start. My conversations with ISPs suggested that given the simplicity of the bulk payment scheme for transit, the benefit of a more complex scheme that related payment to specific packet

6. The various papers in the anthology from which that paper comes (McKnight and Bailey, 1997) give an excellent snapshot of the state of understanding of Internet economics in 1995.

flows was not worth the trouble. At any rate, there was never a way for ISPs to try the idea in the market.

The large ISPs (Tier 1, or long-distance, ISPs) interconnect in order to exchange the traffic of their customers. These interconnections are called peering connections. In contrast to transit service, which gives the subscriber access to all of the Internet, peering allows ISPs to have access only to the peer ISPs and that ISP's customers. With transit service, it is clear that the subscriber pays the provider, but with peering, who should pay whom?

Another story describes the creation myth of revenue-neutral peering. It is said that when the first two commercial ISPs met to negotiate their peering interconnection, one of the engineer-businessmen who was at the meeting was heard to say: "Wait, I thought *you* were going to pay *me* money." They realized that they did not even have a common basis to agree on which direction the money should flow, let alone how to set an amount. Then, as the story goes, these engineers heard the sound of running feet and realized that the lawyers were coming to start a three-year negotiation over the interconnection agreement. They looked at each other and said: "Quick, before the lawyers get here, let's agree that neither of us pays the other; we just interconnect, shake hands." And thus, so the story goes, revenue-neutral peering was born.

Over time, peering became more important in the Internet as smaller ISPs, not just the Tier 1 ISPs, arranged to peer in order to exchange their customers' traffic. Without a peering connection, both smaller ISPs would have to pay a transit provider to deliver their traffic to each other; with peering (if it was revenue-neutral, as it usually was) there was no payment by either party. At the same time, the payment model of the domestic telephone system was moving in this direction, away from a payment scheme ("settlement") where telephone companies paid each other for traffic they carried, to what was called a "bill and keep" model, where the customers paid their local telephone company and each company retained those payments. Calls were exchanged on a revenue-neutral basis. Given that the telephone system was moving toward revenue-neutral interconnection, there was little interest in bringing a more complex scheme into the Internet, especially since it would require a complex set of agreements between providers, which are hard to negotiate.

The next shift in the Internet payment model occurred mostly in the last few years, as content providers negotiated with access providers to have direct interconnections from their content caches into the access networks. Firms like Netflix, which need to deliver high-volume video traffic, have many servers around the globe that cache their content, and these servers have high-speed connections into the networks of broadband access providers. From a routing perspective, these connections resemble peering connections, but after some very public disputes, the payment model that usually emerged was that the content provider paid the access ISP for the high-capacity dedicated interconnections.

The unstated assumption that was embedded deeply in these negotiations was that the value flow matched the packet flow–that is, the sender paid the receiver. One might have argued that when a viewer watched a video, the viewer derived more value than the sender, but the possibility that the receiver might pay seldom came up. There was no question in the minds of any of the negotiating parties I spoke to as to which way the money was flowing—only disagreement about the amount. No packet marking is needed to inform that payment model.

(I believe that there are some cases where a broadband ISP does pay a content provider to interconnect. It is difficult to find out what is actually happening, because all of these interconnection agreements seem to be covered by nondisclosure agreements. However, some smaller ISPs, which do not have the aggregate traffic volumes to justify a content provider or CDN incurring the cost of installing a directly connected cache, may be paying those providers to connect a cache, because it actually saves the ISP money that would otherwise go to its transit provider.)

We now see more complex models emerging for payment between content providers and access ISPs, for example the idea of "zero-rating," where a user with a monthly usage quota can receive certain content without having it count against that quota, presumably because of a business arrangement between the ISP and the content provider. This concept is most common in mobile access networks. This model, again, does not seem to require any sort of architectural support such as new expressive power in the packet. One reason is that zero-rating is a local, bilateral arrangement between two actors.

The Failure of New Services

Augmenting the simple best-effort forwarding service of the Internet with additional services might offer new revenue opportunities for ISPs. Two proposals for a new service, multicast and QoS, received a lot of attention from the research and standardization communities. Both were successful technically but failed as commercial offerings. The failures make informative case studies.

Multicast Multicast (see page 25) enables a packet from a sender to fan out to a number of receivers across the Internet. The designers saw multicast as much more efficient than sending the packet once for each receiver. (It also had other benefits, such as eliminating the need for a sender to keep track of all the receivers.) The ISPs saw the economics in an opposite light. Their view was that they bill the sender based on the capacity of the access link. If a customer sends a multicast packet, the ISP bears the cost of carrying all the copies, some of which might cross interconnection links to other ISPs, but the customer only has to pay for the capacity to send one copy. ISPs saw multicast as generating cost, not revenue, and refused to offer it.

Some of us spent a little time thinking about how to design a money-routing protocol for multicast, but our initial approaches were complicated. A user of multicast could pay a fee based on how many times the packet was replicated, but the technical community had not developed multicast protocols that could provide that information. In fact, it is very hard for a routing scheme to compute the total number of receivers, because the packet may be replicated at several points along the path to the various receivers. Since the packet might be replicated inside each of the ISPs it crosses as it fans out across the Internet, the problem of computing a *multicast fee* would repeat at every interconnection point among ISPs. ISPs were very resistant to the idea of adding a complex inter-ISP payment scheme, especially in the case of revenue-neutral peering, where there was no payment mechanism already in place.

Quality of service The term *quality of service*, or QoS, describes a set of mechanisms that enhance or modify the basic best-effort service of the Internet in ways that better match the needs of certain applications. The best example is a service with low variation in latency (low jitter), which improves multiplayer games and interactive voice and video. During the 1990s, a number of people (including me) worked hard to add this service to the Internet. Again, it was successful technically (usually deployed within a single domain) but failed as a commercial offering.

Here is another conversation I had, sometime around 1995. Once we had the standards in place for enhanced QoS, we spoke to the major router vendors like Cisco and encouraged them to implement it. They replied that they respond to customer demand, so we should go find them a customer. I spoke to the chief technology officer (CTO) of a major ISP, and the conversation went sort of like this:

Me I think the QoS tools are going to be very important and you should deploy them.

CTO No.

Me Why not?

CTO Why should I spend money so that Bill Gates can make money selling Internet telephony for Windows?

And then, after a pause:

CTO And if I *did* install QoS, why do you think I would turn it on for anyone but me?

In that short conversation, I learned two painful things, First, in the commercial Internet it was a waste of my time to come up with a new idea unless I could explain how someone could make money from it, and second, there might be a need for what is now called network neutrality regulation. This was 1995.

The technical research community has argued that the Internet architecture has ossified and resists innovation. The economic norms and business practices among ISPs have ossified just as much as the technical architecture. The barrier is coordination. Multicast and QoS require both technical and business coordination. ISPs interconnect technically, but they interconnect at the business level as well, with agreements about peering and transit. Renegotiation of those agreements is risky, because once a discussion starts, all the terms may come up for reconsideration, not just the terms about a new offering. Negotiation about revenue sharing for a new offering is a massive act of rocking the boat. Firms are also worried that if they get together and discuss cooperation about revenue sharing, it might trigger antitrust concerns.

My conclusion from these fruitless exercises was that a neutral body must not only develop technical protocols but also must work out money-routing protocols. In the 1990s, I discussed with the IETF leadership whether that body would work on the money-routing problem, and they refused. The IETF defined itself as a *technical* standards body, and they were not going to touch what they saw as business practices. One of their (legitimate) fears was that this work would take them out of their area of expertise and might damage their overall reputation and respect, but there is no other neutral venue where this work could be done. Several of us at MIT invited major ISPs to come to MIT and sit at a neutral table to talk about deployment of QoS. They were prepared to talk about the issues around technical coordination but totally unwilling to discuss how revenues could flow. Given that attitude, the failure of QoS in the public Internet is not surprising. For an extended discussion of this, see Claffy and Clark (2015).

The ossification of customary business relationships among ISPs may be more of a barrier to innovation than the ossification of technology. As I observed in my discussion of modularity in chapter 3, interfaces are the constraints that deconstrain. In this case, the stable business practices among the ISPs are constraining what might be fruitful innovation in service offerings and revenue generation.

What has happened in the market is that rather than try to change the stable business interfaces between existing actors, new entrants have created new interfaces in order to provide new services. An example of creating new interfaces is content delivery networks, or CDNs. CDNs are highly distributed content caching and delivery systems. A large CDN today will have servers at thousands of locations across the Internet; by replicating content in those servers, delivery is both more efficient and more robust, since the content can be delivered from a server near the user. Those servers are usually colocated near major broadband access providers, with direct interconnection into those access networks. Some CDNs are built by major content providers such as Netflix for their own use; others offer services to their own customers who have content to deliver.

Technically, ISPs could have implemented and sold a CDN service—they have machine rooms and other technical infrastructure. But CDNs today are mostly implemented by a different set of firms, not ISPs. I think the failure of ISPs to offer a CDN-like service is an illustration of the problems of coordination. If a collection of ISPs attempted to offer a CDN service, a prospective content customer would have to negotiate a hosting agreement with each of them (which might imply very high transaction costs, referring back to Coase), and the ISPs would have had to work out complex, multilateral agreements about technical issues (which server should deliver the content to a particular user) and business issues (what sorts of payments would flow between ISPs). Ronald Coase again has the language to describe this situation: in the absence of customary and routine models for business relationships, transaction costs to work out how to offer a new service, especially if the business relationships are multilateral, may be too high to justify market entry. On the other hand, when a single firm offers a CDN service, the content provider only has a single arrangement to negotiate. The CDN firm has to negotiate an agreement with each ISP where it desires to attach a server, but those negotiations (the business interfaces) are bilateral, not multilateral; they happen once, not for each content customer of the CDN. Today, those agreements typically do involve payment to the ISPs, so the CDN business does generate a little revenue for them, but the ISPs are not part of the CDN innovation, just providers of a simple interconnection to a server machine. Whether or not the original innovators who founded the early CDN firms thought about what they did in this way, they moved business interfaces around to deconstrain innovation.

Architecture and Money Flows

With respect to routing money, as with other objectives, if architecture has any role to play, it is not in defining how the system works but rather in making it possible for a desired range of things to happen. However, the networking community has so little experience with money routing that this area is essentially uncharted territory. In general, architects learn what should or could be built into the core of a system by looking at what application designers tried to do in earlier generations. So, given perhaps two decades of experience with the commercialization of the Internet, what lessons can be learned about money flows? I find few lessons, which is both uninformative and troubling.

Perhaps a new money-routing architecture could be designed as part of a future internet architecture. The designers of ChoiceNet understood that payment flow was a key mechanism, and they provided a framework for it. They did not, however, sketch out what typical business terms might be.

There was a second barrier to the deployment of QoS, beyond the lack of a multilateral business plan. The IETF (in my opinion) did not standardize a key element of the scheme, the end-to-end service model. The best-effort service model is weakly specified, and application designers have learned to work with that uncertainty, but if an ISP is going to sell an enhanced service, the customer needs to have some idea of what it is. The service expectation needs to be somewhat standardized, so that application developers can design their applications to use the service no matter which ISP is providing it. Application developers are not going to redesign their applications for each ISP in the Internet. The IETF standardized what a router should do (the PHB) but would not define how the application would interface to it. Their view was that this standard would define an application programming interface (API) through which an application would initiate the service, and that an API is not a network protocol but part of an operating system. The mission of the IETF did not include standardizing features of an operating system. I think they erred. The specification of a service (its semantics, to use an earlier term) is not the same thing as the details of the interface to that service. The Metanet and FII proposals were clear that a definition of the service (or services) had to be part of the network architecture.

I have asked a number of ISPs whether an enhanced QoS for gamers could be successful, and they said that it could be highly profitable. They cannot offer it because they cannot describe to application designers how to exploit it, and they are not prepared to resolve the issues necessary to bring the service to market. There are massive coordination barriers to such an offering, and they are not primarily technical: the most formidable relate to economics and trust.

Architecture and Incentives to Invest

If the private sector is going to be the investor that continues to build the future internet, what will motivate it to invest, specifically in capital-intensive access infrastructure? At a high level, investment is motivated by anticipation of an adequate return on investment (RoI), and the greater the risk (the uncertainty about the RoI), the higher the anticipated RoI must be to justify investment. This high-level concern with RoI plays out in a number of ways in the communications sector.

Competition Competition can be a powerful driver of investment. In particular, competition between firms that have separate physical access technologies (*facilities*) can stimulate investment in those facilities. The telephone companies and the cable companies, occasionally further goaded by new entrants gleefully pulling fiber to the home, have made significant investments in upgrades to their capabilities. However, their will to compete has been inconsistent. Verizon announced an ambitious investment in fiber to the home (FIOS) in 2005, then suspended new investment in 2010, and then announced some limited fresh investment in 2015.[7] I don't know of a definitive study that relates facilities-based competition to investment, but I once had an operator (from another part of the world) explicitly say that they only invest in upgrades where there is facilities competition; in other locations, they see no reason to improve the service.

Sometimes fear of competitors will motivate investment even when the RoI is uncertain. When the telephone companies were first thinking about whether to move beyond simple deployment of digital subscriber line (DSL) service and contemplating advanced technologies that involve pulling fiber at least partway into their access networks, I asked an executive of one such company whether they had a clear model of what the return on this investment would be. He said more or less the following:

> We actually don't have a clear business model for doing these upgrades. But we have a clear model for what happens if we don't: in 10 years we are out of the wireline business. Better to bet on uncertain success than certain failure.

7. Here is an article from my corner of the world: Brodkin (2016) talks about the reversals of Verizon's commitment to invest in FIOS in Boston.

In other parts of the world, such as Europe, a different pattern of ISP competition has emerged, based on a different sort of regulation. There, regulation has required that owners of access facilities (in particular, the copper wires of the telephone companies) must lease these facilities to other firms to create retail ISP competition over these shared facilities. The term for this approach is *unbundling*; the result is an *unbundled network element*, or UNE. In general, the regulated rates at which these UNEs must be shared have been set low to stimulate entrance into the retail market. The facilities-based providers have complained, with some justification, that they see no motivation to invest in upgrades if they carry all the risk and at the same time gain no competitive advantage and a low RoI.

Since this approach to regulation is prevalent in many parts of the world, it is worth considering whether network architecture can influence how it is implemented. The early model of unbundling was leasing of physical elements, such as the copper circuits from an exchange to a customer's premises. However, unbundling of a physical wire may not be enough to sustain competitive entry. The barrier to entry is still high, since the new entrant would have to install its own switch gear in any exchange where it intended to serve a customer. So, especially in Europe, the model of unbundling has evolved to what is called bitstream service, where the facilities owner (the incumbent) provides to the competitor packet carriage from the customer (the residence) to a point somewhere in the incumbent's network, where the competitor connects.

In this case, the incumbent may be carrying its own traffic and traffic from competitors across the same circuits in its network. This sharing of circuits looks very similar to another thing ISPs are doing today: attempting to increase the diversity of their products (and thus retail revenues) by offering a range of services over their IP network, including telephone service (VoIP), Internet protocol television (IPTV), and other offerings, as well as Internet access. These services are carried in packets defined by the IP protocols, but the packet flows are partitioned from the public network by the use of separate addresses and (in some cases) by separate allocation of capacity.

This sharing of a physical network to provide multiple services is similar to the architectural approach of *virtualization*, which was discussed in chapter 7. The boldest vision of virtualization has global scale, which would imply that many facilities owners would have to interconnect to offer the virtualized platform. At this scale, there would be many issues of interconnection at both the physical and business layers. Virtualization within the span of one ISP seems simpler and more practical.

Since access ISPs today are offering multiple services (e.g., virtualizing the access circuit to the residence), new forms of bit-stream unbundling are possible. Today, unbundling regulations typically require that the owner of a residential access circuit allow a competitor to provide retail service over that circuit, but the consumer picks one retail service provider for all their services. Virtualization of the access circuit would allow a more sophisticated version of competition, where the consumer could pick different providers for different services (phone, TV, Internet, etc.) over the same physical path.

So virtualization can be used in a number of ways. The ISP can virtualize its network for its own use, in which case all the business planning is internal to the firm. The ISP could voluntarily offer access to a share of its virtualized network to an independent third party, in which case the regulator would have to decide whether any regulation of this offering was justified (perhaps to limit unreasonable discrimination in pricing or service). Finally, the regulator could compel the ISP to allow independent service providers to have access, which raises the question I asked earlier: why would an owner of facilities consider investing in the maintenance and upgrades to its infrastructure in this scenario. The ISP might have to be given some assurance about the rate of return, which is a form of regulation very much out of vogue today. For a deeper discussion of the different layers of the system as platforms for services at higher layers, see Claffy and Clark (2014).

The different architectures I discussed in chapter 7 will lead to different outcomes with respect to incentive on the one hand and the potential need to rein in unacceptable behavior by the ISP on the other. With virtualization, the question is whether the owner of the facilities would choose to invest. With schemes like NDN, where the ISP has more options for discriminatory treatment of traffic than it has in the current Internet, the question is whether some even stronger version of neutrality regulation would be required.

Architecture? The factors mentioned can influence the incentive to invest, but what about architecture? Again, in general terms, architecture can increase the incentive to invest by reducing risk and creating opportunity. It can reduce risk if it can make the relationship among the actors clear and stable. (Of course, the fundamental goal of architecture is to produce an internet that is fit for purpose, but that goal is more basic than just improving RoI.) Architecture can potentially improve opportunities via the creative but careful design of features that add to the range of service offerings.

Bad Outcomes in the Future

The next stage in money routing may be schemes that attempt to link payment by content providers to the value of the content being delivered rather than to the cost of delivering it. Value pricing represents an attempt by the access providers to extract revenues based on what is delivered and its perceived value to the sender and receiver. For example, a recent paper (Courcoubetis et al., 2016) describes an analysis of the different services provided by Google and attempts to model the value per byte of the different flows (e.g., search vs. YouTube) so as to set a per-service fee for delivery. Whether one views this as the next stage in the future of access network funding or a throwback to telephone era settlement, the traffic classification is going to be derived not from any simple marking in the header but by looking deeper in the packet. (In the language of chapter 6, this is implicit parameterization and adverse interests.)

Today, the only common example of service-specific billing in the Internet is zero-rating for some content entering cellular networks. However, outside the Internet, an interconnection architecture called Internet protocol eXchange (IPX, not to be confused with Internet eXchange point, or IXP, or with internetwork packet exchange) is used for interconnection of private IP networks (not the global, public Internet) that support services such as carrier grade VoIP. It is used by some cellular providers to interconnect their VoIP services, which are over IP but *not* over the public Internet. IPX contains explicit support for per-service interconnection and cascading payments.[8]

Per-service settlement for Internet interconnection is an example of the pessimistic fears about the future that I articulated earlier in this chapter. We see a number of forces today that might drive the future of the Internet into a darker place, including pressures from those who carry out surveillance and regulation of access to content, but economics is probably the most potent driver. As long as the Internet is a creature of the private sector, profit-seeking behavior will be a natural behavior and what we must expect.

In this respect, it may be that some architects will prefer to push for designs with *less* expressive power in order to prevent the implementation of sophisticated tollbooths in the Internet. Architecture is not value-free, and this is an excellent example of a place where values will be close to the surface as we make design decisions.

8. For a good tutorial on IPX, Wikipedia is a place to start. See https://en.wikipedia.org/wiki/IP_exchange.

If our goal is an open network, it may not be possible for the private sector to recover the costs of building expensive infrastructure, if that infrastructure is only providing an open network. In the long run, we may need to think about networks the way we think about roads—as a public-sector undertaking. Alternatively, if the public sector does not invest in critical parts of the Internet, such as broadband access, then architecture or regulation that restricts the opportunities to reap returns on investment in facilities may limit revenues to the point where buildout of facilities does not occur at a suitable rate. Some degree of vertical integration or discriminatory treatment of traffic (including billing) may be the price of a healthy infrastructure. Any architect must think carefully about the extent to which a proposed design takes an ex ante position on this point. For example, if an access provider supports both a healthy open Internet and its own services that compete with Internet services over the same IP infrastructure, is this a driver of investment, a threat to that Internet service, or both?

Summary—Architecture and Economics

Architecture plays a key role in the economics of an internet ecosystem. Architecture influences the industry structure, and the key interfaces can structure the potential relationships between the actors. It is less clear what role, if any, architecture should play in enabling money-routing protocols. It may be necessary to wait another 20 years to see how the economics plays out before we can see what we should have added to the architecture now. Perhaps even with the current Internet, in 20 years we may have not the ISPs that we have today but a different set of actors trying to work within the existing architecture. If so, it will be economic (and political) forces that will make that future happen.

As I observed in my discussion of modularity in chapter 2, interfaces are the constraints that deconstrain. I concluded that the stable business practices among ISPs are preventing what might be fruitful innovation in service offerings and revenue generation, which suggests that the industry structure needs to be re-engineered. I speculate further on this in chapter 15.

13 Network Management and Control

Introduction

One requirement for a future internet that I listed in chapter 4 was that it do a better job of dealing with network management and control, which has been a weak aspect of the Internet from the beginning. There is less research concerning network management than there is about forwarding of data, so this chapter, while dealing with technology, is a bit more speculative. The most general conclusion I draw about architecture is that a design based on strict layering is not the best approach to modularity from the perspective of management and control. I suspect this conclusion will be controversial.

I will pose two key questions, similar to my starting point in other chapters, that will help to sort out this space:

- What is implied by the terms *management* and *control*?
- What does architecture have to do with these concepts?

I start with the first question.

What Are Management and Control?

Network management is a very general term. It describes all the activities carried out in the background to keep the network running. It is similar to the concept of security in that it is ill-defined. I began the discussion of security by asserting that the term *security* was so general that it was aspirational, not operational. Only when we find a substructure for security can we begin to understand the relationships (and potential conflicts) among the subgoals and thus understand how to improve security in practice. I believe that this same observation will be true about management. While I will identify some common themes that run through different aspects of management, the substructure of management contains distinct objectives that should be considered separately.

One definition of network management is that it is those aspects of network operation that involve a person. Network architects talk about the data plane of the network (the mechanisms that actually forward the packets) and the control plane (the mechanisms, such as routing and congestion control, that provide the information necessary for the data plane to function). Those control mechanisms are automatic; there are no people in the loop. In contrast, management covers those actions that involve a person. However, this definition of management does not mean that all issues that require human intervention are homogeneous with respect to network architecture. The answer is quite the opposite.

In the network community, there is a spectrum of opinion about the role of network management. At one end are engineers who say that a properly designed network should run itself, so the need for management is a signal of failure. At the other end are pragmatists who believe that what they would call policy decisions are not worth trying to codify into algorithms, and having a person in the loop for many decisions is a better (and more realistic) way to go. Part of this debate resolves itself when we look at time constants of human intervention. Many network operators would be skeptical about a network that ran itself so thoroughly that it issued purchase orders for new circuits and routers when it saw itself getting near capacity. Business issues would seem to call for human judgment. On the other hand, a network that requires teams of humans to sit in front of monitors 24 hours a day looking for faults would seem to benefit from some further automation.

Breaking Network Management into Parts

A potential framework for dividing the management problem into parts was developed by the standards unit of the International Telecommunications Union (ITU).[1] An ITU standard (Consultative Committee for International Telephony and Telegraphy, 1992) defines network management and breaks it into the following objectives:

- activities that enable managers to plan, organize, supervise, control, and account for the use of interconnection services;
- the ability to respond to changing requirements;

1. The ITU (part of the United Nations) has a number of responsibilities, including standards. That unit of the ITU, called ITU-T, was previously called the Consultative Committee for International Telephony and Telegraphy (CCITT). Most of the standards defined by ITU-T relate to telephony rather than the Internet, but some are relevant, including standards for audio and video teleconferencing.

- facilities to ensure predictable communications behavior; and
- facilities that provide for information protection and for the authentication of sources of, and destinations for, transmitted data.

They then divide the set of management issues into the following categories:

- fault management;
- configuration management;
- accounting management;
- performance management;
- security management.

This set of categories is called the FCAPS framework for management, based on the first letters of the categories.

What these five categories have in common, as defined by the ITU, is the reporting of data and events from managed devices (such as routers) to a management system. The reported information can be used as input to a simple display or a complex management system that gives the operator a higher-level view of the network. A management system also allows the operator to send configuration/control messages to a managed device—again either from a simple interface or a high-level management application.

The ITU specification makes the assumption that while different categories of management have different objectives and requirements, the protocols for reporting and control can be the same. This assumption superficially matches what we see in the Internet, where a common protocol for reading and writing management variables (the simple network management protocol, or SNMP) is used with a variety of management information bases, or MIBs. (The term *management information base* describes the variables associated with each class of managed device.)[2] So a possible modularity to the problem of network management is as follows. At the

2. In detail, this assumption of a common protocol is flawed. In the early days of work on Internet management, there was a lot of attention paid to the problem of communicating with a managed device when the network was failing and communication was impaired. There was a concern that a protocol like TCP, which insists on reliable, ordered delivery, would not be effective for managing failing communications, so the design of protocols for fault management should target the special case of operation in impaired conditions. This situation is quite different from (say) transmitting the data necessary to configure a complex router, which may involve the installation of thousands of rules, and would seem to call for a reliable, sequenced update.

bottom layer, parameters for a managed device can be read and written; these parameters are specific to the device. Above this layer is a (perhaps common) protocol for communication between the managed device and any higher-layer management system. The management system in turn provides an interface to the people responsible for management. The management system provides the human operator with what the military calls situational awareness: an understanding of what is happening in the network.

Management and Control

Both in the Internet and in the ITU framework, the concept of exposing parameters in a device for reading (and perhaps writing) is typically associated with management functions, not control functions. In the Internet, we have SNMP, the simple network *management* protocol. The Internet design community has not focused on whether the control plane should have similar interfaces (and perhaps, by analogy to MIBs, should have control information bases, or CIBs). I believe that the particular approach to the design of the control plane in the early Internet shaped this way of thinking. In the Internet, the important control protocols (most obviously the routing protocols) run on the devices being controlled (the routers), and each device reads and sets its own control variables as part of executing that protocol. Devices exchange connectivity data as part of the routing protocols, but each device computes and populates its own forwarding table. There is no control interface on a router where it can expose the results of its low-level link measurements, and no interface to set the forwarding table (except for a low-level human command line interface, or CLI, to fill in manual entries from a keyboard).

We are now seeing a trend in the Internet toward a new generation of control schemes that moves the control computation out of the distributed routers and into a more centralized controller. The most obvious example of this is software-defined networking (SDN), where a central controller collects network connectivity data and computes and downloads forwarding data for the different routers in the network. (Later, when I discuss performance management, I will mention another example of a move to a more explicit control algorithm.) As part of the development of SDN, the designers had to specify the interface between the router or switch and the route computation function, including the set of variables the controller could read and set, and the communication protocols used to implement this function.

Given this trend, when we consider the relationship between management and network architecture, I will generalize the consideration to include both management and control functions. While the distinction between management (involving people) and control (automated functions) may be helpful in some contexts, it is not material in other respects. If a device makes certain variables available for reading and manipulation, it is not material whether they are manipulated by a person, an algorithm, or an algorithm that sometimes has a person in the loop.

The addition of control to this analysis will add new categories to the five I listed. Routing and congestion control are the most obvious.

Other Approaches to Management

The conception of management (and control) as a process driven by data gathered from managed entities is a somewhat limited view. Other methods for assessing the state of the network include end-to-end measurement, active probing, and recording packets in the network. The most critical control algorithm in the Internet, congestion control, depends on data explicitly gathered not from each of the routers along the path but from observing end-to-end behavior. There have been any number of papers written on how congestion control might be done better than it is now, given our improved understanding, but while most of these depend on carrying more information in the packet, none that I can think of involve the reporting of state from routers to a central controller.

Active Probing

The most basic tools of active probing in today's Internet are *ping* and *traceroute*. They surely qualify as management tools, since they are used by people, often ordinary users frustrated with the state of the network. Professional network managers also often use these tools. Probing with ping and traceroute serves a number of purposes, including understanding performance (often when it is poor) and fault localization. The original purpose of traceroute was troubleshooting—determining the path a packet was taking on its way to the destination. Ping (or more precisely the ICMP Echo option) detects if some device on the network is sufficiently functional to respond to a message. Ping was specifically designed for this purpose, but traceroute is a clever kludge, a hack that uses a packet crafted with a hop count (the Time to Live, or TTL, field in the IP header) set so that it expires at a point along the path in order to solicit an ICMP control message back from that element. Traceroute does reveal information about the different elements along the path, but implicitly. The ICMP response was not designed for this purpose and does not give clear information.[3]

3. The measurement community has struggled to make use of the ICMP response in this context, dealing with issues such as ambiguous IP addresses on the return

Of course, one problem with tools like ping and traceroute is that sometimes the probed element does not answer. Routers are sometimes configured *not* to answer, for reasons of both performance and security. The Internet is composed of Autonomous Systems operated by different entities that are sometimes not interested in having their insides probed by outsiders. Most operators have come to understand that having their routers respond to a traceroute probe is a useful mutual agreement, but not all operators buy into the agreement at all times. It is important to remember that while measurement is often just troubleshooting, measurement of some other region of the Internet is sometimes an adversarial activity and sometimes an action with political motives.

The architectures I have discussed in this book give little attention to the question of whether there should be tools in the architecture (fields in packets or additional probing functions) to facilitate active measurement of the network, whether to deal with issues of configuration, performance, or fault localization. But the design of any such mechanism would bring into focus an important aspect of tussle, which is that many operators would prefer to keep the details of these issues to themselves. Active probing can be seen as a case of trying to discover from the outside something that the operators on the inside already know but do not reveal.

If operators are (understandably) sometimes reticent about the status of their region of the network, perhaps a way to balance the interests of all parties would be to define, as part of a management architecture, a representation of a region of a network that hides some details (perhaps the exact router topology, for example) but gives an abstracted view of the current state of the region.[4] If there were a uniform agreement on a set of useful abstract parameters to report, this outcome might represent a resolution

packets, and variable processing delay in the control processor of the probed router. An ICMP message explicitly specified to support this sort of probing could provide information that is easier to analyze, for example a unique ID associated with a router, as well as the IP address of the port being probed. The IETF, in an attempt to deal with the issue of variable latency in the reply to a probe, has developed the two way active measurement protocol (TWAMP) mechanism (see RFC 5357). A probing mechanism that combines the features of TWAMP with traceroute might be of great utility to the measurement community, but it must be designed in a way that prevents abuse.

4. Some operators provide an abstract view of their network. For example, AT&T has a Web site where it lists the current latency between all its city pairs; see https:// ipnetwork.bgtmo.ip.att.net/pws/network_delay.html. Whether or not current intercity latency is the most useful measure, it illustrates the idea of an abstract view.

of the tussle issues around the desire of all parties to probe their neighbors. Again, while probing is often motivated by the need to troubleshoot, this sort of agreement sounds more like industrial organization rather than network management.

Packet Sampling

Packet sampling, attaching a device to the network that copies and records a sample of the traffic, is another important tool for network management.[5] Tools such as Internet Protocol Flow Information eXport (IPFIX) and its kin sample the packets passing through a router to identify the flow, reporting data such as source and destination IP addresses, number of packets, and other information. The necessity to sample adds uncertainty to the data, but flow data is a rich source of information that can inform performance and configuration management, and in some cases security. It is a good example of a different sort of data reporting in support of management, going beyond the simple counters normally reported using SNMP. It is also an example of a tool developed without explicit support from the architecture. It is possible that some additional information in the packet header could enrich what is learned from sampling the packets going through a router.

Management of the Management System

Management (and control) systems themselves must be configured and monitored. This sounds painfully recursive, and in some cases is, but the specific cases are usually dealt with in pragmatic ways.

Specifically, how does a management system discover the set of entities to manage? If a managed device can generate alerts (active notifications of a change in status), where should that device send those alerts? These questions will come up as I look at the categories of management and control. Further, there is a significant security aspect to these questions. What entity has the *authority* to manage another entity? Unauthorized reading of management variables seems less harmful than malicious modification, but even reading could lead to an undesirable revelation of system status, and flooding a management interface with queries can potentially lead to overload of a managed system (a form of DoS attack).

5. Given the huge volume of traffic on the Internet, it is not practical to capture every packet. Current monitoring devices copy a sample of the packets, hence the term.

The design of any management mechanism must include an analysis of how that system is managed and controlled, and the security implications of that system.

The Role of Network Architecture

The previous discussion suggests that the ITU model of management (the export of management parameters from a device) is not the only aspect of management to consider. However, in that context, are there any aspects of management that might rise to the level of architecture? One component might be the bottom layer: the definition of the variables that can be read (and written) to monitor and control devices. These parameters are the foundation for building situational awareness and control of the network—different management systems can be built on top of them, but if the basic data is not available, management will be difficult and flawed. In the current Internet, I don't think there has been any organized consideration of which management parameters might be considered part of the architecture. But if some of those parameters achieve the status of "something on which we all need to agree," then they become part of the architecture, even if they earn that status after the fact. In chapter 3, I claimed that a key aspect of architecture is the definition of interfaces. Interfaces modularize the system and specify what is shared among the entities on the different sides of the interface. The set of parameters a managed device makes available defines its management interface. If there are classes of devices (for example, routers) that would benefit from an interface that is globally agreed upon and stable over time, that interface takes on the character of architecture. The exact protocol used to read and set these variables may change, but the existence of those parameters is a more basic and enduring characteristic.

Instrumenting the Data Plane

The previous discussion of packet sampling provides a particular illustration of a more general issue: should the data plane of an architecture contain mechanisms intended to assist in aspects of network management and control? Should the expressive power of packet headers be enhanced with fields (e.g., a flow identifier) that would help with performance analysis? Could the data plane be reengineered to help with fault isolation?

In the early days of the Internet, colleagues from the telephone company (when they were not telling us that packet switching would not work) strongly advised us that the data plane had to include tools for fault diagnosis. The digital version of the telephone system included features (such as management fields in the transmitted data units in their system) used for fault diagnosis. Those colleagues argued that the ability of a network to diagnose faults was just as important as its ability to forward data, and that our failure to appreciate this requirement was just another sign of our inexperience.

Enhancing the data plane to support management as a part of its normal function (for example, better fault localization or detection of performance impairments) would shift tussle in a fundamental way. Today, operators can distort what is learned from using active probing (giving ping probes priority or refusing to respond to them), but it is harder to distort any management data gathered as a side effect of forwarding data without distorting the performance of the data plane itself. Of course, fields in the data packets that only play a role in management (such as the traceback schemes for localizing DDoS attacks described in chapter 10) may be ignored by the routers. The ideal tool for instrumenting the data plane will be a feature that is an inherent part of the forwarding process. This goal, in the presence of misaligned interests, is a significant design challenge that calls for crafty thinking.

State and the Dynamics of the Control Plane

Dynamic control requires that the environment have some ability to provide a control signal to a device—called feedback. Abstractly, a controlled device must include a parameter that regulates its operation (a *state variable*), and the feedback signal adjusts this parameter. This is a very general statement of dynamic control, but the essential point is that a device without controllable parameters, a stateless device, cannot be part of a dynamic control system. In their pursuit of simplicity, the designers of the original Internet strove to minimize state in the routers. Routers do not keep track of the packets they forward other than to count the total number of bytes and packets forwarded in some interval. There is no per-flow state in routers, which simplified the steps for forwarding packets.[6] Since packet forwarding must be highly efficient (routers in the core of the Internet today are required to forward many millions of packets per second out each port), if the Internet could be made to work without keeping state in routers, so much the better.

6. In the early days of router design, engineers explored the idea of keeping a table of the forwarding information for recently forwarded packets to reduce the cost of looking up the destination address in the full forwarding table. The idea was more complex than it was worth.

What we have learned over the years (perhaps in a fragmentary way) is that even if routers do not require state for the actual forwarding, it may be necessary for control, and adequate control of the network and its resources is a prerequisite for the prime requirement of forwarding.

Congestion control Congestion control makes an excellent case study of different approaches that require different sorts of state in different elements, and different dynamic control mechanisms. When, in the late 1980s, Van Jacobson proposed the congestion control algorithm that is still in use in the Internet today (Jacobson, 1988), one of his central challenges was to find a useful control variable. The software that implements the IP layer (not only in the router but in the sending host) does not have any useful state variables. However, the transmission control protocol (TCP), which many applications use and that runs in the end points above the IP layer, has a control variable (the so-called window): the number of bytes that the sender can have in flight across the Internet at any instant. When the sender receives an acknowledgment that one of its packets has reached the receiver, it sends another packet into the system.[7] This simple feedback loop was initially designed to keep a sender of packets from overwhelming a receiver, but what Jacobson realized was that it could be modified to deal with overload in the network as well as overload at the receiver.

The scheme that Jacobson devised made minimal assumptions about the functionality in the routers. It worked as follows. When traffic offered at a router exceeds the outgoing capacity, a queue of packets forms in the router. When the queue of packets exhausts the storage space, the router must discard an incoming packet. TCP keeps track of lost packets and retransmits them, but as part of the congestion control scheme Jacobson proposed, the sender also cuts its sending window (the number of bytes it can have in flight) in half, thus reducing its sending rate and the resulting congestion. The sender then slowly increases its sending window until it again triggers a queue overflow and loss of another packet, and the pattern repeats. This algorithm is a simple control loop: the sender continuously hunts for the acceptable speed at which to send, slowly increasing its sending rate until another dropped packet signals it to slow down.

7. To oversimplify how window-based flow control works, assume that the round-trip time from sender to receiver is constant. A sender should receive an acknowledgment of a packet one round trip later, so the resulting sending rate is the window size divided by the round-trip time. If the window size is 10 packets, and the round trip is 0.1 seconds, then the sending rate is 10 packets every 0.1 seconds, or 100 packets/second.

When Jacobson proposed this scheme, several of us asked why he used the actual dropping of a packet as a congestion signal rather than some explicit control message. His answer was insightful. He said that if he proposed some complex scheme that the router should implement in order to determine when to send a message that congestion had been detected, coders would almost certainly mis-code it. But there is no coding error that can allow a router to avoid dropping a packet if it has run out of memory. Nonetheless, with the goal of better performance, the research community set about trying to design a mechanism called explicit congestion notification, or ECN, to allow the router to signal a sender to slow down before the router had to drop a packet. The design of ECN was complicated by the fact that the packet lacked a field in which to carry the ECN indication (a lack of what I have called expressive power), so most of the design effort went into figuring out how to repurpose an existing field in the packet for this purpose. ECN has not gained wide use in the Internet for a variety of reasons, and the most common signal to a sender to slow down is still a dropped packet.

While Jacobson's scheme has worked very well and has been critical to the success of the Internet, it raises several issues. First, not all applications use TCP as a transport protocol. Other protocols might not react to a dropped packet in the same way. Applications that stream real-time traffic (audio and video) normally send at a constant rate (the encoding rate of the content) and mask the consequence of a dropped packet rather than retransmit it. The idea that real-time applications should slow down in response to a lost packet is not consistent with the need to send the encoded content (e.g., speech) at the rate at which it is encoded. Further, the TCP congestion adaptation algorithm (the cutting of the window size by two on a lost packet) is implemented in the end node. There is no way the network can detect whether the sender has actually slowed down. What if a malicious user just patches his code to omit this step? The end node will continue sending faster and faster, other (more obedient) senders will keep slowing down, and the result will be a very unfair allocation of capacity to the different senders.

More generally, this scheme acts on an individual TCP flow, which makes a single TCP flow the unit of capacity allocation. What if a sender just opens two TCP flows in parallel? He will then be able to transmit twice as fast. Today, we see Web servers opening many TCP flows in parallel, although usually the goal is not to thwart congestion control but rather to deal with other limits to throughput. The more philosophical (or architectural) question is how capacity (when it is scarce) should be allocated, and which actor should be responsible for the allocation decision. Per TCP flow? Per sender, independent of how many TCP flows that sender has? Per sender to a given destination? And so on. Early (and persistent) debates about these questions yielded the following understanding of the situation. First, there is a tussle between users and multiple ISPs over control. An ISP might assert that since it owns the network circuits and has a service agreement with the user, it should determine how resources are allocated among users. But perhaps the congestion is occurring at a distant point in the Internet, where the ISP actually dealing with the congestion has no service agreement with any of the users contributing to the congestion. Users might assert that they should have some control over which of their flows they choose to slow in response to congestion. These debates suggested that the answer to how scarce capacity should be allocated (to flows, to users, to destinations, and so on) might depend on context and thus should *not* be specified by the architecture. These debates suggested as well that we as designers did not understand (or at least did not agree on) how to build a more general mechanism, and TCP-based per-flow congestion feedback is still the practice today.

The research community has developed a number of proposals either to improve the Jacobson algorithm (some of which have seen limited deployment) or to replace it (which are more focused on a future internet). The designers of these schemes are usually striving for better performance, but I will describe several of them through the lens of control state in the router and expressive power in the packet.

Jacobson's scheme uses the queue of packets as a crude tool to trigger a control signal—when the queue overflows, the dropped packet is the control signal. Several proposals use the length of the queue (the number of packets held there at the moment) as a more sophisticated basis for generating a congestion indication. The scheme called RED (which stands for random early detection or random early drop) (Floyd and Jacobson, 1993) dropped packets from the queue even before the queue was full, selecting them at random as they arrived, with increasing probability as the queue grew. With proper setting of the parameters that controlled the probability of dropping, this scheme had many benefits compared to waiting until the queue was actually full, but it proved difficult to set those parameters correctly in an automatic manner, and the need for manual configuration (e.g., congestion management) limited the deployment of RED.

Later schemes tried to improve on this approach and eliminate the need for any manual configuration. These include CoDel (Nichols and Jacobson, 2012) and Pi (proportional-integral controller). There is vast literature (and I use the term *vast* without exaggeration) on congestion and its control, which I do not attempt to review here. These schemes make architectural assumptions about state and expressive power. The developers of CoDel determined that per-flow state in the routers was important to optimize operation, separating packets into different queues on a per-flow basis, so that the router could service the queues according to some proportional scheme. They proposed the idea of per-flow queues as a minor extension to CoDel, but it is a fundamental change in our architectural assumptions about state. One could ask, looking generally at mechanism and function, what the benefits from per-flow state on the various control functions of the Internet are once we accept the cost and complexity of implementing it.[8] For example, once routers support per-flow state, could that state be used for load-balancing as part of a mitigation scheme for DDoS attacks, as I discuss in chapter 10?

There have been more future-oriented proposals to improve congestion control. They require more modification to the architecture (more expressive power in the header) and thus cannot be deployed as described in the current Internet. The eXplicit Control Protocol (XCP) (Katabi et al., 2002) puts per-flow congestion state in the packet to avoid having per-flow state in the router. To oversimplify (as I have done in many places), the sender puts a proposed send window value in each packet it sends, and as the packet passes through the routers on the path to the receiver, those routers, using a suitable algorithm, modify that window. The receiver of the packet then returns that window value to the sender, who uses it to control the next packet sent. The rate control protocol (Dukkipati, 2008) takes a similar approach, putting a rate variable into each packet so that again there is no per-flow control state in the router. Schemes like these require that a router estimate the round trip of traffic flowing out over each of its links—a stable control loop needs an estimate of the delay in the loop. These schemes also do not define what a flow is—they operate on individual packets, based on the assumption that a flow is that set of packets that share a common rate parameter.

Another framework in which to consider congestion control is re-feedback or *re-ecn*, proposed by Briscoe starting in 2005.[9] Re-ecn is a scheme that takes into account the incentives of the different actors (both senders

8. In fact, routers today support this sort of function—it has crept into the implementations without being considered from a more fundamental or architectural perspective.

9. For the reader interested in the re-ecn work, which is not just a protocol but also a proposal to reframe congestion, a good place to start is the Web page maintained by Briscoe at http://www.bobbriscoe.net/projects/refb/.

and ISPs) to deal with congestion and specifies a *policer* that an ISP can use to detect whether the user is responding correctly to the congestion signals received from the network. The policer is a new PHB designed to regulate the potential tussle between sender and ISP.

State and network management Few of the architectural proposals in chapter 7 discuss congestion control in detail, or more specifically what expressive power in the header might be useful for this purpose. They do not discuss the balance of what needs to be specified in the architecture and what needs to be adaptable in different contexts. NDN is an interesting alternative in this respect. NDN is based on an architectural decision to have per-packet state in each router, not just per-flow state. Whenever a router forwards an interest packet, it makes a record of that event. If and when a data packet returns, the router matches that returning data packet to that stored record. The router can log the time when the interest and data packets arrive, so that it can measure the delay. Dynamic control loops can be devised based on the measured delay. Per-packet state, once in place, may enable a variety of control functions, giving NDN a powerful capability that other architectures with less state in the router cannot implement. NDN can implement per-hop forms of congestion control by limiting the number of pending interest packets. It can perform routing experiments by sending interest packets out different paths and seeing whether a data packet returns. While descriptions of NDN often focus on its forwarding function, its ability to accommodate a range of novel control algorithms is an equally important aspect of its fundamentals.

Some of the active network proposals discussed starting on page 140 allow the active packets (packets with code the router executes when the packet arrives at a node) to create temporary state that supports more sophisticated network control. The PLANet scheme (Hicks et al., 1999) discusses the use of active code to create *scout packets* that fan out to seek good routes to a destination, creating transient per-packet state to keep track of the scouting function, somewhat reminiscent of the transient state created by interest packets in NDN. But NDN and PLAN are different; NDN is most emphatically not an active network approach—in NDN, any algorithm that implements (for example) exploration of good routes would be a part of the core router function, not installed by the source. In PLAN, the source node implements the scouting code and sends it to be evaluated by nodes along the path. The latter scheme does raise the question of what happens if different scouting programs from different source nodes are simultaneously exploring the network and perhaps making conflicting decisions that end up having undesirable consequences. But the approach illustrates a form of end-to-end thinking—in principle, only the source knows what sort of service it is seeking and thus what sort of scouting algorithm will best meet its needs. On the other hand, a network could measure its own performance and report that information to the end node as part of a service offering.

Layering and Control

In chapter 2, I described the Internet as having a simple layered structure (see figure 2.1) with a network technology layer at the bottom, the application layer at the top, and an internet layer in the middle that provides the service specification that links the technology to the application. In today's Internet, the network technology layer itself has become layered. We see complex technology layers below the IP layer: multihop subnetworks using their own forwarding elements, often called switches or layer 2 switches to distinguish them from routers.[10] A lower layer based on a different architecture can provide a different set of capabilities for handling traffic not offered by the Internet architecture. The layers are usually considered independently. With the exception of the RNA proposal, I don't know that there has been much consideration of how critical control functions cut across layers. In particular, there is seldom any discussion of what interfaces (definitions of what information will flow between the layers) are needed in the context of control. The RNA scheme provides the most explicit discussion of the question.

Congestion control is an example of the complexity that can arise in control mechanisms with a layered network architecture. What happens when congestion occurs at a switch rather than at a router? How does the switch participate in the congestion control algorithm? In Jacobson's original and very insightful design, the switch just drops a packet. As he said, any device knows how to drop a packet. But with a congestion control scheme that involves expressive power in the packet, both the switch and the router have to understand and manipulate the same fields in the packet, which is not compatible with most conceptions of the independence of different network layers.

Fault control (or management) provides another illustration of the issue. When a physical component fails, the consequence manifests at every layer. If there is no interface to allow coordination among the layers, every

10. I introduced this concept on page 131 in the discussion of RINA in chapter 7. Multiprotocol label switching, or MPLS, is often used at this level. I discuss MPLS in the Appendix. Switched ethernet is another technology for building networks.

layer may independently initiate some sort of corrective action, and these actions may conflict with each other. In a layered system with routers at one layer and switches at a lower layer, the switches will have their own routing protocols, and when a fault occurs, both the switch protocols and the Internet routing protocols may undertake a rerouting effort, while the application might attempt to initiate a connection to a different end point. I know of little discussion in the research community as to how information exchange between the layers might improve network control challenges such as these, perhaps because such a question is inconsistent with the idea of layer independence.

This sort of example suggests that from the point of view of control, the conception of modularity based on layering may be less useful than modularity based on the viewpoint introduced by the RINA proposal, the *scope* of a mechanism. Mechanisms will have different scopes: some can operate within a region of the network, while others must operate globally. We designed Internet routing this way—the Internet allows alternative routing protocols inside individual regions of the Internet and uses the global BGP to hook these routing protocols together. The relationship between regional and global routing protocols is well-understood if implicitly specified. The designers think of these two mechanisms as running not at different layers but at different scopes within the internet layer.

A key question for control and management schemes, and indeed for network architecture, is what defines the scope of a particular mechanism. I think the answer lies in the framing of control as a dynamic mechanism—a mechanism that responds to an input by changing the state of the system. Dynamic systems can be characterized by the rate at which they respond to change—usually called the time constant of the mechanism. My example of fault recovery illustrates how different error recovery schemes at different scopes can work harmoniously—they have to work at different time constants. Within a geographically limited scope where the round-trip times across the scope are small, a routing protocol should be able to find new routes quickly. A local network might reroute within some known time, say 50 ms, if alternative routes are available. If this time bound were known, then a higher-level fault-recovery scheme such as Internet-level routing should wait 50 ms before it reacts.

In contrast, dynamic congestion control of the sort I described operates end-to-end (at a global scope) at the same time constant. The only way that nested congestion control schemes could run at different scopes would be if they had a different time constant. This is what happens in delay tolerant networks (DTNs). Data is transported in a store-and-forward manner between nodes in the DTN. Between nodes in a terrestrial region, the DTN could use a network like the Internet, with its congestion control scheme. Congestion can also occur in the store-and-forward nodes—if it is impossible to forward the data onward, the storage of the node might fill up. But the feedback necessary to deal with that congestion has a very different (much longer) time constant than the internal congestion controls of the Internet, so in a DTN there are nested congestion control schemes running at different time constants.

In chapter 7, I discussed the difference between a layer and a scope. One view is that network architectures define a layer, and scopes exist within a layer. The idea of a time constant may provide a different answer. Perhaps when some control problem (such as error recovery) is solved by nested mechanisms with different time constants, those different mechanisms are in different layers. But that proposition raises a fundamental challenge for network architecture—different mechanisms seem to have a different scope across which different time constants can be used. The layering of error recovery would not be the same as the layering of congestion control. If different mechanisms have a different natural scope of operation, the concept of a layer may be insufficient as an organizing principle for network design. The layering of a network architecture may have a different structure for each mechanism in the design, including data forwarding, control, and management.

There is evidence of this confusion about layering in the evolution of layer 2 switches in the current Internet. As I noted, ISPs sometimes use layer 2 switches instead of routers when building their internal networks. Switches use a separate internal addressing system (with the scope of the ISP) as a basis for routing and forwarding rather than IP addresses. The use of level 2 switching has a number of advantages in practice: a switch is less complex than a router (and often much less expensive), and switches can provide advanced functionality within their scope (for example, providing many virtual networks for the ISP, often called virtual local area networks, or VLANs). In the language of RINA, switches operate at one scope, routers at another, larger scope. Designers of layer 2 switches saw layer 2 as a separate architecture. Layer 2 architectures were developed by design teams different from those that designed the Internet, with the expectation that the two layers (or scopes, to speak the language of RINA) would be almost independent. The level 2 switches, using the mechanisms defined by their architecture, provide an abstraction of the network connecting the Internet routers at the edges of the level 2 network: the illusion of a link, or a virtual circuit. The only link between the layers was this simple abstraction.

The need for more advanced functionality is now causing this independence to break down. Switches are now crossing the architectural boundary and using data in the packet that was specified as part of the Internet architecture, such as the ECN field, and blending together the routing mechanisms at the two layers. In marketing terms, these switches are called Internet aware layer 2 switches, or layer 2.5 switches. In my terms, switches and routers are now jointly using some aspects of the expressive power defined by IP while using separate expressive power for other mechanisms.

People with a strong belief in strict modularity of layering are put off by this layer blending, but the layerists are wrong. Switches are necessarily crossing the layer boundary to deal with issues with different scopes, such as congestion and QoS (which have an end-to-end scope), and management issues related to policy with their own scope. The modularity of routing (with BGP and intra-AS routing) is consistent with the real-world management scopes of ISPs. In fact, nested control schemes associated with administrative boundaries make a lot of sense in some cases. When a failure occurs inside an ISP, wait and see if the ISP fixes it internally. If not, try to route around the ISP. In this respect, layering is an aspect of the architecture that relates to industry structure.

To the extent that different layers are associated with different actors, layering (or control of how expressive power in the packet is used) will now be a means to control tussle. Since actors sometimes have adverse interests, there may be cases where the architecture should prevent the sharing of expressive power as well as cases where it should facilitate it. Here is an example of that tussle. The protocol called Internet Protocol Security, or IPsec, defines a method of encrypting packets sent across the Internet. Since routers depend on the expressive power of IP, that header had to be left unencrypted. However, the part of the packet defined by TCP is supposed to be of interest only to the end points, so the designers specified IPsec to encrypt everything except the IP header. However, ISPs objected to this decision—they asserted that they needed to see one field in the TCP header, the *port* field. They claimed they needed to see this field as part of network management. There was a layerist response: routers should not be designed to violate layer modularity. But the real argument was not about layers, it was about tussle. Many people felt that ISPs should *not* be able to see that information, because the resulting "management" decisions were likely to be traffic discrimination—actions that were adverse to the interests of the end users. If the port numbers were encrypted, that bit of expressive power would be hidden from the ISPs.

I will return to the question of layering as part of architectural design in chapter 15, but, in my view, a careful architectural consideration of management and control will lead to a radical rethinking of layering.

Categories of Management and Control

In this section, I look at categories of management, starting with the ITU FCAPS list, and seek to identify architectural issues, and further explore what aspects of the management interface might rise to the level of architecture.

Fault Management

The challenge of fault management came up in various places earlier in this book, most directly in the discussion of security and availability.

In my discussion of availability in chapter 11, I proposed a high-level framework for understanding availability:

- It must be possible to detect the failure.
- It must be possible to localize the failed parts of the system.
- It must be possible to reconfigure the system to avoid depending on these parts.
- It must be possible to signal to some responsible party that a failure has occurred.

I noted at the time that this framework papered over the huge issue of which actor should carry out each of these tasks.

There are a variety of ways to detect failures, involving different actors. In some cases, an element may be able to tell that it is failing and raise an alert. In this case, the question is where to direct that alert. There are utterly feeble mechanisms to indicate a failure, such as turning on a small red light in the hope that a person will notice (a wonderful example of a horrible management interface).[11]

Sometimes machines interacting on the network can detect that one of their number has failed. Most Internet routing protocols embed some test

11. Early in the era of time-sharing, when I was programming the Multics system, the I/O controller had an interface for reporting errors to the central computer, but if the central computer did not process the error messages, the controller rang a loud alarm bell. One's programming errors took on a somewhat public character. The management of the management system implies a sort of recursion that has to be resolved somehow.

of correct function into the protocol itself (perhaps a simple *keep-alive* or *handshake* probe). The failure of a handshake protocol is a signal of failure, but at which end? Why is it reasonable to trust a machine when it asserts that another has failed? In fact, the first machine may be failing, not the second machine, or the first machine may just be malicious.[12] Any machine that receives a message saying that it is failing must take that input with considerable caution, a case where it may make sense to have a person in the loop, but the need for rapid recovery will create a tension between a quick response and a considered response. One of the benefits of a system with redundancy is that service can be restored quickly by using redundant elements, and the failed element can be isolated and fixed more slowly.

The protocol for forwarding Internet email has a built-in redundancy/resilience mechanism. The DNS can list more than one IP address for a mail transfer agent, so if sending mail to the first one fails, the sender can try another one. However, there is no means for the sender to report the failed transfer agent. If the failure could be reported, it might be repaired more quickly, but a number of issues must be resolved for such a scheme to work, which might rise to the level of architecture. The first is to provide the address where an error report should be sent. The DNS could store a new kind of record that gives the name of the machine to which to report failures. The second issue is to prevent the abuse of this mechanism (malicious reports or DDoS attacks). The third issue is to deal with a possible flood of legitimate error reports when lots of senders simultaneously detect that a receiver has failed. These issues could be mitigated to some extent with an *incast* mechanism (incast is the opposite of broadcast—many to one instead of one to many), which could aggregate multiple error reports (legitimate or otherwise) into one as they flow across the network toward that reporting point.

In simple cases (consider a home network), a standard method of logging errors within that scope should be part of the basic configuration process. For example, when the machine first connects to the network, an extension to the dynamic host configuration protocol could configure the machine with the address of a place to send fault reports. The home router could be the recipient for these messages, and such a framework could be part of a new service that allows for diagnosis of problems in the home network.

12. There was a famous rolling outage of the AT&T phone system that was similar in character to this pattern. One machine self-detected a fault and reset itself, and in recovering from this reset, it sent a sequence to its neighbor machines, which (because of a bug) then reset themselves, and so on. This went on for nine hours (Neumann, 1990).

In the case of email, the two-party nature of the forwarding steps makes localization somewhat straightforward. However, in other cases (most obviously the failure of a packet to reach its destination in a timely manner), localization is much harder. Without the ability to localize the problem, it is much harder to resolve it by avoiding the failing component (a final option is trying other options more or less at random), and there is no way to report the fault. The tools the Internet has today to localize faults along the forwarding path are minimal: usually the only option is *traceroute*, with its many limitations. But as I noted, it may not be in the best interest of a particular region of the network to let outsiders successfully localize faults within that region, and when the fault is caused by the successful blocking of an attack, it is absolutely not in the best interest of the target of the attack that the attacker be able to diagnose the reason the attack failed. I believe that fault localization is a poorly understood but critical aspect of network design, perhaps understudied because it is not purely a technical problem.

Quality of experience, or QoE, is the subjective measure of the degree to which a user is satisfied with the behavior of an application. Many factors can influence the perception of QoE by the user: the expectation of the user, whether the user paid for the application, or the overall mood of the user. However, in this context, the relevant aspects of QoE are those that relate to the character of the network. In this context, QoE raises issues of fault isolation as well as performance and security. When the user encounters an impairment to QoE that is caused by some phenomenon in the network, the steps to resolve the problem very much resemble those I identified for dealing with issues of availability. The issue of localization is thus central to correcting QoE impairments. Lacking localization, the user is reduced to waiting until some other person (presumably the person who manages the relevant entity) notices that something is wrong and fixes it, and, as I noted earlier, localization of a problem in a distant region of the network may be an adversarial action.

I believe that in the future an increasing focus on measurement of QoE and diagnosis of QoE impairments will create a generalized requirement for localization that is not restricted to network faults but also issues such as performance or flaws in higher-level services. As such, a generalized approach to localization of issues in a network would be a major advance.

Configuration Management

Configuration is the process of setting up the elements of a system so that they function properly. As a simple example, the dynamic host configuration protocol (DHCP) allows for the automatic configuration of a host when it first attaches to the Internet. DHCP changed host configuration from a manual and somewhat mysterious management task to an invisible control function hidden from the user. DHCP provides three critical pieces of information: an IP address for the new machine to use, the address of a router that can provide a path to the Internet, and the address of a DNS server that provides access to domain name resolution.

The challenge with configuration is the bootstrapping problem: how does the new device know what existing device on the network is allowed to configure it? With DHCP, a machine broadcasts to find a DHCP server when it is first attached to the network. Lurking inside this broadcast discovery scheme is a potential security problem. If a newly attached host requests configuration information by broadcasting its request and believing whatever machine answers, what if a malicious machine answers? This mechanism is usually not a major vulnerability, but it should serve as a reminder that the initial phase of configuration is a moment of vulnerability in system setup, whether the mechanism is DHCP, bluetooth peering, or configuring devices in a smart home. There may be some necessary first step a person takes, such as typing validation information into the new machine, to add assurance to the initial configuration.

The problem of configuration is not restricted to the network layer. Once a new computer is attached to the network, the next step is to further configure the machine by installing applications. Where do the apps come from? On a smartphone, they come from the app store, but how does the device find the app store? The answer, of course, is that the smartphone comes preconfigured with an initial app that can find the app store—presumably it has some built-in knowledge that lets it make a trustworthy connection to the store. Configuration at every level depends on having some starting knowledge built into the device to bootstrap the process.

The rise of the IoT—small, fixed-function devices—will raise new issues in configuration. These devices are normally preconfigured with the correct application software—a thermostat is programmed to be a thermostat, a security camera is programmed to be a camera, and so on. But to the extent that they need to be configured for a specific installation, how is this done? Devices like this often lack a keyboard or a display. They will need some way of being configured over the network, and perhaps there is a role for a standardized (e.g., architected) way for IoT devices like this to request configuration information when they are first installed. Just as DHCP is standardized as a way to bootstrap device configuration, there should be a standardized protocol to bootstrap configuration at the application level for IoT devices.

Accounting Management

In chapter 12, I discussed a range of schemes for money routing that depended on new fields in packets and presumably also on new sorts of tools in routers to track and report usage of different sorts.

Operators today use fairly simple tools to gather data to inform accounting functions: packet and byte counts, data from samples of the packet stream (such as IPFIX), and others. In 1991, as the first commercial ISPs were being launched, the IETF looked at accounting and published an RFC (Mills et al., 1991) that frames the problem. It discusses methods of reporting based on packet capture, and in many respects the state of the art does not seem to have advanced all that much. The RFC is cautionary with respect to inventing complex tools for accounting, lest they be used. If ISPs had the tools to bill users based on the amount of traffic they sent, that might chill the user's interest in experimenting with new applications. Such a scheme certainly would have chilled the deployment of Internet video streaming.

Performance Management

Performance, as it relates to architecture, is not a simple matter of throughput between two end points. Various proposals I have discussed in this book have implications for performance, but in very different ways. Performance is a multidimensional issue that will have different manifestations in different architectures. Here are three examples:

- ALF was designed to improve host processing performance.
- NDN uses caching to reduce the latency and network load when delivering popular content. In NDN, performance is a function of how the routing protocol finds the closest copy and the cache replacement algorithm in the various routers in the system. It is possible that the cache replacement algorithm needs to be tuned based on the dominant class of content being retrieved, and this tuning may be a management function. If so, what parameters should a router report about its cache to facilitate management?
- MF improves the availability of mobile devices as they move from network to network. The GNS is key to this goal. Does the GNS require management? Should the latency of the GNS be tracked and reported?

For each of these proposals, part of the analysis must be whether the mechanisms related to performance need management, need a control protocol, or function as a natural consequence of the design of the data plane.

With the advent of massive content delivery over the Internet and the use of CDNs with complex caching schemes to improve the delivery of content, new issues related to performance have arisen that seem to call for new interfaces for the management (or control) of these schemes. CDN providers can cache many copies of the same content at different locations around the Internet and can select a specific source for any transfer in order to optimize the delivery. By careful management, CDN providers can operate their interconnection links essentially fully loaded without triggering the negative consequences of actual congestion. However, to achieve this efficiency, they have to detect what the actual instantaneous load on the link is. Today, there is no way to extract that information through a management/control interface; CDN providers must estimate whether the link is fully loaded by looking for transient evidence of congestion. This method is both imprecise and potentially misleading, since monitoring for congestion at the end point cannot localize the congestion to a given link. If routers supported a standardized protocol for reporting link utilization (with the proper access controls to reflect business concerns, of course), this would allow the design of more sophisticated algorithms to manage content delivery from CDNs.

Security Management

In chapter 10, I broke the problem of security into four sub-objectives. Each of them will raise its own requirements for management, some of which I discussed in that chapter. However, I do not yet understand the role of architecture in achieving better security well enough to say much about how architecture might improve the *management* of security mechanisms, but a few thoughts follow.

Attacks on communication With the exception of availability, I argued that end-to-end encryption was the starting point for preventing these attacks. The major issue here is key management. Some architectures (for example, NDN) make key management a part of their design, while others leave the problem to higher levels. The CA system, while not part of the network, has risen to such a level of importance that it is (perhaps like the DNS) trending toward becoming architecture. The CA system has massive issues of management, with organizations such as the CA/Browser Forum[13] meeting to discuss which root authorities are trustworthy and other issues. This situation may justify the view that a system that requires

13. See https://cabforum.org/.

this much management is a poorly designed system. On the other hand, key management is known to be tricky.

The process of configuring a Web server to support TLS has been a manual and complex management task, which has prevented many Web site operators from implementing the security protocols. A recent effort, the Let's Encrypt initiative,[14] has attempted to change the process of configuring TLS from a manual management process to an essentially automated task requiring minimal user intervention. While again this effort seems a bit removed from network architecture, it illustrates that for many problems there are a range of solutions, from the more manual (management) to the more automatic (control).

Attacks on the host PHBs that protect a host from attack need to be configured and managed, but I am not sure about the extent to which the management interface needs to be standardized. More experience is needed.

Attacks on the network itself The most obvious attacks on the network today (aside from DDoS) are attacks on the interdomain routing system. Other architectures with different feature sets will manifest different opportunities for attack. Setting up secure BGP today requires a great deal of manual configuration, including installation of public-private key pairs, and registration of address blocks. As with the Let's Encrypt effort, there is an open question as to how to automate the configuration of secure BGP. Much of security management is the configuration of security parameters such as keys, but it is not clear what parts of this process require architectural support.

Denial of service attacks As I discussed in chapter 10, DoS attacks (and DDoS attacks in particular) are a problem that arises at the that level and must be managed at least to some extent at that level. I described a range of approaches, each of which has its own requirements for management and control interfaces. Routers that participate in traceback logging must make that function available through some interface, which may raise further security issues. The FII SUM requires an association between every sending host and a trusted third party that vouches for its identity, which seems to imply a significant management task. Again, different design approaches may result in schemes with very different degrees of manual intervention.

14. See https://letsencrypt.org/.

Management of Routing

The interior routing protocols in the current Internet are somewhat self-configuring. When two routers discover that there is an active node on the other end of a connected link, they begin to exchange information, with the goal of discovering what is reachable through that other router. The ports on each router need an IP address (manual configuration management), and sometimes a name (for reverse lookup) is assigned to that address, but little more is needed in general. There are many aspects to BGP that reflect policy and require manual configuration.

The emergence of new centralized route computation schemes such as SDN require new management/control interfaces on routers and switches, as I noted.

Conclusions

This chapter is more speculative than some of the earlier chapters. Research on network architecture and design has provided few candidate mechanisms to consider, and our operational experience with the current Internet is based on a set of ad hoc methods that use mechanisms in ways for which they were not intended. I think our operational experience provides a poor basis for thinking about fundamentals. While I believe that I have identified a few potential network features that rise to the level of architecture, and that I have posed some important research challenges, it is not clear how the research community should proceed to learn more about this area. What we need is operational experience with networks at scale, but we cannot easily use the running Internet for experimentation. I fear that this area may remain underdeveloped and feeble.

However, there is an important insight that comes from studying management and control functions: conceiving of the network technology as being built of layers that are largely independent may not be sustainable. Issues of performance cut across these layers, as does fault isolation and recovery, and the assumption of independence (and the resulting simple modularity) may fail.

As a concrete example, people conceptualize routing in the current Internet in two inconsistent ways. The global routing (BGP) and interior routing protocols are thought of not as different levels but as different scopes, as the advocates for recursive scopes would argue. The interactions between global routing and intra-AS routing are well-specified. In contrast, when routing and forwarding mechanisms such as MPLS or switched Ethernet are used, this is conceptualized as a different architecture at a lower layer—we talk about layer 2 switches running beneath IP. The implication of calling these mechanisms a separate layer is a higher degree of independence. A layer encapsulates all its internal behavior and provides an abstraction of its service to the layer above it. Because the protocols of a layer 2 switch are different from those of a layer 3 router, there is no intention to support layer interaction based on the expressive power of the IP packet header.

This degree of independence has worked to this point for the simple best-effort forwarding service of the current Internet because the weak specification of the best-effort service model means that the Internet layer just absorbs whatever behavior the layer 2 architecture produces. The independence of the layers equates to the simplicity of the service abstraction presented to the higher layer. If, in the future, an internet architecture implements delivery services that are more precisely specified or defines mechanisms for fault isolation or other advanced management or control functions, the independence of a lower-layer architecture will not be sustainable. It will be necessary to think of these various layers working in a more integrated way—working within a common architecture.

[illegible] layer 2 switches [illegible]. The implication [illegible]
[illegible] layer is a [illegible] degree of independence [illegible]
[illegible] its [illegible] and provides abstraction [illegible]
services to the layer above [illegible] the protocols of a layer [illegible] which
are different from those [illegible] layer 3 [illegible] there is no [illegible]
[illegible] powered [illegible] IP packet header [illegible]
The degree of independence that [illegible] to this point [illegible]
[illegible]
[illegible] that the internet [illegible]

14 Meeting the Needs of Society

by David Clark and kc claffy

What Do We Want Our Future Internet to Be?

The goal of this chapter is to identify some desirable properties of a future Internet, looking through the lens of societal concerns, and consider what (if anything) network architecture has to do with these goals.

Several years ago, we were moved to try to collect in one paper a list of all the societal aspirations for the future of the Internet that we could find and organize them into categories (Clark and Claffy, 2015). For example, we collected statements from governments and public interest groups. The resulting list of aspirations did not originate with us, nor did we agree with all of them. We cataloged these aspirations in order to subject them to critical analysis and motivate a debate over which of them are desirable, well-specified, realistic, and achievable.

This exercise led us to three high-level conclusions, perhaps obvious but worth stating since they are sometimes neglected. First, not only are many of the aspirations hard to achieve, but some are incompatible with others. Second, many are underspecified and resist operational definition; it is unclear how to translate the aspiration into concrete goals against which to measure progress. Third, most of the current tools society has to shape the future of the Internet seem unequal to the task.

These conclusions, while potentially pessimistic, raise the question of whether a different internet might be a better vehicle for the pursuit of these goals. For this reason, we have taken our list from that paper as a starting point through which to look at this final architectural requirement: a future internet should be designed to meet the needs of society.

This chapter is a revision of a previously unpublished paper written jointly with kc claffy. I appreciate her substantial contribution to this work.

In pursuit of these goals, we encounter again what I called the fundamental tussle. Governments or advocacy groups express many aspirations on this list as societal goals—desirable outcomes for the citizenry and thus in the public interest. However, the Internet's architecture and infrastructure are now primarily under the stewardship of the private sector, driven by profitability and commercial viability, constrained by technological and economic circumstances, and sustained by interconnecting and interoperating with competitors in a multistakeholder ecosystem. Navigating the inherent tension between private-sector objectives and societal aspirations is essential to shaping the future of the Internet.

Catalog of Aspirations

Here is our catalog of aspirations for the future of the Internet:

1. The Internet should reach every person by some means. (Reach)
2. The Internet should be available to us everywhere. (Ubiquity)
3. The Internet should continue to evolve to match the pace and direction of the larger IT sector. (Evolution)
4. The Internet should be used by more of the population. (Uptake)
5. Cost should not be a barrier to the use of the Internet. (Affordable)
6. The Internet should provide experiences that are sufficiently free of frustration, fears, and unpleasant experiences that people are not deterred from using it. (Trustworthy)
7. The Internet should not be an effective space for lawbreakers. (Lawful)
8. The Internet should not raise concerns about national security. (National Security)
9. The Internet should be a platform for vigorous innovation and thus a driver of the economy. (Innovation)
10. The Internet should support a wide range of services and applications. (Generality)
11. Internet content should be accessible to all without blocking or censorship. (Unblocked)
12. The consumer should have choices in their Internet experience. (Choice)
13. The Internet should serve as a mechanism for the distribution of wealth among different sectors and countries. (Redistribution)

14. The Internet (and Internet technology, whether in the public network or not) should become a unified technology platform for communication. (Unification)
15. For any region of the globe, the behavior of the Internet should be consistent with and reflect its core cultural/political values. (Local values)
16. The Internet should be a tool to promote social, cultural, and political values, especially universal ones. (Universal values)
17. The Internet should be a means of communication between citizens of the world. (Global)

As we organized these aspirations, we found that many of them could be clustered into three more general categories:

- Utility
- Economics
- Security

The Economics Cluster

The Internet should reach every person by some means *(Reach)* The *Reach* aspiration is generally uncontentious; almost every country has some form of it. The differences relate to granularity (household or community?), bandwidth (how much?), and methods of achieving reach. Developed countries focus on reaching the yet unserved population, usually rural areas. In developing countries, where most of the population may not have access, the focus may be on wireless access using mobile devices. To achieve *Reach* in rural areas that lack sufficient revenue to justify private investment in infrastructure, some nations have provided subsidies or tax incentives to build or maintain networks. In some cases, the public sector has directly funded construction. In the United States, direct public investment has happened at multiple levels, from federal stimulus money to municipal construction of residential broadband networks.

The Internet should be available to us everywhere *(Ubiquity)* The *Reach* aspiration has a corollary in the age of mobile communications—every person should have access to the Internet approximately everywhere they go, implying the integration of high-performance wireless technology into the Internet.

Cost should not be a barrier to the use of the Internet *(Affordable)* This goal is a component of *Uptake*, since cost is a major barrier cited by nonusers today. The phrase "cost should not be a barrier" could be mapped to the simpler phrase "the Internet should be low-cost." However, we don't expect wine to cost as little as tap water. Low cost might map to lower value, which might be counterproductive. Perhaps an emphasis on value would be more productive as a means to uptake.

The Internet should support a wide range of services and applications *(Generality)* The original Internet architecture embedded this aspiration, since it was designed to support a cooperative network of time-shared general-purpose computers. Benefits that follow from this aspiration include *Innovation* and *Uptake*, since the more value the Internet can deliver, the more users it will attract.

Although there is no obvious way to quantify progress toward *Generality*, the range of Internet applications demonstrates its success at this aspiration. But not all applications work well on the public Internet today—most problematic are those that require very high reliability and availability (e.g., remote surgery or remote control of autonomous vehicles). Does *Generality* imply the need to evolve to support such ambitious services, or should they be segregated to more controlled private networks?

The Internet should evolve to match the pace and direction of the larger IT sector *(Evolution)* The Internet was designed to connect computers together, and this aspiration captures the idea that as computing evolves, so should the Internet. In particular, as computing gets faster and cheaper (e.g., sensors), the network should get faster and access to it cheaper. For decades, Moore's law[1] has characterized how (IT-based) demands on broadband infrastructure change much more rapidly than for other sorts of infrastructure, such as the power grid. In 2013, the forecast growth of U.S. power consumption was 0.9 percent per year (U.S. Energy Information Administration, 2013), while the forecast growth of Internet traffic was 23 percent per year (Cisco Systems, 2013).

National policy statements have often had a dual character (Benkler, 2012): getting some level of broadband to everyone (*Reach*) and pushing for deployment of a next generation of broadband (*Evolution*). The U.S. FCC National Broadband Plan published in 2010 aspired to a 10-year

1. Moore's law, proposed by Gordon Moore (1965) observed that the rate of improvement of computer chip fabrication was leading to a doubling of performance about every two years.

milestone for *Reach* and *Evolution*: *"100 million U.S. homes should have affordable access to actual download speeds of at least 100 Mbps and actual upload speeds of at least 50 Mbps by 2020."* (This admittedly now looks less impressive compared to Google Fiber's gigabit-to-the-home deployments around the country since 2011.)

The Internet should be a platform for innovation and thus a driver of the economy *(Innovation)* As a key component of the IT space, the Internet has contributed to economic growth by promoting innovation and creativity technology development, revolutionizing logistics and service industries, among other ecosystem disruptions. One interpretation of the *Innovation* goal is that the Internet must be *open*, a term used to capture many other aspirations. We believe this word is a red flag for muddy (or at least unfinished) thinking. *Open* is a word with strong positive connotations, useful as a rallying cry but dangerously vague. We prefer to refer to more specific objectives in the ill-defined basket called open: stability, specification (open standards), and freedom from discrimination or from intellectual property restrictions. But even these aspirations are not absolute. For example, some forms of discrimination among uses of a platform can promote innovation, assuming clear and consistent rules (Clark and Claffy, 2014). In fact, many traffic discrimination scenarios may benefit users, the most obvious being protecting latency-sensitive traffic from the consequences of commingling with other traffic.

The deeper and more vexing policy question that is poorly informed by theory or fact relates to causality: what underlying properties (e.g., the aspirations *Generality*, *Uptake*, *Ubiquity*, *Evolution*, and *Unblocked*, or access to capital) are key drivers of *Innovation*? Van Schewick (2012) provides a comprehensive analysis of the relationship between a network that is general, unblocked, and free from discrimination and its value as a platform for innovation.

The Internet should serve as a mechanism for the distribution of wealth among sectors and countries *(Redistribution)* Thousands of independent firms combine to provide the Internet ecosystem, each typically striving to be profitable and competitive, and the flow of money is integral to its structure. Contentious arguments about redistribution of capital, either to cross-subsidize from more profitable to less profitable sectors of the ecosystem (e.g., commercial to residential, urban to rural) or from more developed to less developed countries, have long characterized telecommunications policy debates and legislation.

A recent vivid example is the ongoing tension as to whether high-volume (video) content providers should contribute to the cost of the infrastructure. This tension has led to debates on whether access providers should be able to charge content and/or transit providers for access to their customers, and more generally on whether interconnection arrangements should be left to industry or regulated to allocate money flows according to who induces load on the infrastructure as opposed to who carries it (Frieden, 2011).

In addition to cross-subsidizing across industry sectors within one country, governments also aspire to tap into international revenue flows in the Internet ecosystem. The developing world used to bring in substantial hard-currency payments from settlement fees associated with telephone calls into their countries, a revenue source that is diminishing as communication moves onto the Internet. The global controversy about the role of the ITU in regulating international Internet interconnection reflects a motivation by many parties, including governments, to change the current norms for payment for the global flow of Internet traffic in order to be closer to historical telephony-based norms (Toure, 2012; Huston, 2012).

Architectural relevance *Reach*, *Ubiquity*, and *Evolution* directly relate to the discussion in chapter 12 on the incentives of the private sector to invest. Investment can advance these aspirations, but perhaps not in the proportion that society might want. All three are capital-intensive activities and thus would seem to drive up cost, which would put them in conflict with the aspiration that the Internet be *Affordable*. Architectural support for mobility advances *Ubiquity*.

Innovation and *Redistribution* relate directly to the discussion in chapter 12 on money-routing across the Internet. The *Innovation* aspiration is almost directly an expression of hope that the infrastructure providers will invest so that the innovators on top of that platform can profit. Put that way, it is not obvious why such a hope would come true. The aspiration of *Redistribution* is in some direct sense a response to the pursuit of *Innovation*; it is a call (among other things) for the innovators to shift some of their profits to the infrastructure providers. It is interesting that one can find this aspiration expressed in pretty direct terms by some of the actors.

Again, to the extent that there are architectural implications of this set of aspirations, chapter 12 has tried to address them. They relate to architectural modularity and what interactions among the different actors are facilitated by the expressive power of the module interfaces. Can a network architecture stimulate the creation of new services that innovators benefit from purchasing?

The Internet (and Internet technology) should become a unified technology platform for communication *(Unification)* This aspiration is not directly relevant to society; IP network operators tend to share this aspiration as a source of cost savings, or more generally to maximize return on capital investment. As such, it may facilitate the pursuit of other aspirations discussed here. The *Unification* aspiration differs from *Generality*; the latter is about supporting a wide range of services, while *Unification* reflects the economic efficiency of discontinuing other infrastructure and associated investments.

Historically, telephone calls, cable television, and industrial control networks used independent specialized legacy communications infrastructure. Today, Internet technology can provide a unified platform for any important communications application. Many ISPs today run a fully converged IP backbone for economic efficiency and would resist any regulatory intervention that would cause them to separate infrastructures they have unified or plan to unify.

Note that although *Unification* reduces overall costs in some areas, it also may increase costs in others, since the unified platform must support the performance of the most demanding application in each quality of service. For example, a unified IP-based platform must be reliable enough to support critical phone service and have the capacity to carry large bundles of television channels. Unification may also increase risks to *National Security*, since a less diverse infrastructure has a higher potential for systemic failure (Schneier, 2010; Geer, 2007), although this fear is debatable (Felton, 2004).

Architectural relevance Today, in practice we see a two-level IP platform emerging in which ISPs build an IP platform and then run their part of the IP-based global Internet on top of it. Most of the architectural proposals discussed in this book are related to the creation of a new global Internet, not the creation of a new form of unified platform. Given the current trends in industry, it would seem beneficial to have an architectural exploration of this two-level structure.

Unification can lead to a monoculture of technology, potentially increasing the risk of systemic failure and consequences for *National Security*. Avoiding the risks of monoculture is a valid issue to explore from an architectural perspective.

The Security Cluster

Just as in chapter 10, the overarching concept of security (or an aspiration for better security) is too general as a starting point. All of the aspirations grouped here relate to aspects of security, but they break up the problem in slightly different ways than in chapter 10, since (in the language of that chapter) they focus on harms rather than system components. This focus on harms makes sense in the context of aspirations.

The Internet should provide experiences that are sufficiently free of frustration, fears, and unpleasant experiences so that people are not deterred from using it *(Trustworthy)* Most users hope, expect, or assume that their use of the Internet does not lead to their behavior and data being used against them. Users also need to be able to (but often cannot) assess the safety of a given aspect of their Internet usage. Today, users fear side effects of Internet use (i.e., their activities being monitored, or personal information used in unwelcome ways, such as for behavioral profiling). Users fear identity theft, loss of passwords and credentials, malware corrupting their computer, or losing digital or financial assets by having their accounts compromised. The threats are real (Madden et al., 2012; Ehrenstein, 2012; Sullivan, 2013) and include not just crimes but violations of norms of behavior (e.g., spam or offensive postings).

The Internet should not be an effective space for lawbreakers *(Lawful)* An Internet ecosystem that cannot regulate illegal activities will make the Internet less *Trustworthy* and hinder *Innovation*, impeding the role of the Internet as a *General* and *Unified* platform. Generally, crime is a drag on the economy, and a symptom of erosion of civic character, but much of today's cybercrime is international, and there is significant variation in what different countries consider illegal, and there are inconsistent and in some jurisdictions poor tools for pursuing lawless behavior internationally.

The Internet should not raise concerns about national security *(National Security)* While small-scale intrusions, crimes, and attacks may alarm and deter users, a large-scale attack might disable large portions of the Internet, or critical systems that run over it. There are legitimate fears that the Internet could be a vector for an attack on other critical infrastructure, such as our power or water supply.

The Center for Strategic and International Studies maintains a public list of cyber events with national security implications (Lewis, 2014). A few attacks have risen to the level of national security concerns, but they are hard to categorize.

Finally, of course, improvements to specific aspects of security may be in conflict, such as the tension between surveillance and privacy.

Architectural relevance The reader is referred to chapter 10 for a discussion of the issues relating architecture to security.

The user-centered framing of the *Trustworthy* aspiration brings into focus the issue of privacy, which relates to the confidentiality component of the CIA triad in communications security but is not emphasized in chapter 10. Privacy can either be consistent with or at odds with security, depending on the aspect of security under consideration. It is consistent with the prevention of attacks on communication, makes dealing with attacks by one host on another harder, and may be at odds with some aspects of national security. Decisions as to whether (and to what extent) an architecture should favor privacy over accountability are potentially architectural, and certainly not value-free. There are proposals to reengineer the Internet in a more trustworthy direction, such as ensuring that every user's identity is robustly known at all times (Landwehr, 2009; McConnell, 2010). These proposals raise fundamental concerns, such as loss of privacy, ease of mass surveillance, and repression of speech (Clark and Landau, 2011).

The Utility Cluster

The Internet should be used by more of the population *(Uptake)* *Uptake* is about getting more people to use the Internet services available to them. As more essential social services migrate to the Internet to increase the efficiency of delivering them, nonusers may be increasingly disadvantaged. This goal seems generally laudable but invites the question of whether further policy intervention is appropriate to convert the nonusers.

Consumers should have choices in their Internet experience *(Choice)* There are many possible sorts of *Choice* in the Internet ecosystem (e.g., choice of broadband access providers or of software in an app store).

Freedom of choice seems central to U.S. policy thinking, but the word *choice* is ill-defined; it is often used as a proxy for some other aspiration, for which choice is either a means or a consequence. Choice is described as a positive consequence of a competitive market. The logic is that competition leads to choice, and consumers will choose wisely, so competition disciplines providers toward offering products and services that consumers prefer.

But choice presents tensions with other aspirations. Given choice, consumers might pick a network that was more regulated, curated, and/or more stable than today's Internet (e.g., Apple's app ecosystem), an outcome aligned with the *Trustworthy* aspiration but less aligned with *Innovation* and *Generality*; they might prefer a network that is totally free of accountability and thus rampant with piracy, which governments and rights-holders would find unacceptable and constrain by other means; or they might prefer a network that has zero cost but limits the selection of applications.

Overall, we found that this aspiration was ambiguous, positioned in multiple contexts, and subject to multiple interpretations as we attempted to reduce it to operational terms.

Internet content should be accessible to all without blocking or censorship *(Unblocked)* This aspiration implies that ISPs and other network operators must not block or hinder access to content, the objective called network neutrality. It also implies that those with power to compel the blocking or removal of content (e.g., governments) should refrain from doing so. Of course, many blocking and censorship actions taken by governments and private-sector actors are legally justified.

This aspiration is not equivalent to the ideal that all information be free: some commercial content may require payment for access, and some content may be illegal to transmit. Rather than describing the relationship between content producers and users, this aspiration describes the role of the Internet in connecting them.

For any region of the globe, the behavior of the Internet should be consistent with and reflect the region's core cultural/political values *(Local values)* Because values differ so much across the globe, this aspiration arguably implies some partitioning of the global Internet, at least in terms of user experience. In the United States, the relevant values would include First Amendment freedoms (speech, association and assembly, establishment and exercise of religion, press, and petition to redress grievances) but with limitations on certain types of speech and expression. Other regions prefer an Internet that safeguards social structure or regime stability. Debate about the desirability of this aspiration is a critical aspect of international policy development.

The Internet should promote universal social and political values *(Universal values)* This aspiration implies the existence of universal values, such as those articulated in the United Nations charter or the Universal Declaration of Human Rights (UDHR) (United Nations, 1948), namely peace, freedom, social progress, equal rights, and human dignity (Annan, 2013). Although such values are by no means universally accepted, we can imagine translating them into the Internet, as John Perry Barlow (1996) passionately did, to yield aspirations such as:

- Governments should not restrict the ability of their citizens to interact with people outside their borders, as long as there is no harm to others. The physical world analogue is the universal human right of freedom of movement, either within the state or outside the state, with right of return or to leave permanently (United Nations, 1948).
- People should be allowed to communicate directly with citizens of other states without interference from their government; this is a functional implementation of the global right to free (virtual) assembly and speech.
- The Internet should enable and enhance global interactions (as long as they are not criminal) to foster the exchange of ideas. (But since "criminal" has nation-specific definitions, this aspiration would require a liberal interpretation of acceptable interaction across the globe.)
- The Internet should serve as a forum for an international "marketplace of ideas."

Perhaps as a cyber-manifestation of American exceptionalism, the United States has expressed the view that the technology of cyberspace can be a means to export rather U.S.-centric values we hold as universal (i.e., to change other societies to be more like us). Former secretary of state Hillary Clinton (2011) expressed this policy as follows:

> Two billion people are now online, nearly a third of humankind. We hail from every corner of the world, live under every form of government, and subscribe to every system of beliefs. And increasingly, we are turning to the Internet to conduct important aspects of our lives... the freedoms of expression, assembly, and association online comprise what I've called the freedom to connect. The United States supports this freedom for people everywhere, and we have called on other nations to do the same. Because we want people to have the chance to exercise this freedom.

Other nations take a more inward-facing view of what they want the Internet to do for them.

Architectural relevance Any architecture that defines a general-purpose platform for the creation of services would support this basket of aspirations. The more detailed questions have to do with the degree of generality (e.g., QoS features) and the range of applications. Choice at the ISP level (as opposed to the higher-level service and application layer) seems to relate to the next cluster: economics.

The aspiration that citizens be able to communicate globally does not imply that *all* of the Internet experience needs to be globally available in a consistent form, only that there should be an effective basis for global communication among people (i.e., some tools for discourse and exchange). This aspiration benefits from *Generality*. The alignment of the Internet with *Local Values* has a positive and a negative aspect. The positive aspect is the development of applications that are localized to the language and expectations of different parts of the world, which will drive *Uptake*. Even if the Internet is potentially a platform for global communication, the realistic expectation is that, for most users, most of their experience will be domestic. The negative side of shaping the Internet to local values is censorship. In technical terms, censorship is an attack on a communication between willing parties, but most censors do not describe what they do as a security violation, since they claim the right of law. Mechanisms designed to protect communication from attack will blunt the tools of censorship, whether or not we have sympathy with the motives of a censor.

In the current Internet, this tussle over censorship has played out in a particular way. Rather than trying to examine packet flows and block content in flight, countries have been pressuring major content providers to block delivery at the source, based on the jurisdiction of the recipient. Large content providers have in many cases yielded to this pressure and are providing country-specific filtering of content and search results.

The desire for jurisdiction-specific blocking is not restricted to governments. Providers of commercial content such as music and video usually license such content for consumption on a country-specific basis. They are as anxious as any government to regulate access based on the country of the recipient.

This current state of affairs raises a specific value-laden decision for an internet—should the design make it easy or hard to determine the country (legal jurisdiction) of a particular user? The Internet today supports this capability in an approximate way, since most IP addresses are assigned within a country. Of course, informed clients today are defeating this jurisdictional binding by using VPNs and other sorts of tunnels, which is causing censors to block those tools. Most of the actors concerned with access control have accepted this approximation as adequate, but if a new Internet were proposed, one option would be that addresses *always* be assigned on a per-country basis, which would make country-specific regulation of access to content more robust.

An alternative would be to require that requests for content include some sort of "credential of citizenship." This approach seems highly problematic for a number of reasons, including the obvious evasion, which would be to borrow a credential from a person in a different country. Additionally, a country could revoke the right of a citizen to retrieve content by revoking his credential (sort of like revoking a passport perhaps), which seems like a risky allocation of power to the state. However, architectures such as Nebula, which require a distributed control plane negotiation before initiating a data transfer, might be able to embed a certificate of jurisdiction into the PoC in a nonforgeable way.

On the other hand, an architecture could escalate the tussle by making it *harder* to determine the jurisdiction of origin for a query, and see how the adversaries respond. This is what NDN does, where the interest packet carries the name of the content being sought but not the address of the requester, thus making it impossible for the provider of content to determine the jurisdiction of the sender from the interest packet received, unless the provider requires the requester to include some sort of certificate.

To date, countries have been willing to take rather drastic action, including blocking a whole content provider as a consequence of one unacceptable piece of content hosted there. This is a space where any architectural decision will be heavily value-driven. We have argued, in the context of individual accountability, that identity at the individual level should *not* be a part of the architecture. We are less clear about an architectural binding of an internet end point to a jurisdiction. Embedding jurisdiction into the architecture seems to imply that the ISPs would have a role in validating it. One concern is that there may be other sorts of credentials that services may want from clients (such as their age group). A scheme for embedding credentials into the architecture should allow for arbitrary credentials, but ISPs cannot be expected to issue or validate them.

Political science argues that avoidance of escalation is an important issue in international relations. The sorts of arms races we see today (with encryption blocking of VPNs, tunnels, and whole sites) signals that designers today are in an escalatory frame of mind when they design mechanisms. Perhaps, in meeting the needs of society, we need to think about political compromise and not confrontation and escalation when we make value-laden architectural decisions.

15 Looking to the Future

This chapter might deserve a warning label: "Highly Speculative." I said earlier that all architectures are creations of their time and address different requirements as times change. In this chapter, I am going to offer my own views about future requirements (since, to be durable, an architecture should be forward looking), and then I am going to offer my own thoughts as to what a suitable architecture might be. However, these ideas have never been prototyped or otherwise tested by fleshing them out into a complete design, so this chapter might best be read as a possible research agenda for someone who wants to explore some new ideas.

Drivers of Change

What causes network requirements to change over time? In my view, it is the interplay among three important drivers: new developments in network and computer technology, new approaches to application design, and changing requirements in the larger context in which the Internet sits—all the issues outlined in chapter 14.

Within the space of technology, both network technology and end-node technology can drive change. The Internet was designed to hook end-node computing devices together, so as those devices evolve, so will the network. The emergence of the personal computer changed our perspective as designers—we realized that for the Internet to be successful it must connect not a few hundred thousand time-sharing systems but a few hundred million end nodes. The PC also drove innovation in network technology—the PC and LANs coevolved.

Currently, there are important technology trends both at the high end and at the low end. At the high end, we see massive computing complexes that go under the name of "cloud computing." I think that the word *cloud* is a bit misleading—it tends to imply something amorphous and indefinite in form. Cloud computing, in its physical manifestation, is anything but indefinite in form—cloud computing platforms can be buildings the size of a football field, housing hundreds of thousands of processors, and drawing (and dissipating) megawatts of power. Just as the PC drove the development of LANs, massive data centers are driving the development of new networks suited for interconnection within a data center. But I will be bold and assert that at this point I have not seen the requirements for low-cost, high-performance data-center networking motivating the need for new internet *architecture*. I stand ready to be proven wrong.

At the low end, the current dominant end node is the smartphone, and the future may be the class of device called (infelicitously, in my view) the Internet of Things, or IoT. IoT devices are small, fixed-function devices that operate autonomously, without direct supervision by a person. This class of behavior has also been called embedded computing, sensor nets, and machine to machine (M2M) communication. The range of devices that are lumped together under the heading of IoT is highly diverse, from "things" that are fully capable of being part of a global internet to things that use specialized networks that require a dedicated set of protocols. These devices have requirements for network management that are not addressed by the Internet of today, but I don't think we will see a new, global Internet designed to hook IoT-class devices together; clusters of devices will instead be connected to the global internet by some sort of gateway element. Again, I stand ready to be proven wrong. In particular, a new proposal for an internet that does a better job at network management might be a better basis for the IoT future.

In my view, the most important driver of change is the evolution in how applications are designed. Applications today often have a much more complex structure than applications of even 10 years ago, in part enabled by new technology options, but at the same time, today's applications benefit from a much richer development infrastructure—the ecosystem within which the Internet sits. The early designers of the Internet (at least speaking for myself) had a simple conception of the ecosystem. There was the Internet, there were attached computers, and that was it. All an application developer had as a platform for development was the Internet (often using TCP) and (starting in the mid-1980s) the DNS. Everything else was up to the developer. Then the ecosystem got more complex. The invention of the Web transformed the way applications were built. The browser, together with standards for formatting content, provided the application designer with a prebuilt user interface. Then, about 10 years later (in the late 1990s), CDNs were first deployed, and the application designer now had a prebuilt platform for content replication. Today, adding cloud computing to the ecosystem, the joke (and it is not just a joke) is that a developer can build an application in a day. These advances in the ecosystem are enabled by new technology but are the consequence of innovative thinking about how to turn technology into service capabilities such as CDNs.

As I sketched in chapter 2, the application that has defined the Internet for the user has changed over the decades. In the beginning, to use the Internet was to have email. The next dominant application was the Web. Now two applications seem to define the Internet experience for most users: streaming video and social media—the hybrid of communication and content.

Every so often, someone asks me what the next "killer app" will be. Is there something that will come after video and social media that will redefine what the Internet is to its users? One way to explore the future of applications is to ask what industries remain that the Internet has not disrupted. The list of disrupted industries is long: telephone, music, television, almost every sort of merchandizing, starting with bookselling, and many others. Another way to answer that question is to look to the broader research and development agenda in computer science to see what new capabilities are coming. Advanced machine learning is bringing (for better or worse) autonomous vehicles, speech and image recognition, and behavioral profiling. What else will it bring? Cheap video cameras are bringing us into the era of total surveillance, enhanced by ubiquitous facial recognition. The range of hopes and fears for artificial intelligence spans almost every aspect of society, and then there is the vision of the cyborg, the man-machine hybrid.

My own bet for the next "killer app" (at the risk of being proved totally off base) is immersive, multiperson, interactive virtual reality, or it might be swarms of autonomous vehicles.

Whatever your views about these futures (perhaps they make you want to drop this book and go read science fiction, as I advised in chapter 2), I think the question for network designers is not exactly what the next application will be but how applications will be designed in the future and what demands they will place on the capabilities of the network. Video delivery has pushed the network in the direction of massive capacity, which has required a lot of investment and given application designers a lot of capacity to dream about using in new ways. Video has also pushed the larger ecosystem in particular directions, with widely deployed CDNs using complicated schemes to optimize delivery.

It makes a lot of sense to design complex algorithms to pick the best server from which to deliver video. The efficiency of the resulting transfer can more than compensate for any increase in latency before the transfer itself starts, but imagine if a different class of application emerges in the future, perhaps driven by IoT, where the dominant pattern of communication consists of large numbers of very small data transfers. In this case, the design factor that will be critical to overall performance is not the speed of the transfer but the delay before it can start. Either a future internet (and the ecosystem within which it sits) must offer a single delivery model that has sufficient generality to deal with bulk data transfer (e.g., large video files) as well as single packet exchanges or that future network will need to offer several delivery services suited to different classes of applications.

I will make another prediction. We are at a transition at which the future advances in network capability will be not higher access speeds but instead more diverse delivery services. The next measures of utility will be resilience, security, availability, and consistency of service delivery. Autonomous vehicles might benefit from feeds of information about their environment, but they cannot depend on that kind of information if the network drops out unpredictably. To the extent that we are always online, we will become increasingly dependent on reliable connectivity; the network always needs to be there. What this means for network design (aside from continued massive investment in pervasive wireless connectivity) are developments such as giving the end node multiple paths into the network (*multihoming*), adding more resilience and replication to applications, and including support in the network for application resilience as well as network resilience. More predictable packet delivery services, going beyond best effort, will enable new sorts of applications that depend on low and predictable latency. Networks are stuck with the speed of light but should try to approximate that speed in their delivery.

What the network needs to do is facilitate the growth of a richer application development ecosystem, but it should not try to do the whole job. As the application development ecosystem does more, the network can do less as part of an overall solution to application support.

Lessons from Research on Alternative Network Design

There are good ideas and lessons in many of the projects I have discussed in this book. Here are some that were particularly relevant for me:

- NDN goes against conventional wisdom and argues that routers that maintain per-packet state as part of forwarding are practical, which in turn suggests that designers should think about the most useful ways to exploit that state.

- NDN also argues that a network need not have routable addresses that map to end nodes. Packet delivery can be realized using only high-level names and state in routers. In the language of chapter 6, NDN has radically different expressive power.
- Several proposals illustrate that there can be different classes of packets, with different kinds of addresses forwarded using different delivery services. TRIAD, DONA, and XIA illustrate that not all routers need to participate in all the delivery services. A subset of routers can implement one of the schemes (the more costly scheme in these proposals). These architectures exploit the more complex scheme during a setup phase, when the issues of performance are very different.
- XIA specifically argues that a single packet can carry multiple types of addresses so as to invoke more than one kind of delivery service.
- MF argues that it is practical for a router to query a name resolution server as part of forwarding a packet. This structure shows up in other contexts: routers in SDNs also query a server to obtain forwarding information. What is distinctive about the GNS is that the names are of global scope.
- Metanet and FII (among others) make the case for a clear specification of network services. The FII proposal talks about an application program interface, or API, but the details of the interface are not what matters. What matters is the specification of the service—what the delivery service does. In the Internet, the single best-effort service is so weakly specified and highly variable that there is no reason to describe it explicitly. However, if an internet offers more complex services, the user (or the application builder) will need to know what they do in more detail. The specification of these services needs to be globally consistent (so applications can invoke the same service from different locations).
- Nebula demonstrates that robust source routes are possible (perhaps at a considerable cost in complexity and overhead). These routes are robust in the sense that if the routers are trustworthy, the packets will only go where the sender intended them to go.
- In contrast, NewArch teaches that a network architecture need not include any sort of global identifier. If there is a valid locator, management of higher-level identifiers or names can be moved outside the architecture to higher layers. In that case, there can be multiple naming schemes managed by a different set of actors rather than one managed by the ISPs. Mobility (in general) requires decoupling of locators from identity, but solving mobility does not require that the network understand how identity is managed.

- Role-Based Architecture (RBA), part of the NewArch project, teaches that network protocols do not have to be layered.

Some More General Insights about Architecture

Looking across those lessons, and the material in the previous chapters, I offer a few more general insights about architectural design.

Envision the outcome of success Designers of a new architecture should not just focus on the task of forwarding data. That is too narrow an ambition; a new method for delivering packets will not by itself be transformative (unless perhaps it was orders of magnitude cheaper). Reshaping the larger ecosystem in beneficial ways is transformative—perhaps increasing the incentives for the different actors, creating new actors, or making application design easier, for example. Designers must think broadly about new proposals and envision the overall consequences of their proposal once it is successful and mature. They should describe the ecosystem, not just their own system.

As the ecosystem does more, the network has less to do I conclude, based on my preference for minimality, that an internet architecture should not solve more of the problem than necessary. However, from the point of view of an innovator (think about the research groups in the FIA program), it may seem alarming to develop a partial solution and hope that the ecosystem will emerge to complete it. Such a design puts success in the hands of others.[1] The temptation is to design a system that is complete in itself. Such a solution may be easier to start but harder to grow. To increase the success of a new idea, it may be necessary to prototype both the idea and key parts of the ecosystem within which it is expected to live.

The future is not just the Internet That comment, of course, begs the question of what the Internet is, but my specific point is that in the initial design era, our high-level objective was a single, globally interconnected network for everyone. There were other networks built of Internet technology, for corporations, the military, and others, but the compelling vision was the globally interconnected world, and that required (so we thought)

1. There is a quotation that is attributed to a number of people, including at least two presidents: "To have a child is to give fate a hostage." Innovators have the same feeling about their conceptions.

a globally connected network. I am an optimist, so I see that global vision as still viable, even though it is degraded in parts of the world—for example, China blocks many applications, and other countries limit the Internet experience. But as the Internet ecosystem has expanded, the creation of the global experience no longer depends solely on a single interconnected Internet. Today, there are other global networks that are based on the same Internet technology but not directly interconnected with what we think of as the public Internet. These other networks have become part of the application development ecosystem, along with cloud computing, CDNs, and the like. For example, cloud providers use these networks to reach their enterprise customers, thereby shielding that traffic from various attacks and fluctuations in performance.

In this richer ecosystem with multiple networks, what is the role of that part that we call the global Internet? One role is individual access—the means by which people connect. Another role is peer-to-peer interaction—applications that do not depend on the cloud or centralized services. While most applications today are not peer-to-peer, preserving this capability is critical as an option for continued innovation and diversity in the Internet experience.

The implications of these multiple networks on network architecture are not yet clear to me. Since these networks exist in isolation from each other, an architecture need not specify how they interact. I think the architectural answer is again my minimality argument—perhaps with these multiple networks, architecture can do less. Perhaps some problems related to security, resilience, or system overload can be solved just by isolation rather than by a complex mechanism. (This sounds pretty abstract, but I give a specific example later in the chapter, when I discuss the internal management of CDNs.)

Customary business relationships are as ossifying as network protocols
The lesson I learned from the failure of QoS and multicast in the public Internet was that if a technical innovation requires coordination among multiple parties, the designers of the innovation must design a commercial framework (such as money routing) as well as technology. Individual firms in a commercial ecosystem cannot design a money-routing framework—they are takers rather than makers of the economic system in which they sit.

The best lesson about money routing and business relationships comes from another part of the ecosystem, which I touched on in chapter 8, on naming. The holders of intellectual property rights have a single-minded focus on money routing. The naming schemes I listed in chapter 8 allow them to track items of content and the identity of the various rights-holders. The firms in that industry are aggressive competitors when it comes to the production of content, but they cooperate in the design and enforcement of money routing. They define and enforce the financial structure of their industry through trade groups such as the Recording Industry Association of America (RIAA) and collective rights-management organizations such as the American Society of Composers, Authors and Publishers (ASCAP) and Broadcast Music, Inc. (BMI).

The network industry should take a lesson from the content industry. If ISPs want to offer advanced services such as QoS, anycast, or multicast, the industry needs to organize to manage payment flows. It needs a general framework for the collective selling of services. A collective organization (let me call it a Service Fee Management Collective, or SFMC) would also make it easier for ISPs to sell higher-level services such as content caching and delivery.

There are examples of ISPs collectively operating a common undertaking—the Internet exchanges (IXs). Exchanges, especially in Europe, are often run by a membership organization for the collective benefit of the members, who are the ISPs that interconnect there. Of course, an IX is easier to organize than a global money-routing scheme, but if a group of major ISPs put forward a reasonable proposal for an SFMC, other ISPs would agree to become part of it. But just like a technical architecture, a payment scheme will have to evolve as it matures, and good governance (and legal advice) will be necessary for success.

Consider the trade-off between network performance and architectural complexity This question sounds very low-level compared to the points mentioned earlier, but it illustrates a trade-off in the design of a network layer—doing more or doing less. As a specific example, consider the trade-off of removing the delay caused by a round-trip packet exchange at the cost of adding complexity to an architecture. A network can transmit a packet across the United States and back in a little under a tenth of a second. Given the speed of computers today, a tenth of a second is almost forever. A modern computer can execute millions of instructions in that time, and while computers continue to get faster, that round-trip latency will never improve, because it results in large part from the speed of light.

The designers of several of the architectures I have discussed here have worked hard to remove a round-trip delay from the service they provide, specifically in the setup phase. In TRIAD, the setup packet flows through a series of routers that are specialized to handle the DNS-style lookup packet until it reaches the destination, which replies with a packet to complete a TCP connection; one round trip, more or less. In contrast, the current Internet uses the DNS to convert the DNS-style name into an IP address, which the DNS returns to the end point that made the query. That end point then sends a packet to the IP address to initiate a TCP connection. This sequence adds one round trip to the setup phase but in return removes any need for the routers to know about DNS-style names. At the cost of one round trip, the higher-level name is no longer specified by the architecture, which means that different applications can use alternative mechanisms to find the IP address of a target machine, and those services can be provided by independent providers rather than by ISPs.

Several other architectures have eliminated a round trip at the cost of adding complexity to the architecture. MF uses the GNS to recompute the binding from GUID to network in order to redirect a packet as it is being forwarded. If instead the network sent an error message back to the source and the source looked up the binding, this would add about one round trip to the latency caused by the movement of a mobile host. Sometimes the trade-off is more complex. i3 forwards the setup packet through its DHT and then directly to the final destination, while DOA uses the DHT to look up the address of the final destination, which it sends back to the source, again at a cost of one round trip. However, every packet in i3 then incurs the extra delay (with luck, less than a full round trip) of going to the destination via the i3 node where the trigger was stored, while DOA sends the packets directly from the source to the destination.

My view is that saving one round trip in the case of a setup packet or other special circumstance (for example, the movement of a mobile device) is not justified if the result forces a complex mechanism into the network architecture that could otherwise be separated and performed in the larger ecosystem. My view is shaped in part by looking at the larger context within which connections are set up in the Internet today. Using the Web as an example, there can be many round-trip delays before a page is downloaded and displayed. The browser has to resolve the DNS name in the URL into an address, but after that, there may be many embedded components in the page with additional URLs to resolve, and there may be additional steps related to authorization, behavioral tracking, and the like.

The performance objective for retrieving a Web page should be to reduce the latency of the overall transfer, but the internet architecture should not own the entire task. To reduce the latency associated with a DNS lookup, DNS servers can cache the results of previous queries; in this case, the lookup of a DNS name may still take a round trip, but the server that responds with the answer may only be a few milliseconds away, not across the country. Replication of DNS information, content, or anything else at multiple points around the Internet to shorten the distance between client and server may be the better way to reduce the latency caused by round-trip interactions.

In contrast, to return to my hypothetical future dominated by massive numbers of small transfers among IoT-class devices, the overall performance may depend on the latency before the (small) transfer can start, which implies greater attention to removing excess round-trip interactions at the beginning of the transfer. This goal will have implications for naming, security (is a challenge-response authentication protocol necessary?), resilience, and other areas.

Thinking about Design: Packet Delivery Services

In this section, I am going to draw on lessons and insights from the various proposals I have discussed and weave them into yet another proposal for a network architecture. This is a risky undertaking–I risk offending my friends by using only parts of their ideas, and I risk being foolish if upon further inspection the proposal turns out to be flawed. But I want to go through an exercise that talks about an architecture in the context of its ecosystem, the importance of management and control, and an architecture that shows the role of money routing. At this point, my book will get even more opinionated. The reader is warned. My analysis is also going to get somewhat technical and detailed.

Any proposal for a new network architecture has to start with a discussion of packet delivery services and their relationship to requirements. While the discussion must end up talking about the resulting ecosystem, the essential function of a network is to move data. A running theme throughout this book has been the weakly specified best-effort delivery service of the current Internet. The success of this simple service is indisputable. At the same time, there have been a number of efforts, starting in the 1990s, to develop other services, such as multicast and QoS. These services have failed in the public Internet but are being used in other IP-based networks, which suggests that they might have utility in the public Internet as well. Many of the proposals in chapter 7 also define new packet delivery services. Some, like NDN, offer one radically different service; others, like XIA, offer several services.

My current view is that an internet that offers a richer set of services could better support a wider range of applications and generate more revenues for ISPs, thus providing an increased incentive to invest. To support that view, I have to speculate on what those services might be. The answer can only come from looking at the layers above (learning from what applications do) and the layer below (predicting the future trends in technology and how those trends may create new opportunities across the layers). Multicast and QoS were inspired by real-time teleconferencing. TRIAD, DONA, and the ICN proposals were inspired by data (content) delivery. XIA had several delivery services inspired by different goals—find a service, find some data, or find a specific computer. There is also a negative design goal—services should not only facilitate legitimate applications but should also prevent or hinder bad behavior. The current Internet, which lets any source send a packet to any destination, has facilitated DDoS attacks and scanning of end nodes to detect vulnerable machines. Perhaps a future internet should be "deny by default": a receiver should consent to receive a packet as a precondition for receiving it.

A digression on the word *service* As I read an early draft of this chapter, I realized that I needed to add a caution to the reader. I am going to overuse the word *service*. When I first encountered the word in the context of networking, I was confused by it. My ISP friends used it all the time, but it seemed to have no specific meaning. To me, technology was tangible, service was vague. I finally got it, after some tutoring. A service is something you sell; it is how you make money. Technology is something you buy; it is how you spend money. When I was talking to people at an ISP about technology, I was telling them about cost. They wanted to talk about profit. We were talking past each other.

In this respect, I am using the word *service* in a way that is consistent with its use in the context of trade. For those who are familiar with the negotiations regarding the creation of the agreement called the General Agreement on Trade in Services, the concept of service is well-understood. What was distinctive here is that, in a layered system, services as well as technology are layered. There is technology at every level, and every level provides a service to the layer above it, so the word *service* is reused at every level. That is also true of the word *technology*, but somehow that was not as puzzling. So in this section I am going to talk about a packet delivery service such as anycast being used to support a higher-level service such as cached data delivery. There could be many sorts of packet delivery services used to support many sorts of higher-level services. I will try to be clear.

The next question about packet delivery services is what functions should be designed into an internet and what functions should be delegated to the larger ecosystem in which it sits. As we design packet delivery services, where should the division of responsibility be placed? My preference for architectural minimality shapes my view here.

The Utility of Information-centric Networking

The premise of ICN is that an internet should use the names of data as a basis for packet delivery. This idea has attracted much attention as part of the recent work on alternative future internets. At a high level, the justification for this approach is that what a user (or an application acting on behalf of a user) wants the network to do is move data, so why not make the identity of data part of the architecture?

Currently, I am not convinced that there is strong justification for making names for data part of an internet architecture. A number of considerations, many of which I mentioned in chapter 8, lead me to this (probably contentious) conclusion.

The word *data* describes a wide range of things. Some data is transient, some is persistent, some is generated on demand, some is replicated, and so on. The data that makes up a teleconference originates at the locations of the participants, while the data for a Web site may come from a CDN (in which case there would be lots of potential locations from which to obtain it). Putting names for data into the network layer requires the specification of a single naming scheme and, more importantly, a single scheme for binding those names to locations. While designers could propose a very flexible naming/binding scheme (as NDN has done), I worry that a single scheme will prove too constraining to cover all the different patterns of data transfer.

The issue of scale is a fundamental challenge to forwarding based on names. As I discussed in chapter 8, there might be billions (perhaps trillions) of data objects on an internet. I think it is impractical to use flat names for those billions of data objects and create and manage a binding from each of these names to a current location, as in DONA (see page 115). That seems to imply that the names must be structured or organized in ways that give some guidance regarding the location of the data. However, names that are designed to hint at location are not likely to be useful for other purposes, such as providing long-term identity for an object (independent of where it is) or serving to validate the authenticity of the data. Names that are structured to give guidance about location are perhaps more like locators in disguise. The fundamental conundrum for a system like NDN, which routes an interest packet to a location where the data can be found, is that the data may be at many locations across the network, but the name, to be practical, must include some indication as to the location of the data. I like the idea of using a location hint as part of a packet delivery service, but a hint is not a name.

There has to be an explicit reason to add some mechanism to an internet design. If there is no positive reason to include it, then specify its requirements and let other parts of the system implement it. One justification for making names for data part of an internet architecture is data caching. NDN, as well as some other schemes, puts data names into the network layer so that the network can cache data (or the packets that make up the data, in NDN) and deliver the requested data from the cache. However, I am not convinced of the benefit of this idea in practice. First, much data today is commercial content, protected by rights-management schemes that require authentication and authorization as part of accessing it. Content providers are going to object strongly to a system that allows anyone to retrieve a copy of most content. Second, I have not seen a convincing analysis showing that a cache of reasonable size will prove effective in delivering data from the cache. Third, content delivery is now being done in an effective way by CDNs, which replicate content at many points close to the user and deal with rights-management issues, content customization, and many other advanced services, as well as simple caching. I posed the design rule that the more the ecosystem does, the less the network layer should do. By that rule, content caching as part of a packet delivery service is not important.

Another possible reason to make names for data objects part of an internet architecture is so the network can verify the validity of the named object. When one end point connects to another end point, service, or whatever, the end point must be able to verify that it reached the intended destination rather than being deflected somehow to a malicious substitute. Some sort of verification, probably based on a public-private key pair, should be a standard part of establishing communication, but that does not mean that the network architecture has to specify how this is done. My preference is that identity credentials be visible only to the end points, not visible in the network. Credentials that are designed for certification are likely to be long-lasting and allow for persistent tracking, so I do not think that it is necessary that the actual identifiers used to name services or nodes be self-certifying. That function can be implemented using some other information transmitted between the end points.

A final concern I have with using data names in the network is revelation and balance of power. Putting the names of data into the packet header gives the ISP much power to engage in discriminatory behavior based on exactly what data is being transferred. It also gives much power to third parties (state actors, or ISPs acting as agents of the state) to exercise censorship and blocking. In my view (and design is not value-neutral so values may differ here), revealing the identity of data shifts too much power to the ISP and to third parties with adverse interests.

The Utility of Service-Oriented Networking

In chapter 8, I discussed another packet delivery service, which was delivery to a *service*. XIA includes this concept; one kind of identifier in XIA was a service identifier (SID). A service in this context (remember my warning about the overuse of the word) runs on an end node and performs some higher-level activity or application. A Web server is a service, as is a CDN, a game server, or any other higher-level function that might be implemented. Delivery to a service makes sense when the service is replicated in more than one location and the network takes responsibility for selecting a version of the service that is close or preferable by whatever measure the network specifies. An obvious example of a replicated higher-level service is a CDN or other form of data repository.

I equated this delivery service to *anycast,* but saying "anycast" is not a sufficient specification of what the network-level service should do. The specification must consider issues of scale, performance, function, and cost. With respect to scale, how many higher-level services might exist in the future that could exploit anycast? Providing a Web site is not a service at this level; it is an even higher-level service. (Remember my warning—services are layered.) A Web content provider could subscribe to a content caching and delivery service to distribute its content, so the number of CDNs today is a better estimate than the number of Web sites of how many anycast addresses the system would need to support. Today, there are perhaps hundreds of CDNs, in contrast to millions of Web sites. However, there could also be many other sorts of higher-level services that could exploit anycast. I think a million is a reasonable guess as an upper bound on anycast addresses,[2] so I would propose one million as the design target.

Routing to a million anycast addresses might seem impractical, but BGP is currently computing routes to almost 680,000 different regions of the address space, so maintaining routes to a million anycast addresses would be about the same scale as what the Internet does today. The deeper challenge of anycast routing is not scale but rather designing a routing protocol that is robust to manipulation by malicious regions of the network.

The next question is performance and function. The application designer will want to know the answer to a number of questions in order to utilize

2. I am tired of undershooting when I speculate about the future.

the service. How does the anycast mechanism pick the specific copy of the service? Is it the closest measured in latency? The best based on network load? How consistent will the answer be?

I will briefly defer the question of cost.

Anycast can be used as a building block for a data retrieval service similar to that provided by an ICN. Requesting a data object using anycast would involve sending a packet to an anycast address that identifies an appropriate CDN or similar service, with the name of the desired data object in the body of the packet. The name of the data object no longer has to map to the location, so the name can now be used to realize other functions, such as expressing long-term identity of the data object (independent of which CDN is currently holding it) or self-certification. Different sorts of names can be used to deal with different sorts of data as appropriate. The CDN must locate the data object based on its name, but the CDN does not need to deal with every data name across the internet, just the names for data in the CDN. By dividing the data objects among a number of different CDNs, the problem of binding from name to data object becomes more tractable. A higher-level search tool, which today would return a URL, would return the name of the data object and the anycast address of one or more CDNs that might be able to return the data.

I must challenge myself using my minimality principle. Why put anycast into the network as opposed to using something like the DNS to map a higher-level service name to a nearby location as we do today? Here are some advantages.

First, to pick a copy of a service that is closest to a requester, it is necessary to know where the requester is. The network may have better knowledge of this than the DNS does. (For example, in NDN, where there is no concept of an end-point address, there is no way an independent service like a DNS can know where the requester is.)

Second, services such as CDNs are not simple. Not every server that is part of a CDN will necessarily contain the same content. A CDN may have to forward a content request from one server to another in order to reach the machine that can actually deliver the content. Ideally, the selection of the final server would be based on information at both the network level and the CDN level. The routing information that the network layer assembles in order to implement anycast delivery, which has to take into account network topology and conditions, may be a useful starting point for an additional service that the network could offer to the CDN, in which it provides current information about the paths from different servers to different locations in the network.

Third, anycast mitigates some malicious behavior, specifically a DDoS attack. A DDoS attack directed at a replicated service is thwarted to some extent just by the replication—an attacker must find enough attack machines to flood all the copies or else concentrate on a few, leaving the others unhindered. But if the binding from name to address is in the DNS, an attacker can, by probing the DNS, find the IP address of all of the copies, which could allow a crafted attack against select machines. If the service machines only share an anycast address, there is no way for an attacker to single out a specific target.

I think the biggest barrier to the use of anycast is tussle. Using CDNs to illustrate my concern, today the operators of CDNs have control over how the CDN name (the DNS name) is mapped to an Internet address, because that binding is implemented in the part of the DNS that the CDN operator controls. Replacing this approach with the use of anycast shifts that control to the ISPs. The operators of CDNs will have to be convinced that with anycast they have not lost some crucial aspect of control to an actor that may not have interests fully aligned with theirs.

The Utility of Per-Packet State

An anycast delivery service can be implemented within a variety of internet designs—the current Internet has a limited anycast service, XIA includes anycast as one of its delivery services, and the NDN content retrieval scheme is implicitly a form of anycast. I am personally intrigued by the power of per-packet state that is demonstrated by NDN, so while I am proposing to reject what seems like a key design objective of NDN—forwarding based on names of data—I think there is great potential for a variant of NDN in which per-packet state (and per-flow state, which is not a part of NDN) is used to forward packets to services rather than delivering data. I am going to describe a variant of NDN that differs in two important ways—it delivers packets to services rather than delivering data, and it implements a range of explicit delivery services, rather than using the retrieval model of NDN, which uses a single delivery service. I illustrate the power of per-packet state by showing how to implement a variant of anycast that I call stickycast.

One of the potential problems with anycast is that different packets sent to the same anycast address may go to different places. In a scheme like NDN, where there is no per-flow state and the routers deal with each interest packet independently, the routers might make different forwarding decisions for different packets. Continuity is important for supporting requirements such as authentication, rights management, and congestion control, so anycast could be extended so that while the initial anycast packet might go to one or another copy of the service, subsequent packets of the interchange would go to the same place. I believe that the mechanisms of NDN are excellent for solving this problem. An initial packet would be anycast to a service by using a delivery mechanism similar to the way NDN delivers an interest packet. The data name would be carried inside the interest packet, but the routers would not use that information for forwarding. The data name could be encrypted using the public key of the CDN. The ISP could no longer cache data in this variant, but as I said earlier, I am not convinced of the utility of data caching.

To support continuity of the packet exchange, use the per-packet state supported by NDN in a different way. As NDN currently operates, routers have to record the interface through which an interest packet arrives in order to route the data packet back. They can optionally cache that data packet, which I suggest is not of much use. There is a different kind of record that a router could make, which is to set up state based on the returning data packet, so the routers have a bidirectional record of the path of the exchange. Subsequent packets in the flow, marked by a temporary identifier, would follow the bidirectional records, creating the new sort of service I call stickycast. This is an example of my suggestion to think about creative ways in which per-packet or per-flow state can be used. (This proposal is overly simple in that it does not address the complexity added by a mobile host. Per-packet state would deliver the data from the server back to a fixed point where the client is attached. If the client moves, the per-packet state has to be established at its new attachment point. The server will have to reestablish the state, which implies that the server must have some higher-level name for the client, perhaps similar to the scheme in MF.)

I deferred the issue of cost. ISPs will have to offer an anycast service for a cost that is lower than what it would cost the CDN provider to implement the same function internally, unless the anycast service does something that is very difficult for the CDN to do. Augmenting the anycast service with an interface through which the CDN can learn about network topology and current conditions might make the anycast service more valuable to the CDN. That is why I would consider adding that complexity to the scheme. Second, multiple ISPs will have to cooperate to implement anycast delivery, so some money-routing scheme will be required. My hypothetical Service Fee Management Collective (SFMC) would be a way to sort out the money routing in this sort of multiprovider delivery service.

The Utility of Multiple Address Spaces

The proposal in the previous subsection is actually too simplistic. Since many CDNs do not store all of their content in every server location, they may need to pass the request to a location where the content is stored (and the server is not overloaded, or some other problem). This means that the servers have to talk among themselves. This requirement is somewhat reminiscent of what i3 and DOA do, where they use a DHT to find the right copy of the service. (I am not convinced that a DHT should be defined as part of the network architecture—that feels too restrictive to me.) The challenge here is how the servers in the CDN can talk to each other if all they have is a shared anycast address. It would seem that, at least to talk among themselves, they need individual addresses. This would mean, first, that the architecture would have to support location-level addresses, and second, that those addresses might be a means to launch an attack against the CDN, thus losing the security benefits of the anycast group.

This problem can be solved in a way that again illustrates how the architecture can do less rather than more. In order to allow the members of a CDN to talk to each other, connect them using addresses from a separate address space. This could be either a virtual private network (again organized in the context of my SFMC) or one of the separate global networks that are now available in the current ecosystem.

End-Point Addresses and Balance of Power

Almost all the schemes in chapter 7 use end-point addressing as one delivery mode. NDN is distinctive in rejecting the necessity of a routable end-point locator. I think end-point identifiers are not harmful, provided they are used in specific ways, and implementing them is an easy task for the network layer—that is what the Internet does today. Adding end-point addresses into the stickycast scheme mentioned earlier yields some very useful communication patterns. Consider this elaboration of my service invocation scheme. First, a requester sends a data request to a CDN by using anycast. The packet is sent using an NDN-style delivery where there is no source address in the packet header. Instead, the sender encrypts its source address and includes it in the information in the interest packet, along with the name of the desired data. Next, the various nodes in the CDN communicate internally to select the node from which the data will be served. That final node decrypts the address from which the request came and sends an initial message back to that location, which sets up per-flow state in the routers to indicate where the packet came from. That per-flow state means that the message back to the original requester does not need to include the address of the server node. Once that connection has been established (with bidirectional state on the routers along the paths), encrypted data can be exchanged over the connection.[3]

From the perspective of an adverse third party such as a content censor, almost nothing is revealed by this sequence. The initial packet does not contain the address of the requester in a visible form and is directed to an anycast address. The censor can see the anycast address, so he can learn what higher-level service is being contacted, but that is all. A short time later, a connection to an end point will be made, identified by its address, but there is no indication of what service is initiating this request. There is no direct link between the original request and this subsequent connection. In terms of hiding information, this sequence reveals far less than any scheme based on data names.

Selecting the Delivery Service

If an internet offers multiple delivery services, which actor gets to pick the service? Several of the proposals I have described give that decision to the network, either explicitly or implicitly. For example, in the case of a multihomed host, the RINA scheme said that the end-point address was associated with the end point as a whole, not the interface on the end point, and the network would pick the best path to use. As another example of assigning to the network the task of selecting the delivery service, NDN nominally has only one delivery service, the exchange of an interest and data packet, but lurking inside NDN are a number of delivery services. For example, there might be local broadcasting for a small region of the network (or as part of solving the "desert island" problem), as well as a global routing scheme. It is the network, rather than the end point, that picks among them.

I think that the end point rather than the network should have control over which delivery service it invokes in the network. Allocating responsibility to the end node shifts power to the end point while allowing for more explicit marketing of service options by the network provider. In the data retrieval scenario I sketched, both the provider of the data and the requester of the data have some control over what service is to be used. The provider of the data would create a high-level name for the data, including some identifier and one or more services from which it can be requested. Those

3. This state is not per-packet state as in NDN but rather per-flow state. However, per-packet state can still be used to limit the rate at which requests are sent and data is received.

services might be anycast, or, if the data is available at just one location, would involve unicast to that location. The requester would normally request the data by sending a packet to one of the services indicated by the provider but is free to try requesting the data from some other service if there is reason to think that service might be able to provide it. Here are a few specialized cases that illustrate why the end point needs to control the delivery service.

The desert island scenario I discussed this challenge earlier. Two people on a desert island have computers (or smartphones or whatever). There is no connection to the outside world. One person has a copy of the *New York Times* on his device, and the other wants to read it. NDN solves this by sending an interest packet that contains the name of the desired data; the other device can respond with the desired data if it chooses.

It is equally possible to solve this problem by using services rather than data names. First, define a new high-level data retrieval service called local data sharing, and give it an anycast address. Second, define a version of the anycast service called local anycast, probably implemented by broadcast scoped in some way. Given these services (which would be defined as general service options, not specific to the desert island), the desert island story would proceed as follows. Any device that is prepared to offer data would implement the local data sharing service and listen for packets sent to its anycast address. A request packet would be sent to this anycast address by using the local anycast delivery service. The name of the desired data would be in the request packet. The machine with the data could respond, just as before.

Commercial ISPs might choose not to offer the local anycast service or might block the local data-sharing anycast address to hinder content piracy. People not on a desert island might request the same data (using the same name) from a different delivery service by using a different variant of the anycast service. Going back to chapter 8, the anycast address of the data delivery service is serving as a hint as to how to find the data, but the sender is free to use another strategy, hint, or service.

The filming of the protest rally scenario This scenario was proposed in the NSF Future Internet Architecture project by some of our social science collaborators.[4] In this scenario, an activist at a protest rally wants

4. As part of the project, the NSF supported an activity called the Values in Design Council, which assembled experts from a number of disciplines to comment on

to video what happens without being detected. Can a network delivery service help to hide this activity? Assume that there are security services monitoring network activity (in particular, wireless network transmission) as well as scanning the crowd for unacceptable behavior. An activist might pre-position a discrete camera that transmits what it is capturing to nearby receivers. The activist is nowhere near the camera, so if it is seized and confiscated, all that is lost is the camera. But the receivers, which might be smartphones in the pockets of other activists, want to avoid detection.

A highly specialized delivery service can help with this challenge (from the perspective of the activist, not the security forces—design is not value-neutral). One possible design is one-way transmission, with no acknowledgments sent back from the receivers. Several receivers may silently capture what they can, and then later (in a safer location), with luck, the data can be reassembled to form a complete record. In this mode of delivery, there is an opportunity to exploit a cross-layer optimization. Some wireless systems send messages as part of their control architecture, for example, to announce their presence. Crafty software on the video camera could listen to these messages, which have nothing to do with the reception of the video, to detect when there are potential receivers within range and what the signal quality is likely to be. The delivery service can adjust its behavior accordingly, buffering content until a receiver gets close or until content has been sent enough times that with high probability at least one copy has been received successfully.

While a commercial ISP would not be likely to offer this service, it is a network-level service because it needs access to low-level information about the operation of the radio channel. The service that reassembles the video from whatever was captured is a higher-level service. It has to be managed by humans, since the smartphones holding the captured video fragments cannot determine when it is safe to transmit the fragments for reassembly.

Cross-layer optimizations The activist scenario illustrates that some technologies, such as wireless, may have features such as broadcast that help with certain delivery services. One way to allow these cross-layer optimizations in an internet that strives for generality and the ability to work across a wide range of communication technologies is to package them as specific delivery services. The network has to include, as part of its interface,

---

our technical work. It has two Web sites that describe their work, at http://www.nyu.edu/projects/nissenbaum/vid/vidcouncil.html and https://valuesindesign.wordpress.com/initiatives-2/values-in-design-council-next-phase/.

a way to ask what services are available in a particular context. If a network offers multiple delivery services, it would be ideal if all of them were always available, but to mix generality with local optimization, it is also important that the application be able to ask which are available and how well they are working. The designers of XIA have demonstrated the use of delivery services with only local scope, as well as services that are designed globally.

Multicast and QoS These commercial failures still represent opportunities as enhanced services in an internet. The technologies are already in use in other IP-based networks, so they work and are valuable. They will enter the market in one of two ways. Either ISPs will figure out how to collaborate in providing them over the global Internet (using an SFMC to share revenues) or the business will migrate to networks with global reach that are operated by one provider.

Source-controlled multicast The original conception of multicast, as proposed by Deering and Cheriton (1990) (see page 24 and the Appendix), was that receivers of a multicast stream initiated the connection to the stream by sending a message requesting that the packets being multicast be sent to them. In this conception, anyone could join a multicast group, and the sender did not know what set of end points were actually receiving the packets. This approach made sense in a simple model of data delivery where there were no access controls on who could receive what.[5]

An alternative conception of how to implement multicast is that the sender initiates the attachment of a new receiver to the multicast distribution tree. I described how a CDN node could initiate a connection to a receiver by using an NDN-like state setup in the routers along the path from the CDN node to the location of the receiver. That mechanism can be used to set up multicast forwarding state as well as unicast forwarding state. Allowing the sender rather than the receiver to add a new receiver to a multicast gives the sender more control over authorization to receive, for example.[6]

5. Deering believed that encryption was the right way to prevent unauthorized receivers from exploiting content, rather than controlling which nodes could join the multicast receiver group, but this approach requires that all authorized receivers share a decryption key, which is not a robust way to handle keys.
6. Assuming that the routers are trustworthy and do not themselves replicate the packets to unauthorized receivers.

Business risk Based on my experience with the Internet today, the barrier to the introduction of multiple services is not technical—that is what XIA is demonstrating. The barrier will be the inability of the ISPs to work out among themselves how to manage and collect revenues for these services. The management problem arises because as the services become better specified, the user will expect the ISPs to meet the specified service commitment, which will require a degree of cooperation among the ISPs that they, as competitors, will find difficult. Consider the case where there is a failure to deliver the expected service and the ISPs must localize among themselves which of them caused the failure. The issue of collecting and sharing revenues is the challenge that I proposed to address with my Service Fee Management Collective. Without some collective institution like that, I fear that the coordination problems associated with offering advanced services will prove insurmountable.

NDN has a single service that is very clever and flexible. It can provide unicast communication, a form of anycast, and a form of multicast, depending only on how the names of data are assigned and how the end nodes use those names. Given the complexities of offering multiple services, ISPs might prefer an architecture that has a single service, even though the resulting network might be less flexible and not suited for as broad a range of applications.

The Dark Side

Network delivery services must have the dual goals of supporting desirable behavior and hindering undesirable behavior. Chapter 10 discussed a number of security problems and how architecture might contribute to their solution, but the problem that most required a solution at the network level was DDos attacks.

Blocking DDoS attacks A successful DDoS attack depends on the ability of the attacker to launch traffic from many sources toward a target so as to overwhelm it. Chapter 10 has a discussion of a number of ways to impair DDoS attacks, from traceback to the SUM. However, to the extent that delivery services can hinder DDoS attacks by design, so much the better. The anycast scheme, where individual servers do not have distinguishable addresses, helps to diffuse any DDoS attack across the servers. End-node addresses that are not routed globally limit the range of machines that can attack them. NDN-style communication, with state in the routers, can better provide a "deny by default" semantics where the receiver has to provide some indication that it is prepared to receive traffic.

Not all machines are equally likely to suffer from DDoS attacks. Today, servers are much more likely targets than individual end points. A useful threat analysis would consider whether end nodes (like devices in the home) need to be protected from DDoS attacks. To facilitate innovation in new applications and services, it may be a good idea to have two sorts of locations: those that are protected in some way (perhaps by the use of anycast addresses and a server-initiated setup for stickycast) and those that are prepared to risk receiving whatever is sent. If we accept that individual end points do not accrue a high risk by having a known location, this allows end points to experiment with peer-to-peer protocols and allow state setup from a server back to an end-point location. One way an end node can reduce the probability of attack is to change its address from time to time. Of course, when a machine changes its address, it needs to tell other nodes what the new address is so they can continue to find it, but this is the same problem faced by the mobile host.

Incentives If ISPs are going to offer multiple services, there is a risk that they will abuse the capability rather than use it to support a wider range of applications. If various services differ in their quality but have the same general pattern of delivery, ISPs may be tempted to let the basic delivery service degrade to the point where all the users are forced to pay for the enhanced service. One of the reasons for resistance to mechanisms for QoS is the fear of this outcome. However, many of the services I have proposed here are differentiated not by quality but by a different delivery function. Unicast, anycast, multicast and so on are different, not better or worse. There is less risk of abuse if ISPs are offering services that differ in their basic characteristics rather than on the basis of quality.

Thinking about Design: Control and Management

The previous discussion focused on data transfer. Issues related to management and control are equally important in shaping architectural decisions, but there are fewer lessons to draw on from the designs I have described, which reflects a bias in the research community toward focusing on moving data and not on the necessary supporting services. The most important insight from chapter 13 is that different control and management functions have different natural scopes, and an architectural design based on strict layering will not cope with the need for flexible scoping—flexibility to define how different aspects of the expressive power can be used (or protected from use) by different mechanisms across different scopes.

The proposal that comes closest to addressing this challenge is the component of the NewArch project called Role-Based Architecture (RBA), which was discussed on page 94 in chapter 7. In RBA, the different components of the expressive power in the packet were associated with roles, and the relationship among roles was not nested or layered but instead arbitrary. The role-specific headers (RSHs) could be processed in any order and encrypted or deleted as appropriate, for example.

The second insight about management and control is that the per-packet (or per-flow) state of the sort exploited in NDN provides a new set of tools for rethinking key management challenges. Going back to the discussion about management in chapter 13, here are some thoughts about architectural support for the different challenges.

Performance The congestion control schemes by Katabi et al. (2002) and Dukkipati (2008) that I described in chapter 13 went to great effort to avoid creating per-flow state in the router. They added new expressive power to the packet (a window or rate parameter) but continued to assume that routers had no per-flow or per-packet state. If routers have more state, this is an opportunity to rethink congestion control yet again.

Fault isolation In my view, this objective is one of the major open questions in network design. How can an architecture add expressive power (and enhanced PHBs) such that the normal flow of data packets can be used as a diagnostic tool to detect and localize failures and impairments? Again, I speculate that the use of per-packet and per-flow state in routers can be a building block for new schemes, but the design remains a challenge.

Configuration There is very little support in today's Internet for device configuration. The dynamic host configuration protocol (DHCP) is the exception—when a device first attaches to the Internet, it requests key configuration parameters by sending a DCHP request. It gets in response a message with its IP address, the address of a nearby router, and the address of a nearby DNS server. These parameters are sufficient to get a device started as an Internet end node. With today's devices, the user then continues with the manual installation and configuration of applications. This pattern will have to be rethought in the era of IoT devices. IoT devices are not general-purpose computing platforms but fixed-function devices. They come with their intended application code preinstalled, but that code will, in general, still need configuration. However, these devices may lack a display or keyboard, and there may be no straightforward method to configure them.

The right way of thinking about this is that configuration is a service. DHCP was designed as a special-purpose protocol but is just a simple query being sent using some sort of local anycast or multicast packet delivery service. Application-level configuration can similarly be structured as a query sent to a standard service address requesting the initial information to configure the application-level aspects of an IoT device. Device configuration is an example of a case where there needs to be more support in the larger ecosystem, and the network can provide an enabling service.

Accounting It is an open question as to whether there is any useful expressive power to add to a packet to enhance any sort of accounting and traffic monitoring (such as packet sampling), given the possibility of malicious end points inserting false information into the packet.

The Role of the End Node

Minimality gives more responsibility to the end node, which reflects my bias in the tussle between customer and provider. My preference is to give the end node control over what delivery service is selected, how multihoming is managed, and other aspects of service configuration. The designers of the Internet tended to ignore what happens in the end node, since it was not a part of the network, but what happens in the network needs to be complemented by what happens in the host. We accept this with TCP but have not generalized that understanding. In the overall ecosystem, services are composed to form services, in a repeating pattern. Giving the end node more control over this composition gives it more responsibility but also more power. The end node can be given tools to make the task of service composition easier. Today, most of this happens inside applications, so the activity is more or less invisible. For example, the browser starts with a URL, extracts the DNS name, uses the DNS to resolve the name to an address, and establishes a connection to that address. It is the application in the end node that decides whether to use the DNS; a different application might use a different mechanism for name binding. If the DNS-style names are embedded in the packet transfer service, as in TRIAD and DONA, the two services could not be separated.

I am also calling for the network service to give the end node a more complicated service invocation interface. This provides more opportunity for revenues for the network and again gives the end node more control. The end node can select services that are available in specific circumstances (such as a service that exploits cross-layer optimization with specific technology) or query the network to determine what sort of performance a particular delivery service can offer.

Conclusion

This final chapter is speculative. While I have stated some opinions, I would encourage readers to take these as starting points for discussion and debate. Here is a sample of topics for further discussion:

- Can my concerns about a packet delivery service based on names for data objects be mitigated? These concerns were well-understood by the advocates for that approach.
- How can better specifications be developed for candidate network delivery services? This is both a technical challenge and an organizational challenge.
- Can an anycast packet delivery service be devised that is a success in the market? How should it be designed to provide utility and mitigate fears of tussle?
- Is design of money-routing methods an appropriate activity? Can ISPs overcome the business issues associated with bringing advanced services to an internet?
- How can per-flow and per-packet state be exploited to solve key internet challenges such as mitigation of DDoS attacks?
- How can an internet architecture address the challenge of fault localization and other aspects of network diagnosis?
- Can end nodes be given more control over network services (e.g., routing) without creating new attack vectors?
- Can the community reach a better understanding of layers, scopes, overlays, and tunnels?
- Is it appropriate to build into an internet architecture a strong resistance to censorship, as I proposed in my data delivery scenario, or will this trigger a response by actors with adverse interests that will distort the architecture too much?

My goal with this book is to capture and structure some of the thinking in the field over the last few decades but also to encourage more architectural thinking. As I said in the beginning of the book, architecture is a design discipline, and we learn by experimenting and by studying past experiments. I welcome discussion and debate. I can already see a second edition of this book in the future, perhaps with some very different opinions.

Appendix: Addressing and Forwarding

Introduction

The basic function of a network is to move data. In packet networks (the focus of this book), packets move across the network through a series of routers or switches, which make a decision as to how to forward the packet based on some information in it. That information may take several forms—it can refer to a specific end point, a service, a piece of data, a multicast or anycast group, or anything else specified by the network architecture. Generically, that information is the *address* in the packet.

In chapter 7, I discussed a number of alternative proposals for an internet architecture. However, that chapter by no means discussed all the alternative schemes that have been published. Since the Internet was first designed, there have been any number of proposals to improve or redesign it, and one of the most popular topics of study has been addressing. Here I review some of the literature on alternative addressing schemes and attempts to put these proposals into an organizing framework.

Defining Some Terms—Mechanisms for Forwarding

I defined the term *address* in operational terms—it is that information in the packet that allows the packet to be delivered. This very operational definition masks a history of philosophical discussion on the difference between a name, which provides some form of identity; an address, which provides some sort of guidance to a location; and a route or path, which describes the series of routers through which the packet should pass to reach that location. I refer the interested reader to chapter 8 and to the classic papers in this area (Shoch, 1978; Saltzer, 1982) and a slightly later discussion (Francis, 1994a). I will continue with an operational definition of address.

I will discuss addressing and forwarding, and only peripherally routing. While most architectural proposals define an addressing and forwarding scheme, most leave the development of a routing scheme to a later stage, where the architecture is fleshed out to become a complete system. Routing (as opposed to forwarding) is the process of computing the path to a destination. The routing computation produces forwarding information, which the router uses to direct each packet on its way to the destination. In the case of the Internet, the routing computation runs in each of the routers and produces a forwarding table. When the packet arrives, the router looks up the destination address from the packet in the forwarding table to determine how to forward the packet.

With this information as background, here is a quick review of some general classes of addressing and forwarding schemes.

Destination-based forwarding In the basic Internet architecture, there is a (sort of) global, (usually) distributed routing algorithm that runs in the background and computes, for each router, the correct next hop for every set of destination addresses. When a packet arrives at a router, the forwarding algorithm searches in the forwarding table for an entry that matches that destination address and extracts from that entry the stored information about the next hop the packet must take. Within this general framework, different addressing schemes may lead to more or less efficient routing and forwarding algorithms. By way of example, a flat address space, as in DONA, requires that the routing algorithm keep track of every address, while in an address scheme where the structure of the address matches the topology of the network (provider-based addressing), the routing algorithm need only keep track of groups of addresses that map to different parts of the network. For a very detailed analysis of different forms of addressing, including flat, hierarchical, and many others, see Francis (1994a).

Source routing This alternative has the forwarding information in the packet rather than in the router. In general terms, the packet lists the desired sequence of routers through which the packet should flow. Each router in turn removes its address from the list and then sends the packet onward to the router at the next address. Of course, this oversimplified description begs many questions, such as how the sender gets the source route in the first place. Source routing was proposed and discussed by Farber and Vittal (1973), Sunshine (1977), and Saltzer et al. (1980), and the definition of IP includes source routing options, but the idea did not survive as an operational capability. Variants on source routing form the basis of many alternative proposals for forwarding. The reasons for the practical failure of source routing (so far) make an interesting case study of the set of requirements that addressing must meet. In chapter 7, I mentioned that Nebula and Scion (part of XIA) use source routes, and I will discuss more here.

Label switching Destination-based forwarding, as I described it, has a potentially costly step: searching the forwarding table to find the entry that matches the destination address in the packet. An alternative that avoids the search is to set up state for each flow of packets in every router through which the packet will go. In each router, the forwarding table records the outgoing link on which each flow of packets should be sent, and the index (usually called the label) into the forwarding table in the next switch. At each switch, the incoming label is used to look up the correct forwarding entry and is then rewritten with the label for the next router. This process repeats at each router along the preestablished path. This mechanism is called label rewriting, label switching, or label swapping.

Given that the data in the packet is sometimes a destination address, sometimes a sequence of addresses, and sometimes a label (and sometimes other things), the term *address* may not be the best term for that information. The term *locator* captures a somewhat more general idea—that thing that is in the packet. NewArch used the term *forwarding directive* in this way.

The remainder of this appendix starts with a quick review of addressing in the Internet, since that is a system that many people understand. It then looks at an alternative tradition, which is addressing in virtual circuit networks. With this background, it then catalogs a list of objectives that an addressing and forwarding scheme should meet and describes a number of alternative proposals using this set of objectives.

A History of Internet Addressing

Before the Internet, there was the ARPAnet. The ARPAnet was the first packet technology deployed by DARPA and was the first technology used to carry Internet packets. The addressing/forwarding mechanism was a very simple form of destination-based address. In the first version, each packet carried an 8 bit switch number (the so-called interface message processor, or IMP, number) and a 2 bit host number (so there could be four hosts per IMP). As the ARPAnet grew, more addresses were needed, and the next generation of addresses had 16 bits to select the IMP and 8 bits to select the host. Near the end of life for the ARPAnet, logical (or flat) addressing was added (see RFC 878),[1] consisting of a flat, 16 bit field (two of which are flags), so 2^{14} hosts could be addressed without the sender having to know where the receiving host is. (When a host attached to the ARPAnet, it notified the IMP about what logical names it was going to use, and the IMPs propagated this information around the network in an early form of forwarding on a flat address space.)

The 32 bit addresses in the current Internet were not the first option considered in the early design. The initial paper on TCP (Cerf and Kahn, 1974) proposed an 8 bit network field and a 16 bit host field (called the TCP field), with the comment: "The choice for network identification (8 bits) allows up to 256 distinct networks. This size seems sufficient for the foreseeable future. Similarly, the TCP identifier field permits up to 65 536 distinct TCP's to be addressed, which seems more than sufficient for any given network." The option of a variable-length address field was considered in the subsequent early design discussions in order to allow for growth. Regrettably, the team implementing the first router said that the overhead of processing variable-length header fields would lead to unacceptable performance, and the idea was rejected. The result was the 32 bit fixed-length address. Paul Francis, in an epilogue to his PhD thesis (Francis, 1994a), provided a very thoughtful review of the early Internet design literature and pointed out that the Internet almost had a source-routing scheme and variable-length addresses.

During the 1980s, there were a series of proposals to deal with growth. The original structure of the 32 bit address was 8 bits for the network number and 24 bits for the rest, specified in RFC 760 in 1980. A more flexible *class* structure, with class A addresses having 8 bits for the network, class B having 16 bits, and class C having 24 bits, was defined in 1981 in RFC 791. Even this scheme proved inadequate, and in January 1991, the Internet Activities Board (IAB) held a retreat, the results of which are documented in RFC 1287, which contains the following statements:

> This [addressing and routing] is the most urgent architectural problem, as it is directly involved in the ability of the Internet to continue to grow successfully. (1)
>
> The Internet architecture needs to be able to scale to 10^9 networks. (3)
>
> We should not plan a series of "small" changes to the architecture. We should embark now on a plan that will take us past the exhaustion of the address space. This is a more long-range act of planning than the Internet community has undertaken recently, but the problems of migration will require a long lead time, and

1. Here, I refer to a rather large number of Internet RFCs. I do not include citations for all of them. They can be found at https://www.ietf.org/rfc.html.

> it is hard to see an effective way of dealing with some of the more immediate problems, such as class B exhaustion, in a way that does not by itself take a long time. So, once we embark on a plan of change, it should take us all the way to replacing the current 32-bit global address space. (6)
>
> There will be a need for more than one route from a source to a destination, to permit variation in TOS and policy conformance. This need will be driven both by new applications and by diverse transit services. The source, or an agent acting for the source, must control the selection of the route options. (5)

This report initiated the effort to create IPng, which eventually led to IPv6. Two short-term ideas emerged at that time to bridge the gap to IPng. One was CIDR, or classless addressing, which is described in a series of RFCs published around 1993. The other approach to preservation of IP addresses was to let large enterprises with many internal hosts use private address spaces to address these machines rather than using globally routed addresses. The IESG commissioned a subgroup, called the ROAD group, whose deliberations were documented in RFC 1380 in November 1992. They wrote:

> The following general approaches have been suggested for dealing with the possible exhaustion of the IP address space:
>
> … Addresses which are not globally unique. Several proposed schemes have emerged whereby a host's domain name is globally unique, but its IP address would be unique only within its local routing domain. These schemes usually involve address translating.

This idea is the starting point for the idea of network address translation (NAT) devices, introduced by Tsuchiya and Eng (1993). Private address spaces are further documented in RFC 1597, issued in March 1994.

Multicast

The major intellectual advance in addressing during the 1980s was the specification of IP multicast by Steve Deering. Deering's PhD thesis did not appear until 1992 (Deering, 1992), but the initial proposal had emerged earlier (Cheriton and Deering, 1985; Deering and Cheriton, 1990). In multicast, the destination address is not the location of the destination but rather a handle or pointer, called the multicast ID. Routers use this handle to look up a list of next hops that the packet should take as it fans out to all the destinations. Every router on the path from the source to (all of) the destinations must have the proper state information to map the handle to the correct set of next hops. This requirement led to a large research agenda in the design of multicast routing protocols.

Multicast is important for two reasons. First, of course, it was a proposal for a new class of delivery semantics. But multicast also demonstrated three more general points: that the locator need not be a literal address of a destination but can be an arbitrary bit sequence, that different regions of an address space can be associated with different forwarding mechanisms, and that a router can run more than one routing algorithm at the same time. All of these generalizations are important.

While Deering is generally credited with establishing the concept of multicast in the Internet architecture, there are much earlier mentions of the idea. The Stream protocol ST (Forgie, 1979) introduced the concepts of conference connections, multiaddressing, and special forwarders called replicators. The idea of delivery to multiple destinations was present in the Internet research community even before the standardization of IP.

So in the early part of the 1990s, multicast was proposed as a significant enhancement to the delivery semantics of the original Internet; a long-term effort was launched to replace IP with a new protocol with a larger address field; as a stopgap the format of the 32 bit IP address was changed again from the class structure to a scheme called classless interdomain routing (CIDR); and NAT implied a major deviation from the original addressing architecture. Address translation was supposed to be a short-term fix, but it has persisted, with long-term implications.

The Quest for IPng

The call for proposals to replace IPv4 produced two initial contributions. One was PIP (Francis, 1994b), which represented a significant shift from IP in that it used source routing as its basic forwarding mechanism. The addresses in the source route were not global but instead more compact locators that were only unique within a region of the network. To realize this scheme, PIP organized the nodes of the Internet into a hierarchy, and node names were defined within the hierarchy. The other proposal was SIP (Deering, 1993),[2] a more conservative design that included the concept of source routes but used globally routed names for the intermediate nodes. Discussion within the IPng community led to a compromise called SIPP (SIP plus), which used the syntax of SIP but incorporated some of the advanced semantics of PIP. Hinden (1994) describes SIPP, and Francis (1994a) provides a very detailed comparison of these alternatives.

2. This acronym has nothing to do with session initiation protocol, which came along later and reused the three-letter acronym (TLA).

The size of the address field was, of course, a major motivation for the replacement of IPv4. SIP and SIPP used an address field of 64 bits, twice that of IPv4. The final version of IPv6 has an even bigger address field, of 128 bits, not because 2^{64} was too small to address all the nodes the Internet might someday see but rather to facilitate address space management.

A Parallel Universe—Virtual Circuit Networks

One of the key design principles of the ARPAnet was that it was connection-oriented: the ARPAnet established and maintained paths between all pairs of hosts so that every IMP had state for every possible path from source host to destination host. This state was used to manage resource allocation and congestion. This connection-oriented design philosophy contrasted with the connectionless, or datagram, approach of the Internet. This split in approaches defined two different lines of evolution for packet-switched data networks.

X.25 is a connection-oriented outgrowth of a number of early projects, including ARPAnet and the early work in England on packet networks. The design of X.25 included the concept of virtual circuits between sender and receiver and a signaling protocol to set up these circuits. A user on an X.25 network requested a circuit before sending data and received back (if the setup was successful) a short identifier for this circuit, which is used to tag each packet. The full address used in an X.25 packet thus only showed up in the signaling protocol. The full X.25 address somewhat resembled a phone number. It was a sequence of ASCII digits: 3 for country, 1 for network within the country, and 10 for the end point within the country. The country-based assignment reflects the telephony roots of X.25. X.25 was an interface protocol that described how a host (a piece of data terminal equipment, or DTE) talked to its attachment point in the network (the data communication equipment, or DCE). The X.25 specification did not describe how the switches inside the network talked to each other. There were some interesting demonstrations, such as the use of X.25 as an interface to a datagram network, but there was also a need for vendor standards to design interoperable switches. In most X.25 networks, the representation of an address inside the network was the same as the representation across the interface—a short identifier for the circuit.

X.25 contained rich mechanisms for error detection and recovery. Much of the state established inside the switches had to do with the detection and retransmission of corrupted packets. This function seemed important in the beginning, when circuits were noisy. However, as reliability and performance improved, a simpler connection-oriented scheme, frame relay, was proposed, which traded the occasional loss of a packet for a simplification of the switch. Forwarding and addressing in frame relay was similar to that of X.25. Frame relay networks identified virtual circuits using a 10 bit data link connection identifier (DLCI). This compact coding of the virtual circuit would not work if DLCIs had global meeting—there are not enough values. The DLCI had only local meaning, between each switch and the next. The DLCI was a label, and frame relay is an example of forwarding based on label switching.

Label Switching and Cells

The Internet uses the packet as the unit of statistical multiplexing. Packets from different flows share the capacity of any link over which they pass, and variable arrival patterns mean that the instantaneous load on the link will fluctuate. Transient overloads will lead to queuing, which can potentially affect service quality. Even if the link is not overloaded, if a small voice packet arrives just after a large data packet, the voice packet will be delayed. In contrast, the circuit-switched telephone system committed capacity to each call at the time it was placed. A call might not complete if there was insufficient capacity, but once the call was connected, the voice traffic would not encounter any statistical fluctuations in service quality. However, making this static commitment of capacity wasted bandwidth, since capacity remained dedicated to a specific call even if one or both parties were not talking. Telephone company engineers were interested in the possibility that statistical sharing of bandwidth among calls could be practical, but given the goal of minimizing total latency, they thought that multiplexing packets of variable size would introduce an unacceptable level of variation in delay. So when the designers of the telephone system considered replacing their circuit-switched architecture with a scheme more like packet switching, they devised an alternative multiplexing model in which the unit of multiplexing was not a packet of variable size but a small cell of fixed size.

There is no fundamental reason why a cell could not carry a globally routed destination address, but the engineers felt that the overhead of putting a large header on a small cell was unacceptable. Their design approach was to establish a virtual circuit with state in the switch, and the address in each cell was just a simple index into a table rather than a search for a matching destination address.

Perhaps the most extreme form of cell switching was the Cambridge Ring, developed at the Computer Laboratory at the University of Cambridge in the late 1970s, about the same time that Ethernet was being developed at Xerox Parc. The unit of transfer across the Cambridge Ring, which they initially called a packet (Wilkes and Wheeler, 1979) but then more suggestively called a mini-packet (Needham, 1979), has a 2 byte payload and source and destination addresses of 1 byte each. Needless to say, the forwarding decision was of necessity simple, but it is also clear that putting an IP header on a 2 byte payload would represent an unacceptable waste of bandwidth. This system did not have label rewriting, however, since the number of hosts was so small that it was acceptable to have a global ID for each.

A less extreme and more relevant example of cell switching is the Datakit architecture, developed at AT&T Bell Labs (Fraser, 1980; Luderer et al., 1981). The Datakit cell had a 16 byte payload and a 2 byte header, with one byte to identify the outgoing link on the switch and one byte to identify the virtual circuit on the link. These bytes were rewritten at each switch.

The most mature form of cell switching, which to a considerable extent descends from Datakit, is asynchronous transfer mode (ATM). There is extensive literature on ATM, which I do not attempt to catalog here, but the core idea is similar to that of Datakit. The ATM cell contains a 48 byte payload, and a somewhat complex virtual circuit identifier (VCI) in the header. The scheme depends on virtual circuit setup, and the VCI is a label that is rewritten at each hop.

There are actually two motivations tangled up in the design of cell switching (just as there are multiple motivations behind the design of the Internet packet). One is the switching efficiency and jitter control of fixed-size cells. The other is a preference for virtual circuit setup and per-flow state in the switch. The Internet design took the extreme view that there was never any per-flow setup, and no per-flow state in the router. The goal was to reduce the overhead and complexity of sending data—there is no required setup phase; to send a packet, you just send it. But this lack of state and resource reservation meant that the network was making no service commitment to the sender. In the Internet, there is no idea of a "call," no commitment of resources to the call, and no effort to provide a different quality of service to different flows. Indeed, the Internet community spent much of the 1990s figuring out how to add QoS to the Internet, while QoS was always a central tenet of the virtual circuit community. So the per-flow state in the router is not just a necessary consequence of the small cell size and the need for a small header but has a virtue in its own right, allowing better capacity management. In this respect, the Internet and the cell switching/virtual circuit worlds are distinguished as much by their views on circuit setup and per-flow state as they are about fixed-size versus variable-size multiplexing units.

Label Switching Meets Internet 1: The Stream Protocol ST

From essentially the beginning of the Internet, there was an effort to define an alternate forwarding scheme, called the stream protocol or ST, which set up a flow and had per-flow forwarding state in the router. It was first documented in IEN 119 (Forgie, 1979) and provided explicit QoS for speech packets; the design specification discussed in some detail the concept of a flow spec, and the need to set up state in the routers using an explicit setup protocol. ST also supported multicast. It took advantage of the state to use a small packet header, in which the destination address was replaced by a connection identifier (CID), which was only locally meaningful between routers and was rewritten on each hop. This mechanism is thus an example of label switching, perhaps the first in the Internet. The ST evolved over the next decade, and a new protocol, called ST-2, was described in RFC 1190 in 1990. This version of ST was still based on a local label, now called the hop identifier (HID), a 16 bit field. The specification in RFC 1190 contains details on the process of setting up a sequence of HIDs that are unique across each link. Interestingly, in the final specification of ST-2, in RFC 1819 in 1996, the HID is replaced with a stream ID, which is globally unique (a 16 bit nonce[3] combined with the 32 bit source address). This implies a slightly more complex lookup process. RFC 1819 says: "HIDs added much complexity to the protocol and was found to be a major impediment to interoperability." So the final version of ST abandoned the idea of label switching but still depended on a full, per-flow connection setup and state in the packet.

Label Switching Meets Internet 2: Remove the Cells

As the previous discussion suggests, there are actually a bundle of motivations behind the label switching approach, such as the need for low-overhead headers on cells, and the desire for flow setup. There were both packet-based (frame relay) and cell-based (ATM) connection-oriented networks using label switching as the basis of forwarding. In the early networks, the idea of flow setup was that a virtual circuit was equivalent to an end-to-end flow, but it became clear that another use for a virtual circuit was to set up state (and perhaps to allocate resources) to a path that carries aggregated traffic from a set of sources to a set of destinations, such as city pairs

3. A *nonce* is an arbitrary value that is intended to be used only once.

in a large national network. This sort of undertaking is often called traffic engineering: the allocation of traffic aggregates to physical circuits in such a way that, for example, the overall link loads are balanced and there is spare capacity for outages. The goal of traffic engineering, together with the goal of simplifying the forwarding process, led to the proposal to use label switching on variable-size packets rather than cells. Cisco Systems put forward a proposal called tag switching, building to some extent on the ideas of both frame relay and ATM. Tag switching was turned over to the IETF for standardization, where it was first called label switching and then, in its full glory, Multi-protocol label switching (MPLS). As with ATM, there is a wealth of literature on MPLS, including complete books. Wikipedia is a reasonable place to start.

MPLS, like many mature ideas, has become increasingly complex. It supports the idea of nested flows—that is, a set of virtual circuits carried inside another. So the header of a packet can have a series of labels, not just one. When a packet reaches the router at the beginning of an MPLS path, the router adds a label; when it passes along the path, each node rewrites the label; and when it reaches the end of the path, the label is "popped," which may reveal yet another label or may leave no further labels to process, in which case the router processes the packet using the native packet header, for example the traditional IP header.

An MPLS header is 32 bits long, which is a very efficient representation of forwarding state. It consists of a 20 bit label along with some other control information. The IPv4 header is five times bigger.

Label Switching Meets Internet 3: NAT—The Loss of the Global Address Space

The essence of the original Internet forwarding scheme was the existence of global addresses, meaningful to a router anywhere in the Internet. The idea of network address translation devices (or "NAT boxes"), introduced in the early 1990s, was on the one hand a very clever way to conserve scarce Internet addresses and on the other hand a total violation of the global addressing assumption. NAT is based on label switching, except that in this case it is IP addresses themselves that are being rewritten. The IP address field in the header, which was previously static and immutable, is now rewritten inside the NAT box by using stored state. There has been a great deal of work to try to patch up the rift in the Internet architecture created by NAT, and the necessity of establishing and maintaining the right state in the NAT box, given that the Internet lacks any sort of signaling protocol. (Many of the solutions somewhat resemble a signaling protocol, though most would not be so bold as to call them circuit setup protocols.)

The first idea for NAT was simple. When a host behind a NAT box sends a packet, the NAT box rewrites the source address of the packet with its own source address and remembers the internal IP address and port number of the packet. If an incoming packet arrives for that port number, the NAT box uses that remembered state to rewrite the destination address of this incoming packet with the correct local address. (Some NAT boxes also do port remapping.) In other words, the outgoing packet triggers the setup of state for the subsequent incoming packet.

This idea is fine as far as it goes, but what about an incoming packet without a prior outgoing packet—what about a server behind a NAT box? The current solution for most consumer-grade NAT boxes is primitive. The user manually configures static state in the NAT box to allow the remapping of the incoming packet, but there are a variety of more complex solutions for setting up this state dynamically. One approach is to set up the state by using a message from the machine behind the NAT box—the machine that is acting as the server. The IETFT has worked extensively on this problem; for example, see RFC 3303 and the related work on middleboxes.

A more complex scheme is found in IP next layer (IPNL) (Francis and Gummadi, 2001), which discusses the pros and cons of multiple address spaces and (in Internet terms) NAT. Among the advantages are expansion of the address space, changing providers without renumbering, and multihoming. Francis and Gummadi propose an architecture that attempts to reproduce the existing Internet functionality, where all hosts have long-lived globally routed addresses if they choose, routers (including the elements that link the address spaces) are as stateless as they are today, and only a router on the path of the packet can disrupt the forwarding process. Further, the scheme allows a region with a private address space to have its public address reassigned at will without disrupting the flow.

The approach used by Francis and Gummadi involves a source-routing *shim* header, a new header put between IP and the transport protocol. The scheme uses fully qualified Internet domain names (FQDNs) in the first packet of a flow to identify the destination. As the packet reaches an element where address translation is required, the element uses the FQDN to look up the IP address in the next addressing region. This sequence of IP addresses is stored in the packet (not in the router—this is a stateless source-routing scheme). Subsequent packets use this sequence of addresses as a source route. The scheme supports failure and rerouting to alternative entry points into the private address spaces.

The FQDNs in IPNL must be globally meaningful, so the scheme depends on having higher-level global names. The extreme point in this space would be a system in which there are no shared names of any sort, either locators or higher-level names. Examples of such a system include Sirpent and Plutarch, discussed in chapter 7.

A more recent scheme for dealing with NAT is NUTSS (Guha et al., 2004), which uses SIP to set up rewriting state (in NAT boxes), so the setup is associated with a per-flow/per-application signaling phase. This scheme uses an application-level signaling protocol (SIP) to set up forwarding state in NAT routers. It is stateful, as opposed to the stateless character of IPNL.

Another source-routing scheme for dealing with NAT is 4 + 4 (Turányi et al., 2003), a two-stage source route, again made up of traditional IP addresses. The 4 + 4 scheme uses the DNS in a different manner than IPNL. In IPNL, different parts of the address are looked up as the packet crosses different addressing regions (that is, the DNS returns different values in different regions), whereas in 4 + 4, the DNS stores the whole two-part address. The sender looks up the address and puts it on the packet at the source. Storing the complete address in the DNS has implications for what information is visible where and what information can change dynamically. (In IPNL, public addresses have no meaning in private address regions, whereas in 4 + 4 they are meaningful and routed.)

Comparing Mechanisms

Earlier, I divided forwarding schemes into two broad camps: forwarding state in the router and a forwarding directive in the packet. I discussed two forms of forwarding state in the router: forwarding based on globally known locators and label rewriting, which implies per-flow state. There are two general forms of forwarding directives in the packet, source routing and encapsulation, which is discussed later.

Another way to think about these schemes has to do with their relative expressive power. Traditional IP forwarding based on a globally known destination address allows the sender to name a destination. Source routing and label switching both allow the naming of a path to a destination. If there is more than one possible path to a destination, then the ability to name paths is more expressive. Naming a path rather than a destination allows for more control over multipath routing, can support quality of service routing, and other actions. The motivation behind most Internet-centric source-routing proposals is control of the path, so there is recognition in the Internet community that this expressive power is of some value. To oversimplify, these options can be organized as a two-by-two matrix.

	Destination-based	Path-based
Forwarding directive in packet	Encapsulation	Source route
Forwarding state in router	Classic Internet	Label switching

Source Routing

There are two high-level motivations for source routing. One is to simplify what the router does, both by removing the routing computation from the router and by simplifying the forwarding process. The other motivation is to give the end point control over the path of the packet, perhaps to allow the end points to pick their providers or to implement more general sorts of policy routing. Different schemes can be positioned in different parts of this landscape.

The source routes in SIP and SIPP do not simplify the forwarding process in the router. If a source address is included in a SIPP packet, each such address is a globally routed address, so the lookup in each router requires a search of a forwarding table of the same complexity as in the case of a simple locator. PIP had a slightly more compact representation of a source route, in which the intermediate elements in the source route (called route sequence elements) were only unique and meaningful within a hierarchically organized region of the network, not global addresses.

The simpler version of source routing makes the individual elements of the source route only locally meaningful to each router. For example, a router could label its ports using a local index (1, 2, 3,...), and a source route could just be a sequence of the small numbers. This idea leads to compact source routes (though they are still of variable length) but means that the sender has to have a lot of router-specific information to construct the source route, so this idea is only practical in small regions of the network (a LAN), where the issues of scale are not so daunting. Examples of this idea include Paris, an early network designed at IBM (Cidon and Gopal, 1988), and the link-ID option in the Bananas scheme (Kaur et al., 2003), which is otherwise a label-rewriting scheme.

The other use of source routing is to give the sender control over the path the packet takes. In terms of policy and end-node control, SIPP contains a special form of anycast address called a cluster address, which can be used to identify a region (e.g., an AS or provider). Cluster addresses allow a sender to select a sequence of providers without picking the specific entry point into the provider's network. This feature was called source selection policy, and while SIPP did not restrict the use of source routing to this purpose, cluster addresses were the only special form of locator defined by SIPP. Hinden (1994), describing SIPP, lists these other examples of the use of a source route: host mobility (route to current location), auto-readdressing (route to new address), and extended addressing (route to sub-cloud).

Argyraki and Cheriton (2004) propose the wide-area relay addressing protocol (WRAP), a loose source-routing scheme that differs in detail from the IP option. The source route is carried in a shim layer between the IP and next layer, so the source route does not have to be processed by every router along the path but only those to which the packet has been intentionally addressed. The authors claim that the scheme is useful in DoS filtering and QoS routing. The processing of the shim layer is done by a different PHB than for the forwarding function, which could implement functions beyond rewriting the address, but the authors do not discuss this.

Source routing and fault tolerance One of the issues with source routing is that if an element along the specified path has failed, there is no way for the network to remedy this problem and send the packet by an alternate path—the path has been specified by the user. In time, the user may be able to detect that the path has failed and construct a new source route, but the process of discovery, localization, and constructing a new source route may take much longer than the recomputation of routes done inside the network. A scheme called slick packets (Nguyen et al., 2011) proposed a solution to this—the source route is actually a directed acyclic graph (DAG), which gives each router along the specified source route a set of options for forwarding the packet. This scheme faces a number of challenges, of course, including constructing the graph and encoding it in the packet header in a sufficiently efficient manner. Compensating for these issues, the option of alternative routes to deal with failures means that information about short-term failures need not be propagated across the network to sources constructing source routes, since the alternative routes in the header will deal with these failures. XIA is also based on the use of a DAG as an address.

Label Switching

As the previous discussion suggests, there are also a number of motivations for label switching, and divided schools of thought about different mechanisms and their merits. The debate between connection-oriented and connectionless (datagram) networks is as old as the Internet. Two important distinctions in the design philosophies are the unit of multiplexing and the value (and cost) of per-flow state. The mechanism of label switching is often a consequence of these other considerations, but it has taken on a life of its own.

An argument favoring label switching over destination-based forwarding is that label switching (such as MPLS) can provide a more precise allocation of aggregates to circuits (for traffic engineering purposes) than using link weights in OSPF, but Fortz and Thorup (2000) argue that a global computation of OSPF weights can reproduce any desired pattern of traffic allocation so it is claimed that destination-based forwarding and label switching are equally effective in this case (again, so long as there is one path).

One of the objections to label switching is that it seems to imply high overhead for circuit state setup. If the paths are in the core of the network and used for carrying stable aggregates of traffic (as is the usual case with MPLS), then the overhead of path setup is not serious, since the paths are set up fairly statically as part of network management. But if label switching were used end-to-end per source-destination flow, the use of label switching would imply a connection-oriented design with per-flow setup.

The ideas of label switching and state setup can be separated. It is possible to use label switching with Internet-style route computation, datagram forwarding, and no per-flow state in the routers. Bananas (Kaur et al., 2003) provides a clever solution to this problem of how to use label switching and a constant-size packet header without doing per-flow state setup in the network. The goal of Bananas is to allow multipath routing, but it could be used in any context in which a set of global or well-known routes can be computed. Assume that every node has some sort of global address. Conceptually, trace each path backward from the destination toward the source, computing at each stage a hash that represents the path (the sequence of addresses) to the destination. These paths and the associated hashes can be precomputed. For multipath routing, use any well-known multi-path computation algorithm. At the source, compute the hash for the first node of the desired path. Put that and the destination into the packet. At each router, look up the destination, using some sort of prefix match together with an exact match with the hash. Rewrite the hash with the value (stored locally) of the hash of the subpath starting at the next router. In this way, the hash values are a specialized form of label, the label rewriting is done based on the stored information in the forwarding table, and no flow setup is required. All that is required is that all parties agree on what subset of valid paths have been encoded. Some scheme is required for this (the authors suggest one), but this depends on the purpose of the paths being computed. Kaur et al. discuss the operation of Bananas in a number of contexts, such as BGP. This scheme is useful because label switching may be more efficient, and different path-setup schemes might be used at the same time; the source could select among them by selecting which label to put on the packet when it is launched.

Since both label switching and source routing can be used to specify a path, and both can be used with virtual circuit or datagram networks, one might ask whether they are fundamentally different in some way. The distinction here is not one of expressivity but one of control (and packet size). When source routes are used, it is the source that determines the path. When label switching is used, different parties can install the path, and the source only has the choice of selecting which path to use. In some cases, the source may not know the details of the path but only the starting label, so label switching gives more options for which parties can configure paths. On the other hand, source routing allows a path to be established without any sort of distributed path setup. In a label-switching network, every path must either be set up on demand or precomputed. With source routing, a path can be specified by the source at the time it is needed.

New Requirements

The previous discussion cataloged schemes based on a set of criteria, which include expressivity, efficiency (both forwarding and header size), and control over path determination. In the early days of the Internet, these were the primary considerations. Francis (1994a) offered the following criteria for selecting among candidates for IPng:

- Cost: hardware processing cost, address assignment complexity, control protocol (routing) complexity, header size.
- Functional capability—necessary: big enough hierarchical unicast address, multicast shared-tree group address, multicast source-tree group address, scoped multicast group address, well-known multicast group address, mobility, multicast two-phase group address, domain-level policy route, host auto-address assignment.
- Functional capability—useful: ToS field, embedded link-layer address, node-level source route, anycast group addressing, anycast two-phase group address.

All of these have to do with the expressive power of the forwarding scheme and its cost, but in more recent times, there has been a recognition that addressing and forwarding must be responsive to a broader set of requirements.

Addressing and Security

Addressing and security have a complex relationship. It is hard to attack a machine if the attacker cannot find and use the machine's address, so some addressing schemes allow machines to hide their addresses. NAT is viewed as a useful, if incomplete, tool in securing a host, since it does hide the local address of a machine.

The i3 scheme (Stoica et al., 2004), described in chapter 7, protects a node from attack by keeping its address secret. In this scheme, a receiver controls who can send to it by installing a trigger, which is a form of label, into a forwarding overlay. The sender is given the trigger but not the actual destination. When the packet reaches the right forwarding point, the trigger is rewritten with the final address, and then destination-based addressing is used for the remainder of the path. However, hiding addresses and similar tactics cannot fully protect a public site from attack, since to be used a machine must reveal itself in some way. Once it reveals itself, a DDoS attack can mimic normal behavior in all respects and still attempt to overrun a server.

One of the challenges in designing an indirection scheme such as i3 is whether the source and destination are equally trying to protect themselves from attack. Schemes such as i3 attempt to hide or protect the destination. If the destination is truly hidden, then when a packet goes in the reverse direction, the source of that packet must be hidden. This implies that the original destination (perhaps a server) has some measure of anonymity in the distant region. If i3 is used in a symmetric way to protect both ends from each other, then the identity of each end is not easily known by the other. This raises the question of when a machine can hide its address (to protect itself) and when it must reveal its address (for reasons of accountability).

The TVA scheme, described in Yang et al. (2005), is an alternative way of protecting the destination from attack. Instead of hiding the destination address, packets must carry a specific authorization, a *capability*, to be allowed to pass through the network. The forwarding scheme is the basic destination-based Internet mechanism, but the router (in particular, the router at trust boundaries between ISPs) is responsible for checking the capabilities. (This design approach is what I called intentional delivery in chapter 6.) The scheme actually uses a rather complex mix of mechanisms to implement its function. The packet carries a variable-length set of capabilities (which has something in common with a source route in that it does not require state in the router to validate it) but also uses soft state in the router and a nonce in the packet to avoid the variable-length address in most cases. It computes an incremental hash of the source path to assist in tracking sources and allocating capacity among different sources.

It uses fair queuing to limit the congestion that one flow can cause to another.

Yet another indirection scheme is secure overlay services (SOS) (Keromytis et al., 2002), which is again designed to protect servers from being attacked. SOS restricts the problem to protecting servers that have a known and predetermined set of clients—the authors do not offer SOS as a means to protect public servers. They use a three-tier system of defense. The server is protected by a filter that is topologically placed so that all packets to the server must go through it. The authors assume the filter can run at line speed and cannot be flooded except as part of a general link-flooding scheme. This means that the filtering must be simple, so they only filter on the source address. To allow for more complex filtering, they require that all legitimate traffic to the filter first pass through an overlay mesh, where one of the overlay nodes knows the location of the filter. The address of this node is secret, and the overlay uses DHT routing to get the packet to the correct overlay node. To protect this layer, the authors have a set of secure overlay access points (SOAPs), which perform the first line of checking and perform a hash on the destination address to get the identifier used to drive the DHT. The paper contains a discussion of the justification for this rather complex set of mechanisms, and an analysis of the various attacks that can be mounted against it.

Yaar et al. (2004) propose a scheme called stateless internet flow filter (SIFF), which allows a receiver to give a capability (permit to send) to a sender; the routers check these capabilities and reject them if forged, but otherwise give them priority over unmarked traffic. In this way, traffic without a permit to send (including malicious traffic) is disadvantaged relative to favored traffic and presumably preferentially dropped as the network becomes fully loaded with attack traffic. Portcullis (Parno et al., 2007) is concerned with preventing attacks on the blocking system itself. Systems using capabilities to provide preferential service to selected flows offer strong protection for established network flows. The denial-of-capability (DoC) attack, which prevents new capability-setup packets from reaching their destination, limits the value of these systems. Portcullis mitigates DoC attacks by allocating scarce link bandwidth for connection establishment, and Parno et al. argue that their approach is optimal in that no algorithm of this sort can improve on their assurance.

All of these schemes, TVA and SOS in particular, have rather complex and rich sets of mechanisms, which arise when the full range of attacks are contemplated, defenses are selected for these attacks, and then these defenses in turn must be defended. This does beg the question of whether there is a different way, perhaps more simple, of factoring the security problem.

Tussle and Economics

The simple model of the Internet was that the network computed the routing and everyone used the result, but both senders and receivers may want to have some control over where the traffic goes. Senders and receivers may want to pick a path through the network as part of picking a service provider, obtaining a specific QoS, or avoiding a particular part of the network, for example. Different third parties may also want to have some control over routing, which may cause them to invent a separate address space. This set of considerations has also been understood for some time. Francis (1994a) says:

> There are several advantages to using a source routing approach for policy routing. First, every source may have its own policy constraints (for instance, certain acceptable use or billing policies). It is most efficient to limit distribution of this policy information to the sources themselves. Second, it may not be feasible to globally distribute policy information about transit networks. Further, some sources may have less need for detailed transit policy information than others. With a source routing approach, it is possible for sources to cache only the information they need, and from that information calculate the appropriate routes. (55)

Yang (2003) proposed the new internet routing architecture (NIRA). NIRA is primarily about routing and providing the ability for users to select routes. This objective is proposed in order to create a competitive market for packet forwarding and impose the discipline of competition on ISPs. As a part of this, NIRA proposes an efficient scheme for encoding explicit routes in a packet. It uses addresses extravagantly, specifically to assign a separate address to each valid route in a region of the network. To control the cross-product explosion of sources and destinations, NIRA breaks the routes into three parts: a *source part*, a *middle part*, and a *destination part*. A packet carries (as usual) a source address and the destination address. For the first part of the path, the source address guides the packet, following the reverse of the route to that source. Across the middle part (large global ISPs), traditional routing is used. For the final part, the destination address is used. So any node only has a separate address for each path to and from it and the middle part of the network, not all the way to the destination.

It is interesting to contrast NIRA and Bananas in this context. Bananas computes routes all the way from the source to the destination. As a result, there are a huge number of routes, and there is no plausible way to assign a distinct global ID to each such route. Instead, NIRA uses a clever trick to rewrite the pathID at each node. NIRA computes IDs for "route halves" and asserts that each one of these can have a unique ID (an address) that is valid within that region of the network, so no rewriting is needed. In exchange for this simplicity, paths that do not follow a simple "up, across, and down" pattern require explicit source routing. Bananas can use, with equal efficiency, any route that has been precomputed.

Relation of Addressing to Identity and Naming

There has been a long-standing set of arguments in favor of separating the notions of location (address) and identity. The Internet uses the IP address for both, which hinders mobility. But using the IP address for identity in the packet provides a weak form of security, and separating the two requires an analysis of the resulting issues. It is possible that if identity is separated from location, there is no form of identity weaker than strong encryption that is of any real value.

Many schemes that separate identities from location do use the identity as a way to "look up" the location. IPNL uses the DNS to look up the addresses to deal with NAT, as does 4 + 4. The unmanaged network protocol (Ford, 2004) used "flat" identifiers that are public keys. That is, any node can make its own identifier and later prove that it is the entity with which this identifier goes by using the private key associated with the identifier. The scheme uses a DHT to allow nodes to incrementally find each other and establish paths across the DHT overlay among all the nodes. Turfnet (Pujol et al., 2005), another scheme for tying independent addressing regions together by using a common naming scheme, uses flat identifiers that are flooded up the routing tree, which raises interesting performance issues with regard to finding an entity.

In contrast, FARA (Clark et al., 2003), discussed in chapter 7, argues that it is not necessary for the packet forwarding layer of an architecture to convert from any sort of higher-level identity to a network-level locator. The hypothesis in FARA is that any end-point identity (EID) scheme can be private among the end nodes, that higher-level naming schemes like a DNS can be used to find the location of entities, and entities can manage their locations (e.g., they can move) without having to provide a means to "look them up" using the EID.

Jonsson et al. (2003) propose a split between names and locators, called SNF, for split naming forwarding. They suggest that locators need not be global, and that source routing or state in translation gateways can be used to bridge addressing regimes, but offer few details. They propose that naming is a least common denominator, so naming schemes must be well-specified and global. But there can be more than one, and this is good in that different schemes can compete. Names map to things at locations, so it appears that Jonsson et al. name machines, not higher-level entities. They describe naming as an overlay that can route but at low performance, somewhat like IPNL. They also propose an ephemeral correspondent identifier (ECI) that is used by the transport layer. This is visible in the packet and becomes a short-term identifier that does not change if the locator changes.

Enhanced Label Switching

Gold et al. (2004) propose a scheme called SelNet, a virtualized link layer. It is a label-based forwarding scheme (somewhat like MPLS) with the feature that each label includes a next-hop destination and a selector, which is a generalization of a label that can trigger not just a rewriting but a range of services and actions. Actions might include, for example, forwarding, local delivery, or multicast. Further actions can include removing or replacing a label so a packet can be sent with a series of labels, which produces a variant of source routing, or the labels can trigger rewriting, which is more stateful and more resembles MPLS. In this respect, SelNet is an interesting generalization with rich expressive power.

The SelNet design does not constrain how actions (the things that selectors refer to) are established. They can be static and long-lasting, or they can be dynamically set up. SelNet includes a protocol somewhat like ARP, called XRP, for extensible resolution protocol, which allows a sender to broadcast to find a receiver and get back an address/selector pair in reply. Gold et al. observe that validation or verification can and should be done before returning this information (in contrast to ARP, which always answers), which gives a measure of protection somewhat like a dynamic NAT. This idea of a security check before answering is a clever idea that allows for a range of checks, including application-level checks, but it begs the question of what info should be in the request packet, which the authors do not elaborate.

The form this would take in a new architecture is not clear. Gold et al. describe it as a link layer, or layer 2.5 scheme, but this seems to derive from the desire to interwork with IP. In a new scheme, this might be the way that IP worked. The novelty seems to be the idea of a selector with generalized and unspecified semantics, the separation of the forwarding from the (multiple) ways in which selector state is set up, and the idea of a security check at label setup time. I believe that by defining a few selectors with global meaning (well-known selectors need a security analysis), this system can emulate several prior indirection schemes.

Mobility

Work on mobility seems to be particularly constrained by the current Internet architecture, in particular the overloading of an address with identity information. I do not attempt to discuss the large number of schemes for mobility, since this is as much about routing as about addressing.

In general, schemes can be divided into end-to-end and network aware. In the end-to-end schemes, a host that moves gets a new address that reflects its new location, part of an address block that is already routed across the network, so the routers see nothing special about a mobile host. In the network-aware scheme, there is some sort of indirection, either in the routers or in a special node (e.g., a home server), so that the sending host does not have to be told about the move. There are issues of complexity, scale, and speed of response.

Mysore and Bharghavan (1997) make the point that multicast and mobility have many similarities. They explore the option of using current multicast as a way to track a mobile host. They note the major problem: how does the mobile host find the multicast tree to join, since flooding will be very expensive. They summarize the other problems, all of which arise from current details. This paper might be a nice input to a fresh study of mobility in a new architecture.

Lilith (Untz et al., 2004) is an addressing/routing scheme for ad hoc networks of limited scope, where broadcast and flooding can be used. It uses flooding to set up flows and MPLS to utilize them. The authors note the interesting point that if you discover topology and routes at the same time, for example by using flooding, then you need a lower-level set of addresses that scope the flooding. So the authors don't use IP addresses for the labels, because IP broadcast only works within a subnetwork, and they are trying to build a subnetwork at the IP level. Because of the state in the routers, they call this a connection-oriented approach, but this is a particular use of the term. They say that they prefer connections to allowing each forwarder to decide what to do, but it is not clear exactly what the dynamics of their route setup scheme are. It is not clear how this scheme would differ if the path setup message from the destination back toward the source set up an IP forwarding entry rather than an MPLS label rewrite entry.

Making Source Routing Robust

As I discussed, there are a number of problems raised by source routing. One is that it seems to take control of resources away from the network operator and give it to the user. There is no reason to believe that an ISP will want to carry a packet for a user unless the ISP is going to be compensated, or at least is party to an agreement to carry that class of traffic. Also, perhaps giving control of the routing to the user creates a new and massive attack mechanism, where the routes can be used to implement a DoS attack against some part of the network. Another problem is that in a simple source-routing scheme, there is no guarantee that the packet will actually follow the specified path. Schemes have been proposed to try to address some of these issues.

Platypus (Raghavan et al., 2009) is an authenticated source-routing system built around the concept of network capabilities. Platypus defines source routes at the level of the ISP—it defines a route as a series of "waypoints" that link ISPs. Inside an ISP, default routing is used. A Platypus header is thus a sequence of capabilities, each specifying a waypoint. The process of obtaining a capability allows an ISP to maintain control over which traffic it agrees to carry.

Another scheme for creating robust and policy-compliant source routes is ICING (Naous et al., 2011). ICING is described in chapter 7; it is essentially the forwarding scheme for the Nebula proposal. I have now followed the history of addressing and forwarding up to the era of the NSF Future Internet Architecture Program.

Acknowledgments

This book has been a long time in the making, and many individuals and groups of people have helped to shape it. I owe a debt to the many people who have worked on the Internet over the last 40-plus years, to people who have shaped my career, and to those who have contributed in many ways to the book.

In the beginning, Robert Kahn and Vinton Cerf wrote the initial paper that proposed the design of the Internet. They kicked off the whole venture, and both continued to shape the Internet as it grew up. Without them, none of this would have happened as it did. Vint chaired the initial design group, and it was because of him that I took over the chairmanship of the IAB during the 1980s. His support and encouragement were invaluable.

I cannot now recall the names of all the folks who made up that initial design team, but they were wonderful collaborators in a glorious time. We were setting out into uncharted territory with few guideposts, working out the basic design decisions that now define the core of the Internet. I acknowledge all they did to make the Internet. Those were enjoyable times.

In the 1990s, a group of us undertook to extend the service model of the Internet to provide explicit quality of service assurances. That effort, while successful only in part, taught me a great deal. I had the benefit of working with some very smart people, including Scott Shenker, Lixia Zhang, Deborah Estrin, and Sally Floyd.

A number of the ideas in this book first took shape in the DARPA-funded NewArch project. The lead collaborators on that project included John Wroclawski, Karen Sollins, Robert Braden, Ted Faber, Aaron Falk, Mark Handley, and Noel Chiappa. We had some wonderful conversations, both enjoyable and insightful.

I started my career with a typical focus on technology. It was in the 1990s that I realized that the technologists were no longer in charge of the Internet's future. My own career took a turn at that point, as I realized the value of working with people from other disciplines, who brought their own valuable perspectives on the shaping of the Internet. Marjory Blumenthal showed me the value of interdisciplinary work and transformed how I undertake research. Her influence on this book is indirect but profound.

Much credit must go to the lead investigators for the various NSF Future Internet Architecture projects, including Lixia Zhang, Van Jacobson, Peter Steenkiste, David Andersen, Jonathan Smith, Dipankar (Ray) Raychaudhuri, Arun Venkataramni, Tilman Wolf, and Ken Calvert. Many of the ideas in this book were shaped by discussions in the FIA investigator meetings. As a part of the Future Internet Architecture project, Helen Nissenbaum organized a group of social scientists from different disciplines to join our meetings and add a nontechnical dimension to our discussion. That was a wonderful contribution to the effort.

Research cannot happen without the efforts of dedicated program officers at funding agencies, who oversee research grants and set a vision for the research community. Darleen Fisher at NSF must get a special acknowledgment for what she has done to support the network research community, specifically with respect to the Future Internet Architecture project but also for her work over many years to support and encourage the field of networking. She has supported and encouraged my work on FIA, and I have greatly enjoyed working with her. Guru Parulkar was also instrumental in launching this program. He had a vision early on about the potential of the Future Internet program, and he reached out to get me involved, for which I am grateful. In its early stages, Victor Frost did a great deal to make the program happen. Those in leadership positions at NSF, including Peter Freeman and Taieb Znati, were also critical to the early success of the project. Many others at NSF have helped to nurture this program.

In addition to my funding from NSF, Ralph Wachter, then at the Office of Naval Research (ONR), provided funding for me during the critical early stage of this project. The first drafts of several chapters of this book were written with funding from ONR. Mari Maida at DARPA provided funding for the NewArch project, which was a highly speculative bit of grant-making. Her willingness to take this risk was one of the starting points for this work. The support and encouragement from those who serve as program officers is much appreciated.

I greatly benefited from conversations with John Wroclawski in the early stages of writing this book, and I have noted in the text places where his ideas were instrumental in shaping my thinking. Karen Sollins, Josephine Wolff, Shirley Hung, Steve Bauer, Willian Lehr, and Nazli Choucri were also sources of valuable insights.

I received very helpful comments and suggestions on this book from a number of reviewers, including Jon Crowcroft, Vint Cerf, and Marjory Blumenthal. A number of people were willing to read drafts of the book, including Karen Sollins, John Wroclawski, Steve Bauer, Butler Lampson, and Lixia Zhang. kc claffy gets special thanks—not only was she a joint author of chapter 14, but she read the manuscript *twice* and gave me very valuable comments. Among her many virtues is the ability to look at something I have written and tell me what I was trying to say. The book is much better for her efforts. The series editor for the MIT Press, Sandra Braman, also gave me extensive and detailed comments that greatly improved the book. I thank all of these people for their efforts.

In a perfect world, every time someone said something wonderful and insightful, I would take out my notebook and write down what they said and who said it. The real world is not like that, and I am distressed at all the important thoughts I can no longer trace back to their origin. Most of what I know somebody else told me. I know that there are names that should be highlighted here that are missing.

The research reported here and the preparation of this book were supported by the National Science Foundation under agreement 0836555, and by the Office of Naval Research under contract N00014-08-1-0898. As always, the opinions contained herein are those of the author and do not reflect the opinions of the supporting agencies.

Glossary

address (Internet) An Internet address is associated with each end node on the Internet or (more precisely) with each interface or connection point that an end node has on the Internet. The addresses now most commonly used on the Internet are called IPv4 addresses (versions 1–3 were experimental). These addresses are 32 bits long, which allows the Internet to address a little over 4 billion end points. The Internet is running out of addresses as it gets bigger, and the next version of Internet addresses (called IPv6) will allow the Internet to grow almost without bound.

anycast A packet delivery service in which there are several potential destinations for a packet and the network picks one of them based on some set of rules (such as proximity).

autonomous system (AS) A region of the network, typically operated by a single Internet service provider. The word *autonomous* reflects the independence of the various ISPs and their ability to implement their regions of the network as they choose.

best effort The packet delivery service of the current Internet, in which the network is expected to do its best to deliver all the packets sent, but there is no specific level of commitment as to speed, loss rate, and other service parameters.

bitstream access A term related to the regulatory requirement that an incumbent network operator unbundle its facilities to a competitor. In contrast to a physical element that is unbundled, bitstream access is a virtual data path representing a share of some physical assets.

border gateway protocol (BGP) The global routing protocol used among autonomous systems on the Internet. An AS can use one of a number of routing protocols internally but must communicate with other ASs using BGP.

capability As used in computer security, a capability is a hard-to-forge data item that indicates that the possessor of that item is authorized to take some action. In contrast to an access control list, which gives permission based on identity, a capability gives permission based on possession, independent of what actor holds it.

certificate An assertion, cryptographically signed by a certificate authority, that confirms the public key of some entity and that the entity has the right to use the domain name in the certificate.

certificate authority (CA) An authority, assumed to be trustworthy, that issues certificates that confirm the public key of some entity. Higher-level certificate authorities vouch for lower-level certificate authorities, and *root certificate authorities* provide the starting point for sequences of certificate authorities.

checksum A value that results from a computation performed on all the bits in a data item, with the property that if any bit in the data item is changed, the checksum will change. A checksum is thus a means to detect that a data item has been modified. Different checksum algorithms provide different degrees of assurance. A *cryptographic checksum* should provide an indication that the data item has been modified, no matter how many bits in the item have been changed.

cleartext In cryptography, the data item in its readable form, before it has been encrypted.

cloud computing A general term for a service that provides large-scale computing and storage capabilities on the Internet. A cloud service can be physically distributed or centralized, but the user does not need to be concerned with the exact location of the service. Cloud computing is typically provided in whatever quantity is needed, so applications that use cloud computing can easily increase their performance as demand grows.

content delivery network (CDN) A service, typically implemented as a set of servers widely replicated across the Internet, that hosts content and delivers it to end nodes on demand. By replicating the content at many points across the Internet, a CDN can increase the performance and availability of the delivery. Large firms may build their own CDNs, and CDN providers sell their service to third parties that want efficient delivery of content.

control plane In network architecture, those mechanisms that relate to the control of the network, as opposed to the mechanisms in the data plane that relate to the actual forwarding of packets.

conversion architecture A class of internet architecture that connects different sorts of networks together by converting the service of one into the service of the other. The success of this approach depends on whether the various networks have a native service that is sufficiently similar and the ability to specify the resulting end-to-end service. Contrast with spanning architecture.

cyphertext In cryptography, the data item after it has been converted from the cleartext by the encryption algorithm.

datagram A packet in a class of internet architecture where the routers do not maintain per-flow state, so that each packet is processed independently of the others.

data mule A physical entity that can move packets from one place to another. An example is a bus on a rural route equipped with a WiFi base station, which picks up and drops off packets when it is in range of a specific end point. Another example is a smartphone that picks up data as it is carried near some end node and sends it on when it has further connectivity.

data plane In network architecture, those mechanisms in a network that support the actual forwarding of data.

debug To debug a system is to track down and remove flaws (programming errors, hardware faults, and the like) that prevent the correct operation of the system.

deep packet inspection (DPI) A term that describes a class of per-hop behavior in which a device (e.g., a router) examines all of the packet, including the data, rather than just the header. DPI serves a number of purposes, including more complex rules for forwarding the packet and observation of what the sender is doing.

domain name system (DNS) A service provided as part of the Internet that converts a *domain name* (names such as www.example.mit.edu) into an Internet address. It is implemented as a set of servers distributed across the Internet, each being responsible for certain domain names.

end node A computing device that is attached to the Internet with the intention of sending and receiving packets. Also called a host.

facilities A term in telecommunications that describes the physical network assets of a provider (e,g., the actual circuits and routers). A *facilities-based* broadband provider is a firm that owns its own circuits to the home as opposed to renting them from some other provider.

flow A sequence of packets that implement the delivery of a data item. Packets in a flow typically share a common source and destination address and other fields in the packet header.

gossip scheme A scheme that spreads some piece of information by passing it between directly connected devices so that it eventually reaches all parts of the network.

hash Somewhat similar to a checksum. A hash is a value (typically of fixed length) that results from a computation across all the bits in a data item, such that the value of the hash depends on the value of every bit in the item. A *cryptographic hash* has the feature that while it is relatively easy to compute the hash of a data item, it is essentially impossible to do the reverse—find a data item that yields a given hash. A hash algorithm with this strength makes it impractical to forge a data item that matches a given hash.

header The first part of a packet, which contains the control information that specifies how the packet is to be forwarded. It includes information such as the destination address, packet length, what service the packet is to receive, and the like.

hypertext markup language (HTML) Another key protocol that defines the Internet. It specifies the format of a basic Web page.

hypertext transfer protocol (HTTP) One of the key protocols that defines the operation of the Web. It specifies the messages used to request and retrieve a Web page.

incast A form of packet delivery in which packets (or messages) from multiple sources to a given destination are combined into one packet/message. A typical use would be when some network element fails and many monitoring points want to report the error. To avoid flooding the recipient of these reports, the different reports of the same error can be combined as they flow toward the recipient.

interface A very general term that describes how a module is accessed from outside. An interface can manifest as hardware (a USB socket is an interface) or as software. The specification of an interface describes the functions that the module provides and how they are to be invoked.

Internet The term (with a capital I) used in this book to refer to the existing global, general-purpose packet-switched internetwork.

internet The term (with a lowercase i) used in this book to refer to any of a class of internetworks, based on one or another architecture, with the same general goals as the existing Internet—an internetwork with global reach, connecting a range of computing devices, and supporting a range of applications.

Internet protocol (IP) The standard that specifies the basic packet transport service of the Internet: how the relevant part of the header of packets is formatted, and the required operation of routers.

Internet protocol television (IPTV) describes the delivery of television content over networks using the Internet protocol. IPTV is in contrast to other, earlier protocols (both analog and digital) that have been used to deliver television over cable systems, satellite, and other platforms.

Internet service provider (ISP) An organization that provides packet forwarding service in a region of the Internet.

jitter Variation in latency experienced by different packets in a flow.

kludge A term in computer science that describes a solution to a problem that is ad hoc, often using elements in ways they were not intended to be used but at the same time having a degree of cleverness. To describe something as a kludge signals that the implementor did not go to the effort of solving the problem in a clean way but also signals respect.

latency The delay (typically measured as a two-way or round trip) between a sender and receiver. High latency can degrade the utility of some applications, such as teleconferencing and multiplayer games.

layer A layered system is specified as a sequence of modules with asymmetric dependency. Upper layers build on the services provided by the lower layers and are dependent on them, but the correct operation of the lower layers should not depend on the correct functioning of the upper layers. One strict interpretation of a layered system is that a module at layer N can only invoke services from layer $N-1$ (but not below that) and provide services only to layer $N+1$. Network architectures are often specified in layers.

layer 2 While there are many layered models for an internet architecture, layer 2 often refers to a particular network technology, on top of which (at layer 3) an internet is specified.

metadata Data about other data. The term is used in many contexts. The fields in a packet header, which describe where the packet is to go and how to deliver it, are metadata with respect to the data in the packet. Information about the creator of a data item, who is allowed to read or write it, and other similar information are metadata about that item.

middlebox A general term for a device that is interposed in a flow of packets between the source and the destination and performs some function (implements a PHB) that somehow modifies or controls the data being transferred.

module A component of a system. Large computer systems are typically specified and constructed as a number of modules, which connect to each other through their interfaces.

multicast A mode of packet delivery in which the destination refers to a set of end points and the network attempts to deliver the packet to all the end points to the best of its ability.

multihoming A situation in which an end node is connected to an internet at multiple points. The points may be on the same network or different ones. For example, a smartphone might be connected to the Internet over both a WiFi connection and a cellular connection. Different internet architectures will give either the network or the end node control over which point of connection is used.

outage A failure of availability that affects some region of a network. An outage is typically characterized by how many users are affected and for how long.

overlay A situation in which one network scheme is used to support another network scheme running on top of it. For example, an IP packet could be used to carry another IP packet as its data. The second IP internet would be an overlay on the first. An overlay is similar to a spanning architecture, but the term *spanning architecture* has the implication of heterogeneous lower-level network architectures, while an overlay network can be two versions of the same architecture, one running on top of the other.

packet The unit of data that is transmitted across the Internet (or any *packet-switched* network). A packet has a *header* on the front with information that controls the delivery, such as the destination address, and a *payload*, which is the data to be delivered.

path vector A kind of message used as part of a routing protocol that contains the sequence of elements on the path to a given destination.

peering A mode of interconnection between two autonomous systems on the Internet in which each agrees to receive traffic intended for it and its customers but not for other parts of the Internet. Contrast with *transit.*

ping A network management message that can be sent into the Internet. It is sent to a particular destination with the expectation that the destination will respond, thus proving that it is reachable and functioning.

port A field in the TCP packet header that indicates which application at a particular computer sent or received a packet. For example, a Web server is typically on port (80), and other services are on other well-known ports.

principal In computer security, the actor that is actually responsible for some action. When a computer sends a message on behalf of a person, it is the person, not the computer, that is the principal.

protocol The term used to describe the specification of how elements on the Internet interact to implement some service. The specification typically includes the format and meaning of messages exchanged among elements, the acceptable sequences of messages, and other information. A protocol is often declared to be a *standard,* which means that the designers have declared that the protocol is the accepted (or perhaps mandatory) way to realize the particular service.

quality of service (QoS) A term that can describe in general terms the key performance metrics of a packet delivery service (e.g., throughput, latency, jitter, packet loss, and the like) or more specifically a set of mechanisms and standards intended to provide an improved or enhanced QoS for some particular flows of packets. A typical enhancement might be a service with reduced jitter, which is beneficial for real-time teleconferencing and multiplayer games.

resilience The characteristic of a network that it continues to provide service even as elements are failing or have fallen to an attacker. A system that is *resistant* to attack should not allow an element to be penetrated successfully; a resilient system will keep running up to a point even if parts are penetrated.

router A device that receives and forwards packets at a point in the Internet. It uses the information in the header of the packet to determine the necessary treatment of the packet. In the early days of the Internet, a router was called a gateway.

sandbox In an operating system, an enclosed execution environment for a program that prevents misbehavior of the program from having any consequences outside the scope of the enclosure.

socket In an operating system, the interface provided by the transport protocol (e.g., TCP) to an application.

soft state With respect to a particular sort of state (e.g., per-flow or per-packet state in a router), the feature that if the state is discarded, it can be reconstituted as part of the normal system operation. If one device sets up state in another device and then crashes, that state has to be cleaned up somehow. Soft state has the advantage over hard state that it is easier to clean up old state; it can just be deleted if it seems stale.

software-defined networking (SDN) A modern approach to operating a region of the Internet, in which the routers do not implement a traditional distributed routing computation but instead receive the information they use to forward packets from a centralized controller. SDN requires the definition of new protocols to specify how the forwarding element (router or switch) communicates with the controller.

source route A form of destination address that is a sequence of addresses that identify intermediate points through which the packet is to flow on its way to the destination.

spanning architecture A class of internet architecture that connects different sorts of networks together by using the native service of each to carry the elements of a higher-level service end-to-end. Contrast with *conversion architecture*.

state An element that can be in more than one state can respond to the same input in different ways. A device that is stateless with respect to some input will always respond in the same way. A device that can be in more than one state records that state in *state variables*. One use for state variables is to record recent events, so a stateless device is sometimes called memoryless. A router that does not maintain per-flow state must process each packet independently—a *datagram* design. A router that maintains per-flow state can treat the packets in a flow in a consistent way.

Time to Live (TTL) A field in the IP packet header that indicates how many routers the packet is allowed to pass through before being discarded. The TTL field prevents a packet from looping forever if it encounters a routing inconsistency that sends the packet into a loop.

traceroute A sequence of network management packets sent into the Internet to map the routers between the source and a specified destination. The different packets in the sequence are sent to the same destination but with a different value in the TTL field, so that the packet is discarded at different points along the path. Ideally, the router where the packet is discarded will send an error message to the source, thus revealing its address.

traffic analysis A form of surveillance in which the goal is not to see what data is being sent but rather which parties are communicating and the patterns of communication. Traffic analysis involves looking at the metadata of the flow, not the data.

transaction cost In addition to the actual cost of a transaction, there can be costs arising from a search for the best price, negotiation over the price, and other associated tasks. Collectively, these are called transaction costs.

transit A mode of interconnection between two autonomous systems on the Internet in which one agrees to provide the other with access to all of the Internet. Contrast with *peering*.

transmission control protocol (TCP) The standard that specifies how an application-level data unit is broken up into packets that are transmitted across the Internet. It specifies how packets (or more properly the bytes in a packet) are numbered, how packets that are lost in transit are resent, and how the data unit is reassembled at the receiver out of these packets.

transport layer In the layered model of the Internet, the layer that is responsible for the end-to-end data processing. It defines an end-to-end service model that applications can use. The usual transport layer of the Internet is the transmission control protocol (TCP).

tussle A term my coauthors and I introduced to describe the interactions among actors with misaligned interests as they try to shape a system (such as the Internet ecosystem) to their benefit.

unbundle In telecommunications regulation, an element of a system is unbundled if it is made separately available to competitors. For example, a telephone company might be required to unbundle the wires to residences and allow competitors to use them to offer retail services.

unicast The basic packet delivery service of the Internet (and most networks), in which a packet is delivered to a single specified destination.

virtual This term describes the general case where a physical element (a processor, link, or the like) is divided into shares, each of which provides the same function as the physical element, but with lower performance since the physical element is being shared. It can also describe an abstract service that is not literally made by dividing a physical element but instead by combining various elements in a way that results in a service resembling a physical element.

virtual circuit A virtual circuit can be a share of a physical circuit, or an abstract service (such as the service provided by TCP) that resembles a circuit in its behavior. In the case of a circuit, the expected behavior is that the data that goes in one end comes out the other end in order.

virus (computer) A form of malicious software that has been crafted so that it can propagate from one computer to another by some means, thus infecting a sequence of computers over time.

Voice over IP (VoIP) The delivery of telephone service over networks based on Internet protocols.

Acronyms

ACM	Association for Computing Machinery
ADU	application data unit
ALF	application layer framing
ANTS	active node transfer system
API	application programming interface
ARPA	Advanced Research Projects Agency (now DARPA)
AS	autonomous system
ATM	asynchronous transfer mode
BGP	border gateway protocol
CA	certificate authority
CABO	Concurrent Architectures Are Better than One
CDN	content delivery network
CIA	confidentiality, integrity, and availability
CID	content identifier—a class of identifier in the XIA scheme
CIDR	classless interdomain routing
CN	ChoiceNet
CSPP	Computer Systems Policy Project—now known as the Technology CEO Council
DARPA	Defense Advanced Research Projects Agency
DDoS	distributed denial of service
DHCP	dynamic host configuration protocol
DHT	distributed hash table
DLCI	data link connection identifier
DNS	domain name system
DOA	delegation-oriented architecture
DoC	denial of capability
DOI	digital object identifier
DONA	data-oriented network architecture

DoS	denial of service
DPI	deep packet inspection
DTN	delay/disruption tolerant network
ECN	explicit congestion notification
EGP	exterior gateway protocol
EID	end-point identity
FARA	forwarding, association, and rendezvous architecture
FD	forwarding directive
FIA	Future Internet Architecture (project)
FII	framework for internet innovation
FQDN	fully qualified domain name
GENI	Global Environment for Network Innovations
GNS	global name service (in MobilityFirst)
GUID	global unique identifier (in MobilityFirst)
HID	host identifier—a class of identifier in the XIA scheme
HTML	hypertext markup language
HTTP	hypertext transfer protocol
i3	Internet indirection infrastructure
ICANN	Internet Corporation for Assigned Names and Numbers
ICCB	Internet Configuration Control Board
ICN	information-centric networking
IETF	Internet Engineering Task Force
IMP	interface message processor
INID	in-network identifier
IoT	Internet of Things
IP	Internet protocol
IPFIX	Internet protocol flow information eXport
IPX	Internet protocol eXchange
ISP	Internet service provider
ITU	International Telecommunications Union
LAN	local area network
MF	MobilityFirst
MIB	management information base
MPLS	multiprotocol label switching
NA	network address (in MobilityFirst)
NAT	network address translation
NDN	named data networking
Netinf	network of information
NII	national information infrastructure
NRS	name resolution server (in Netinf)
NSF	National Science Foundation

OSI	open systems interconnection
PHB	per-hop behavior
PoC	proof of consent
PoP	proof of path (a concept from Nebula) or point of presence (referring to a telephone exchange)
PSIRP	publish/subscribe Internet routing paradigm
QoE	quality of experience
QoS	quality of service
RED	random early detection or drop
RFC	"Request for Comment," publications of the Internet Engineering Task Force
RID	rendezvous identifier
RINA	recursive internetwork architecture
RNA	recursive network architecture
RoI	return on investment
RS	rendezvous system (part of NewArch)
RSH	role-specific header
SDN	software-defined networking
SID	server identifier–a class of identifier in the XIA scheme
SIP	session initiation protocol
SNMP	simple network management protocol
SUM	shut up message
TCP	transmission control protocol
TLS	transport layer security
TOR	The Onion Router
ToS	Type of Service (IP header field)
TTL	Time to Live (IP header field)
TWAMP	two way active measurement protocol
UNE	unbundled network element
URL	uniform resource locator
VINI	virtual network infrastructure
VoIP	voice over Internet protocol
VPN	virtual private network
XCP	eXplicit Control Protocol
XIA	expressive Internet architecture
XID	XIA identifier
XIWT	Cross-Industry Working Team

References

Abbate, Janet. 2000. *Inventing the internet.* Cambridge, MA: MIT Press.

Adhatarao, S. S., J. Chen, M. Arumaithurai, X. Fu, and K. K. Ramakrishnan. 2016. Comparison of naming schema in ICN. In *2016 IEEE international symposium on local and metropolitan area networks (LANMAN),* 1–6. doi:10.1109/LANMAN.2016.7548856.

Alexander, D. Scott, Marianne Shaw, Scott M. Nettles, and Jonathan M. Smith. 1997. Active bridging. In *Proceedings of the ACM SIGCOMM '97 conference on applications, technologies, architectures, and protocols for computer communication. SIGCOMM '97,* 101–111. New York: ACM. doi:10.1145/263105.263149.

Andersen, David, Hari Balakrishnan, Frans Kaashoek, and Robert Morris. 2001. Resilient overlay networks. In *Proceedings of the eighteenth ACM symposium on operating system principles: SOSP Sosp '01,* 131–145. New York: ACM. doi:10.1145/502034.502048. http://doi.acm.org.libproxy.mit.edu/10.1145/502034.502048.

Andersen, David G. 2003. Mayday: Distributed filtering for internet services. In *Proceedings of the 4th conference on USENIX symposium on internet technologies and systems USITS'03,* 3-3. Berkeley, CA: USENIX Association. http://dl.acm.org/citation.cfm?id=1251460.1251463.

Anderson, Ross, and Roger Needham. 2004. Programming satan's computer. In *Computer science today,* 426–440. New York: Springer.

Anderson, Thomas, Larry Peterson, Scott Shenker, and Jonathan Turner. 2005. Overcoming the internet impasse through virtualization. *Computer* 38 (4): 34–41. doi:10.1109/MC.2005.136. http://dx.doi.org/10.1109/MC.2005.136.

Anderson, Thomas, Timothy Roscoe, and David Wetherall. 2004. Preventing internet denial-of-service with capabilities. *SIGCOMM Computer Communication Review* 34 (1): 39–44. doi:10.1145/972374.972382. http://doi.acm.org/10.1145/972374.972382.

Annan, Kofi. 2013. Universal values—peace, freedom, social progress, equal rights, human dignity acutely needed, Secretary-General says at Tubingen University, Germany. http://www.un.org/press/en/2003/sgsm9076.doc.htm.

Argyraki, Katerina, and David R. Cheriton. 2004. Loose source routing as a mechanism for traffic policies. In *Proceedings of the ACM SIGCOMM workshop on future directions in network architecture: FDNA '04*, 57–64. New York: ACM. doi:10.1145/1016707.1016718. http://doi.acm.org/10.1145/1016707.1016718.

Balakrishnan, Hari, Karthik Lakshminarayanan, Sylvia Ratnasamy, Scott Shenker, Ion Stoica, and Michael Walfish. 2004. A layered naming architecture for the internet. *SIGCOMM Computer Communication Review* 34 (4): 343–352. doi:10.1145/1030194.1015505. http://doi.acm.org/10.1145/1030194.1015505.

Barlow, John Perry. 1996. A Declaration of the Independence of Cyberspace. https://projects.eff.org/ barlow/Declaration-Final.html.

Bavier, Andy, Nick Feamster, Mark Huang, Larry Peterson, and Jennifer Rexford. 2006. In vini veritas: Realistic and controlled network experimentation. In *Proceedings of the 2006 conference on applications, technologies, architectures, and protocols for computer communications: SIGCOMM '06*, 3–14. New York: ACM. doi:10.1145/1159913.1159916. http://doi.acm.org.libproxy.mit.edu/10.1145/1159913.1159916.

Belady, Laszlo A., and Meir M. Lehman. 1976. A model of large program development. *IBM Systems Journal* 15 (3): 225–252.

Benkler, Yochai. 2012. Next Generation Connectivity: A review of broadband Internet transitions and policy from around the world. http://www.fcc.gov/stage/pdf/Berkman_Center_Broadband_Study_13Oct09.pdf.

Bogost, Ian. 2007. *Persuasive games: The expressive power of videogames*. Cambridge, MA: MIT Press.

Braden, Robert, Ted Faber, and Mark Handley. 2003. From protocol stack to protocol heap: Role-based architecture. *SIGCOMM Computer Communication Review* 33 (1): 17–22.

Briscoe, Bob, Arnaud Jacquet, Carla Di Cairano-Gilfedder, Alessandro Salvatori, Andrea Soppera, and Martin Koyabe. 2005. Policing congestion response in an internetwork using re-feedback. *Proceedings of ACM SIGCOMM'05, Computer Communication Review* 35 (4): 277–288. doi:http://doi.acm.org/10.1145/1080091.1080124. http://www.cs.ucl.ac.uk/staff/B.Briscoe/projects/2020comms/refb/refb_sigcomm05.pdf.

Brodkin, Jon. 2016. Verizon is actually expanding FIOS again, with new fiber in Boston. *Arstechnica*. https://arstechnica.com/information-technology/2016/04/verizon-is-actually-expanding-fios-again-with-new-fiber-in-boston/.

Brunner, John. 1975. *Shockwave rider*. New York: Harper & Row.

Caesar, Matthew, Tyson Condie, Jayanthkumar Kannan, Karthik Lakshminarayanan, and Ion Stoica. 2006. Rofl: Routing on flat labels. *SIGCOMM Computer Communication Review* 36 (4): 363–374.

Cerf, V., and R. Kahn. 1974. A protocol for packet network intercommunication. *IEEE Transactions on Communications* 22 (5): 637–648. doi:10.1109/TCOM.1974.1092259.

Chen, Shuo, Rui Wang, XiaoFeng Wang, and Kehuan Zhang. 2010. Side-channel leaks in web applications: A reality today, a challenge tomorrow. In *Proceedings of the 2010 IEEE symposium on security and privacy: SP '10,* 191–206. Washington, DC: IEEE Computer Society. doi:10.1109/SP.2010.20. http://dx.doi.org/10.1109/SP.2010.20.

Cheriton, David. 2000. Triad. *SIGOPS Operating Systems Review* 34 (2): 34. doi:10.1145/346152.346236. http://doi.acm.org/10.1145/346152.346236.

Cheriton, David R., and Stephen E. Deering. 1985. Host groups: A multicast extension for datagram internetworks. In *Proceedings of the ninth symposium on data communications: SIGCOMM '85,* 172–179. New York: ACM. doi:10.1145/319056.319039. http://doi.acm.org/10.1145/319056.319039.

Cheriton, D. R. 1989. Sirpent: A high-performance internetworking approach. *SIGCOMM Computer Communication Review* 19 (4): 158–169. doi:10.1145/75247.75263. http://doi.acm.org/10.1145/75247.75263.

Chiang, M., S. H. Low, A. R. Calderbank, and J. C. Doyle. 2007. Layering as optimization decomposition: A mathematical theory of network architectures. *Proceedings of the IEEE* 95 (1): 255–312. doi:10.1109/JPROC.2006.887322.

Chirgwin, Richard. 2015. Spud? the IETF's anti-snooping protocol that will never be used. *The Register.* http://www.theregister.co.uk/2015/07/30/understanding_spud_the_ietfs_burnafterreading_protocol/.

Cidon, Israel, and Inder S. Gopal. 1988. Paris: An approach to integrated high-speed private networks. *International Journal of Digital & Analog Cabled Systems* 1 (2): 77–85. doi:10.1002/dac.4520010208. http://dx.doi.org/10.1002/dac.4520010208.

Cisco Systems, Inc. 2013. Cisco Visual Networking Index: Forecast and Methodology, 2012–2017. http://www.cisco.com/en/US/solutions/collateral/ns341/ns525/ns537/ns705/ns827/white_paper_c11-481360.pdf.

Claffy, KC., and David D. Clark. 2015. Adding Enhanced Services to the Internet: Lessons from History. Social Science Research Network Working Paper Series. http://ssrn.com/abstract=2587262.

Claffy, KC., and D. Clark. 2014. Platform models for sustainable internet regulation. *Journal of Information Policy* 4:463–488.

Clark, D., and KC. Claffy. 2015. An Inventory of Aspirations for the Internet's Future, Technical report, Center for Applied Internet Data Analysis (CAIDA), University of California, San Diego.

Clark, David, Robert Braden, Aaron Falk, and Venkata Pingali. 2003. FARA: Reorganizing the addressing architecture. *SIGCOMM Computer Communication Review* 33 (4): 313–321. doi:10.1145/972426.944770. http://doi.acm.org/10.1145/972426.944770.

Clark, David, Lyman Chapin, Vint Cerf, Robert Braden, and Russ Hobby. 1991. Towards the Future Internet Architecture: RFC 1287 Network Working Group. https://www.ietf.org/rfc/rfc1287.txt.

Clark, David, and KC. Claffy. 2014. Approaches to Transparency Aimed at Minimizing Harm and Maximizing Investment. http://www.caida.org/publications/papers/2014/approaches_to_transparency_aimed/.

Clark, David, and Danny Cohen. 1978. A Proposal for Addressing and Routing in the Internet, IEN 46. http://www.postel.org/ien/pdf/ien046.pdf.

Clark, David, and Susan Landau. 2011. Untangling attribution. *Harvard National Security Journal* 2. http://harvardnsj.org/wp-content/uploads/2011/03/Vol.-2_Clark-Landau_Final-Version.pdf.

Clark, David, Karen Sollins, John Wroclawski, Dina Katabi, Joanna Kulik, Xiaowei Yang, Robert Braden, Aaron Falk, Venkata Pingali, Mark Handley, and Noel Chiappa. 2004. New Arch: Future Generation Internet Architecture. http://www.isi.edu/newarch/iDOCS/final.finalreport.pdf.

Clark, David D. 1988. The design philosophy of the darpa internet protocols. In *Symposium proceedings on communications architectures and protocols: SIGCOMM '88*, 106–114. New York: ACM. doi:10.1145/52324.52336. http://doi.acm.org/10.1145/52324.52336.

Clark, David D. 1997. Internet economics. In *Internet economics*, eds. Lee McKnight and Joseph Bailey, 215–252. Cambridge, MA: MIT Press.

Clark, David D., and Marjory S. Blumenthal. 2011. The end-to-end argument and application design: The role of trust. *Federal Communications Law Journal* 63 (2): 357–390.

Clark, David D., and David R. Wilson. 1987. A comparison of commercial and military computer security policies. In *Proceedings of the 1987 IEEE symposium on research in security and privacy (SP'87)*, 184–193. New York: IEEE Press.

Clark, David D., John Wroclawski, Karen R. Sollins, and Robert Braden. 2005. Tussle in cyberspace: Defining tomorrow's internet. *IEEE/ACM Transactions on Networking* 13 (3): 462–475. doi:10.1109/TNET.2005.850224. http://dx.doi.org/10.1109/TNET.2005.850224.

Clark, D. D., and D. L. Tennenhouse. 1990. Architectural considerations for a new generation of protocols. In *Proceedings of the ACM symposium on communications architectures and protocols: SIGCOMM '90*, 200–208. New York: ACM. doi:10.1145/99508.99553. http://doi.acm.org/10.1145/99508.99553.

Clinton, Hillary. 2011. Remarks: Internet Rights and Wrongs: Choices and Challenges in a Networked World. http://www.state.gov/secretary/rm/2011/02/156619.htm.

Coase, R. H. 1937. The nature of the firm. *Economica* 4 (16): 386–405. doi:10.1111/j.1468-0335.1937.tb00002.x. http://dx.doi.org/10.1111/j.1468-0335.1937.tb00002.x.

Computer Systems Policy Project. 1994. Perspectives on the National Information Infrastructure: Ensuring Interoperability.

Consultative Committee for International Telephony and Telegraphy (CCITT). 1992. *Management framework for open systems interconnection (OSI) for CCITT applications: X.700*. International Telecommunications Union. https://www.itu.int/rec/T-REC-X.700-199209-I/en .

Courcoubetis, Costas, Laszlo Gyarmati, Nikolaos Laoutaris, Pablo Rodriguez, and Kostas Sdrolias. 2016. Negotiating premium peering prices: A quantitative model with applications. *ACM Transactions on Internet Technology* 16 (2): 14–11422. doi:10.1145/2883610. http://doi.acm.org/10.1145/2883610.

Cross-Industry Working Team. 1994. An Architectural Framework for the National Information Infrastructure "Corporation for National Research Initiatives." http://www.xiwt.org/documents/ArchFrame.pdf.

Crowcroft, Jon, Steven Hand, Richard Mortier, Timothy Roscoe, and Andrew Warfield. 2003. Plutarch: An argument for network pluralism. *SIGCOMM Computer Communication Review* 33 (4): 258–266. doi:10.1145/972426.944763. http://doi.acm.org/10.1145/972426.944763.

Dannewitz, Christian, Dirk Kutscher, Börje Ohlman, Stephen Farrell, Bengt Ahlgren, and Holger Karl. 2013. Network of information (Netinf)—An information-centric networking architecture. *Computer Communication* 36 (7): 721–735. doi:10.1016/j.comcom.2013.01.009. http://dx.doi.org/10.1016/j.comcom.2013.01.009.

Day, John. 2008. *Patterns in network architecture: A return to fundamentals*. Upper Saddle River, NJ: Prentice Hall.

Decasper, Dan, Zubin Dittia, Guru Parulkar, and Bernhard Plattner. 1998. Router plugins: A software architecture for next generation routers. In *Proceedings of the ACM SIGCOMM '98 conference on applications, technologies, architectures, and protocols for computer communication: SIGCOMM '98*, 229–240. New York: ACM. doi:10.1145/285237.285285. http://doi.acm.org/10.1145/285237.285285.

Deering, S. E. 1993. SIP: Simple internet protocol. *IEEE Network* 7 (3): 16–28. doi:10.1109/65.224022.

Deering, Stephen E., and David R. Cheriton. 1990. Multicast routing in datagram internetworks and extended LANs. *ACM Transactions on Computer Systems* 8 (2): 85–110. doi:10.1145/78952.78953. http://doi.acm.org/10.1145/78952.78953.

Deering, Stephen Edward. 1992. Multicast Routing in a Datagram Internetwork. PhD diss. Stanford University. UMI (GAX92-21608).

DeNardis, Laura. 2015. *The global war for internet governance.* New Haven, CT: Yale University Press.

Dukkipati, Nandita. 2008. Rate Control Protocol (RCP): Congestion Control to Make Flows Complete Quickly. PhD diss. Stanford University. http://yuba.stanford.edu/~nanditad/thesis-NanditaD.pdf.

Ehrenstein, Claudia. 2012. New Study in Germany Finds Fears of the Internet Are Much Higher Than Expected. *Die Welt.* http://www.worldcrunch.com/tech-science/new-study-in-germany-finds-fears-of-the-internet-are-much-higher-than-expected/c4s4780/.

Fall, Kevin. 2003. A delay-tolerant network architecture for challenged internets. In *Proceedings of the 2003 conference on applications, technologies, architectures, and protocols for computer communications*: *SIGCOMM '03*, 27–34. New York: ACM. doi: 10.1145/863955.863960. http://doi.acm.org/10.1145/863955.863960.

Farber, D., and J. J. Vittal. 1973. Extendability considerations in the design of the distributed computer system (DCS). In *Proceedings of the national telecommunications conference*. Atlanta, Georgia.

Feamster, Nicholas Greer. 2006. Proactive Techniques for Correct and Predictable Internet Routing. PhD diss., MIT.

Feamster, Nick, Lixin Gao, and Jennifer Rexford. 2007. How to lease the internet in your spare time. *SIGCOMM Computer Communication Review* 37 (1): 61–64. doi:10.1145/1198255.1198265. http://doi.acm.org.libproxy.mit.edu/10.1145/1198255.1198265.

Federal Communications Commission. 2005. FCC 05-151, Policy Statement Regarding Broadband Access to the Internet. https://apps.fcc.gov/edocs_public/attachmatch/FCC-05-151A1.pdf.

Federal Communications Commission. 2010. The National Broadband Plan: Connecting America. http://download.broadband.gov/plan/national-broadband-plan.pdf.

Federal Communications Commission. 2015. Protecting and Promoting the Open Internet, GN Docket No. 14-28. https://apps.fcc.gov/edocs_public/attachmatch/FCC-15-24A1.pdf.

Felton, Ed. 2004. Monoculture Debate: Geer vs. Charney. https://freedom-to-tinker.com/blog/felten/monoculture-debate-geer-vs-charney/.

Floyd, Sally, and Van Jacobson. 1993. Random early detection gateways for congestion avoidance. *IEEE/ACM Transactions on Networking* 1 (4): 397–413. doi:10.1109/90.251892. http://dx.doi.org/10.1109/90.251892.

Ford, Bryan. 2004. Unmanaged internet protocol: Taming the edge network management crisis. *SIGCOMM Computer Communication Review* 34 (1): 93–98. doi:10.1145/972374.972391. http://doi.acm.org/10.1145/972374.972391.

Forgie, James. 1979. ST—A Proposed Internet Stream Protocol: IEN 119. https://www.rfc-editor.org/ien/ien119.txt.

Fortz, B., and M. Thorup. 2000. Internet traffic engineering by optimizing ospf weights. In *Infocom 2000. nineteenth annual joint conference of the IEEE computer and communications societies,* Vol. 2, 519–5282. doi:10.1109/INFCOM.2000.832225.

Francis, Paul. 1994a. Addressing in Internet Protocols. PhD diss. University College London. http://www.cs.cornell.edu/people/francis/thesis.pdf.

Francis, Paul. 1994b. PIP Near-Term Architecture: RFC 1621 Network Working Group. https://tools.ietf.org/html/rfc1621.

Francis, Paul, and Ramakrishna Gummadi. 2001. IPNL: A Nat-extended internet architecture. *SIGCOMM Computer Communication Review* 31 (4): 69–80. doi:10.1145/964723.383065. http://doi.acm.org.libproxy.mit.edu/10.1145/964723.383065.

Fraser, Anthony G. 1980. Datakit—a modular network for synchronous and asynchronous traffic. In *Proceedings of the international conference on communications,* Boston, Massachusetts.

Frieden, Rob. 2011. Rationales For and Against FCC Involvement in Resolving Internet Service Provider Interconnection Disputes. Telecommunications Policy Research Conference. http://papers.ssrn.com/sol3/papers.cfm?abstract_id=1838655.

Gaynor, M., and S. Bradner. 2001. The real options approach to standardization. In *Proceedings of the 34th annual Hawaii international conference on system sciences,* 10. doi:10.1109/HICSS.2001.926526.

Geer, Daniel E. 2007. The evolution of security. *Queue* 5 (3): 30–35. doi:10.1145/1242489.1242500. http://doi.acm.org/10.1145/1242489.1242500.

Gibson, William. 1984. *Neuromancer.* New York: Ace.

Godfrey, P. Brighten, Igor Ganichev, Scott Shenker, and Ion Stoica. 2009. Pathlet routing. In *Proceedings of the ACM SIGCOMM 2009 conference on data communication: SIGCOMM '09,* 111–122. New York: ACM. doi:10.1145/1592568.1592583. http://doi.acm.org/10.1145/1592568.1592583.

Gold, Richard, Per Gunningberg, and Christian Tschudin. 2004. A virtualized link layer with support for indirection. In *Proceedings of the ACM SIGCOMM workshop on future directions in network architecture: FDNA '04*, 28–34. New York: ACM. doi:10.1145/1016707.1016713. http://doi.acm.org/10.1145/1016707.1016713.

Greenberg, Andy. 2015. Hackers remotely kill a jeep on the highway—With me in it. *Wired*. https://www.wired.com/2015/07/hackers-remotely-kill-jeep-highway/.

Guha, Saikat, Yutaka Takeda, and Paul Francis. 2004. NUTSS: A sip-based approach to UDP and TCP network connectivity. In *Proceedings of the ACM SIGCOMM workshop on future directions in network architecture: FDNA '04*, 43–48. New York: ACM. doi:10.1145/1016707.1016715. http://doi.acm.org/10.1145/1016707.1016715.

Hafner, Katie. 1998. *Where wizards stay up late: The origins of the internet*. New York: Simon & Schuster.

Hicks, M., J. T. Moore, D. S. Alexander, C. A. Gunter, and S. M. Nettles. 1999. Planet: An active internetwork. In *Infocom '99: eighteenth annual joint conference of the IEEE computer and communications societies*, Vol. 3, 1124–1133. doi:10.1109/INFCOM .1999.751668.

Hinden, Robert. 1994. RFC 1710: Simple Internet Protocol Plus White Paper. https://tools.ietf.org/html/rfc1710.

Horrigan, John B. 2000. New Internet Users. Pew Research Center. http://www .pewinternet.org/2000/09/25/new-internet-users/.

Huston, Geoff. 2012. It's Just Not Cricket: Number Misuse, WCIT and ITRs. http://www.circleid.com/posts/number_misuse_telecommunications_regulations _and_wcit/.

International Telecommunications Union. 2016. Key ICT Indicators for Developed and Developing Countries and the World (Totals and Penetration Rates). Data provided by ITU, extracted from their ITU World Telecommunication/ICT Indicators database. http://www.itu.int/en/ITU-D/Statistics/Documents/statistics/2016 /ITU_Key_2005-2016_ICT_data.xls.

Jacobson, V. 1988. Congestion avoidance and control. In *Symposium proceedings on communications architectures and protocols: SIGCOMM '88*, 314–329. New York: ACM. doi:10.1145/52324.52356. http://doi.acm.org/10.1145/52324.52356.

Jonsson, Andreas, Mats Folke, and Bengt Ahlgren. 2003. The split naming/forwarding network architecture. In *First Swedish national computer networking workshop (SNCNW 2003)*. Arlandastad, Sweden.

Katabi, Dina, Mark Handley, and Charlie Rohrs. 2002. Congestion control for high bandwidth-delay product networks. In *Proceedings of the 2002 conference on applications, technologies, architectures, and protocols for computer communications: SIGCOMM '02*, 89–102. New York: ACM. doi:10.1145/633025.633035. http://doi.acm.org/10.1145/633025.633035.

Kaur, H. Tahilramani, S. Kalyanaraman, A. Weiss, S. Kanwar, and A. Gandhi. 2003. Bananas: An evolutionary framework for explicit and multipath routing in the internet. In *Proceedings of the ACM SIGCOMM workshop on future directions in network architecture: FDNA '03*, 277–288. New York: ACM. doi:10.1145/944759.944766. http://doi.acm.org/10.1145/944759.944766.

Keromytis, Angelos D., Vishal Misra, and Dan Rubenstein. 2002. SOS: Secure overlay services. In *Proceedings of the 2002 conference on applications, technologies, architectures, and protocols for computer communications: SIGCOMM '02*, 61–72. New York: ACM. doi:10.1145/633025.633032. http://doi.acm.org/10.1145/633025.633032.

Kirschner, Marc, and John Gerhart. 1998. Evolvability. *Proceedings of the National Academy of Sciences* 95:8420–8427.

Koponen, Teemu, Mohit Chawla, Byung-Gon Chun, Andrey Ermolinskiy, Kye Hyun Kim, Scott Shenker, and Ion Stoica. 2007. A data-oriented (and beyond) network architecture. In *Proceedings of the 2007 conference on applications, technologies, architectures, and protocols for computer communications: SIGCOMM '07*, 181–192. New York: ACM. doi:10.1145/1282380.1282402. http://doi.acm.org/10.1145/1282380.1282402.

Koponen, Teemu, Scott Shenker, Hari Balakrishnan, Nick Feamster, Igor Ganichev, Ali Ghodsi, P. Brighten Godfrey, Nick McKeown, Guru Parulkar, Barath Raghavan, Jennifer Rexford, Somaya Arianfar, and Dmitriy Kuptsov. 2011. Architecting for innovation. *SIGCOMM Computer Communication Review* 41 (3): 24–36. doi:10.1145/2002250.2002256. http://doi.acm.org/10.1145/2002250.2002256.

Kuhn, Thomas S. 1962. *The structure of scientific revolutions*. Chicago: University of Chicago Press.

Kushman, Nate, Srikanth Kandula, and Dina Katabi. 2007. Can you hear me now?!: It must be BGP. *SIGCOMM Computer Communication Review* 37 (2): 75–84. doi:10.1145/1232919.1232927. http://doi.acm.org/10.1145/1232919.1232927.

Lamport, Leslie, Robert Shostak, and Marshall Pease. 1982. The Byzantine generals problem. *ACM Transactions on Programming Languages and Systems* 4 (3): 382–401. doi:10.1145/357172.357176. http://doi.acm.org.libproxy.mit.edu/10.1145/357172.357176.

Lampson, Butler W. 1973. A note on the confinement problem. *Communications of the ACM* 16 (10): 613–615. doi:10.1145/362375.362389. http://doi.acm.org/10.1145/362375.362389.

Landwehr, Carl E. 2009. A national goal for cyberspace: Create an open, accountable internet. *IEEE Security and Privacy* 7 (3): 3–4. doi:10.1109/MSP.2009.58. http://owens.mit.edu:8888/sfx_local?__char_set=utf8&id=doi:10.1109/MSP.2009.58%7D,&sid=libx%3Amit&genre=article.

Lewis, James. 2014. Significant Cyber Events. Center for Strategic and International Studies. http://csis.org/program/significant-cyber-events.

Licklider, J. C. R., and Robert W. Taylor. 1968. The computer as a communication device. *Science and Technology*. April. Reprinted at http://memex.org/licklider.pdf.

Luderer, G. W. R., H. Che, and W. T. Marshall. 1981. A virtual circuit switch as the basis for distributed systems. In *Proceedings of the seventh symposium on data communications: SIGCOMM '81*, 164–179. New York: ACM. doi:10.1145/800081.802670. http://doi.acm.org/10.1145/800081.802670.

MacKie-Mason, Jeffrey K., and Hal R. Varian. 1996. Some economics of the internet. *In Networks, Infrastructure and the New Task for Regulation*, eds. Werner Sichel and Donald L. Alexander, 107–136. Ann Arbor: University of Michigan Press. http://deepblue.lib.umich.edu/handle/2027.42/50461.

Madden, Mary, Sandra Cortesi, Urs Gasser, Amanda Lenhart, and Maeve Duggan. 2012. Parents, Teens and Online Privacy. Pew Internet and American Life Project. http://www.pewinternet.org/Reports/2012/Teens-and-Privacy.aspx.

Masinter, Larry, and Karen Sollins. 1994. Functional Requirements for Uniform Resource Names. http://www.ietf.org/rfc/rfc1737.txt.

Matni, Nikolai, Ao Tang, and John C. Doyle. 2015. A case study in network architecture tradeoffs. In *Proceedings of the 1st ACM SIGCOMM symposium on software defined networking research: SOSR'15*, 181–187. New York: ACM. doi:10.1145/2774993.2775011. http://doi.acm.org.libproxy.mit.edu/10.1145/2774993.2775011.

McConnell, Mike. 2010. Mike McConnell on how to win the cyber-war we're losing. *Washington Post*, February 28.

McKnight, Lee, and Joseph Bailey, eds. 1997. *Internet economics*. Cambridge, MA: MIT Press.

Mills, C., D. Hirsh, and G. Ruth. 1991. Internet Accounting: Background. https://tools.ietf.org/html/rfc1272.

Monge, Peter R., and Noshir S. Contractor. 2003. *Theories of communication networks*. Oxford: Oxford University Press.

Moore, Gordon E. 1965. Cramming more components onto integrated circuits. *Electronics* 38(8): 114 ff.

Mysore, Jayanth, and Vaduvur Bharghavan. 1997. A new multicasting-based architecture for internet host mobility. In *Proceedings of the 3rd annual ACM/IEEE international conference on mobile computing and networking. Mobicom '97*, 161–172. New York: ACM. doi:10.1145/262116.262144. http://doi.acm.org.libproxy.mit.edu/10.1145/262116.262144.

Naous, Jad, Michael Walfish, Antonio Nicolosi, David Mazières, Michael Miller, and Arun Seehra. 2011. Verifying and enforcing network paths with ICING. In *Proceedings of the seventh conference on emerging networking experiments and technologies: CONEXT '11*, 30–13012. New York: ACM. doi:10.1145/2079296.2079326. http://doi.acm.org/10.1145/2079296.2079326.

National Research Council. 1994. *Realizing the information future: The internet and beyond.* Washington, DC: National Academies Press. doi:10.17226/4755. https://www.nap.edu/catalog/4755/realizing-the-information-future-the-internet-and-beyond.

National Research Council. 1996. *The unpredictable certainty: Information infrastructure through 2000.* Washington, DC: National Academies Press. doi:10.17226/5130. https://www.nap.edu/catalog/5130/the-unpredictable-certainty-information-infrastructure-through-2000.

Needham, R. M. 1979. Systems aspects of the Cambridge ring. In *Proceedings of the seventh ACM symposium on operating systems principles: SOSP '79*, 82–85. New York: ACM. doi:10.1145/800215.806573. http://doi.acm.org/10.1145/800215.806573.

Neumann, Peter G. 1990. Cause of AT&T network failure. *RISKS-FORUM Digest* 9 (62). https://catless.ncl.ac.uk/Risks/.

Nguyen, Giang T. K., Rachit Agarwal, Junda Liu, Matthew Caesar, P. Brighten Godfrey, and Scott Shenker. 2011. Slick packets. In *Proceedings of the ACM sigmetrics joint international conference on measurement and modeling of computer systems: Sigmetrics '11*, 245–256. New York: ACM. doi:10.1145/1993744.1993769. http://doi.acm.org/10.1145/1993744.1993769.

Nichols, K., and B. Carpenter. 1998. Definition of Differentiated Services per Domain Behaviors and Rules for Their Specification. http://www.ietf.org/rfc/rfc3086.txt.

Nichols, Kathleen, and Van Jacobson. 2012. Controlling queue delay. *ACM Queue* 10 (5). http://queue.acm.org/detail.cfm?id=2209336.

Nixon, Ron. 2013. Postal service confirms photographing all U.S. mail. *New York Times,* August 3. http://www.nytimes.com/2013/08/03/us/postal-service-confirms-photographing-all-us-mail.html.

Nygren, E. L., S. J. Garland, and M. F. Kaashoek. 1999: PAN A high-performance active network node supporting multiple mobile code systems. In *Proceedings of the IEEE second conference on Open architectures and network programming proceedings: Openarch '99*, 78–89. New York: IEEE. doi:10.1109/OPNARC.1999.758497.

Open Interconnect Consortium. 2010. Internet Gateway Device (IGD) V 2.0. http://upnp.org/specs/gw/igd2/.

Parno, Bryan, Dan Wendlandt, Elaine Shi, Adrian Perrig, Bruce Maggs, and Yih-Chun Hu. 2007. Portcullis: Protecting connection setup from denial-of-capability attacks. *SIGCOMM Computer Communication Review* 37 (4): 289–300. doi:10.1145/1282427.1282413. http://doi.acm.org/10.1145/1282427.1282413.

Perlman, Radia. 1988. Network Layer Protocols with Byzantine Robustness. PhD diss. MIT. http://publications.csail.mit.edu/lcs/pubs/pdf/MIT-LCS-TR-429.pdf.

Postel, Jon. 1981a. Internet Protocol: Request for Comments 791. http://www.ietf.org/rfc/rfc791.txt.

Postel, Jon. 1981b. Service Mappings. http://www.ietf.org/rfc/rfc795.txt.

Pujol, Jordi, Stefan Schmid, Lars Eggert, Marcus Brunner, and Jürgen Quittek. 2005. Scalability analysis of the turfnet naming and routing architecture. In *Proceedings of the 1st ACM workshop on dynamic interconnection of networks: DIN '05*, 28–32. New York: ACM. doi:10.1145/1080776.1080787. http://doi.acm.org/10.1145 1080776.1080787.

Raghavan, B., P. Verkaik, and A. C. Snoeren. 2009. Secure and policy-compliant source routing. *IEEE/ACM Transactions on Networking* 17 (3): 764–777. doi:10.1109/TNET.2008.2007949.

Rosen, Eric. 1982. Exterior Gateway Protocol (EGP). https://tools.ietf.org/html/rfc827.

Saltzer, Jerome. 1982. On the naming and binding of network destinations. In *Local computer networks*, ed. Piercarlo Ravasio, Greg Hopkins, and Najah Naffah, 311–317. North-Holland. Reprinted as RFC 1498.

Saltzer, Jerome H., David P. Reed, and David D. Clark. 1980. Source routing for campus-wide internet transport. In *Local networks for computer communications*, eds. Anthony West and Phillippe Janson. http://groups.csail.mit.edu/ana/Publications/PubPDFs/SourceRouting.html.

Saltzer, J. H., D. P. Reed, and D. D. Clark. 1984. End-to-end arguments in system design. *ACM Transactions on Computer Systems* 2 (4): 277–288.

Sandvine. 2016. Global Internet Phenomena Report. https://www.sandvine.com/hubfs/downloads/archive/2016-global-internet-phenomena-report-latin-america-and-north-america.pdf.

Savage, Stefan, David Wetherall, Anna Karlin, and Tom Anderson. 2000. Practical network support for IP traceback. In *Proceedings of the conference on applications, technologies, architectures, and protocols for computer communication: SIGCOMM '00*, 295–306. New York: ACM. doi:10.1145/347059.347560. http://doi.acm.org/10.1145/347059.347560.

Schauer, Frederick. 1991. *Playing by the rules: A philosophical examination of rule-based decision-making*. Oxford: Oxford University Press.

Schneier, Bruce. 2010. The Dangers of a Software Monoculture. https://www.schneier.com/essays/archives/2010/11/the_dangers_of_a_sof.html.

Schwartz, B., A. W. Jackson, W. T. Strayer, Wenyi Zhou, R. D. Rockwell, and C. Partridge. 1999. Smart packets for active networks. In *Proceedings of the IEEE second conference on open architectures and network programming: Openarch '99*. 90–97. New York: IEEE. doi:10.1109/OPNARC.1999.758557.

Shoch, John F. 1978. Inter-network naming, addressing, and routing. In *Proceedings of IEEE COMPCON, fall 1978*. https://www.rfc-editor.org/ien/ien19.txt. Reprinted in *Tutorial: Distributed processor communication architecture*, ed. K. Thurber, IEEE Publ. EHO 152-9, 1979, 280–287. New York: IEEE.

Smith, Aaron. 2010. Home Broadband 2010. http://www.pewinternet.org/files/old-media/Files/Reports/2010/Home%20broadband%202010.pdf.

Snoeren, A. C., C. Partridge, L. A. Sanchez, C. E. Jones, F. Tchakountio, B. Schwartz, S. T. Kent, and W. T. Strayer. 2002. Single-packet IP traceback. *IEEE/ACM Transactions on Networking* 10 (6): 721–734. doi:10.1109/TNET.2002.804827.

Sollins, Karen R. 2002. Recursively Invoking Linnaeus:A Taxonomy for Naming Systems. Technical Report MIT-CSAIL-TR-2008-064, Massachusetts Institute of Technology, Computer Science and Artificial Intelligence Lab. http://hdl.handle.net/1721.1/42898.

Song, Dawn Xiaodong, and A. Perrig. 2001. Advanced and authenticated marking schemes for ip traceback. In *Infocom 2001 twentieth annual joint conference of the IEEE computer and communications societies*. Vol. 2, 878–886. doi:10.1109/INFCOM.2001.916279.

Star, S. Leigh. 1998. The structure of ill-structured solutions: Boundary objects and heterogeneous distributed problem solving. In *Distributed artificial intelligence*, Vol. 2, eds. Les Gasser and Michael N. Huhns, 37–54. Amsterdam: Morgan Kaufmann/Elsevier.

Star, S. Leigh, and Karen Ruhleder. 1996. Steps toward an ecology of infrastructure: Design and access for large information spaces. *Information Systems Research* 7 (1): 111–134.

Stoica, Ion, Daniel Adkins, Shelley Zhuang, Scott Shenker, and Sonesh Surana. 2004. Internet indirection infrastructure. *IEEE/ACM Transactions on Networking* 12 (2): 205–218. doi:10.1109/TNET.2004.826279. http://dx.doi.org/10.1109/TNET.2004.826279.

Strauss, Neil. 1994. Rolling stones live on internet: Both a big deal and a little deal. *The New York Times*, November 22. http://www.nytimes.com/1994/11/22/arts/rolling-stones-live-on-internet-both-a-big-deal-and-a-little-deal.html.

Sullivan, Bob. 2013. Online Privacy Fears Are Real. NBCNews. http://www.nbcnews.com/id/3078835/t/online-privacy-fears-are-real.

Sunshine, Carl A. 1977. Source routing in computer networks. *SIGCOMM Computer Communication Review* 7 (1): 29–33. doi:10.1145/1024853.1024855. http://doi.acm.org/10.1145/1024853.1024855.

Tennenhouse, David L., and David J. Wetherall. 1996. Towards an active network architecture. *SIGCOMM Computer Communication Review* 26 (2): 5–17. doi:10.1145/231699.231701. http://doi.acm.org/10.1145/231699.231701.

Touch, Joe, Ilia Baldine, Rudra Dutta, Gregory G. Finn, Bryan Ford, Scott Jordan, Dan Massey, Abraham Matta, Christos Papadopoulos, Peter Reiher, and George Rouskas. 2011. A dynamic recursive unified internet design (DRUID). *Computer Networks* 55 (4): 919–935. doi:http://dx.doi.org/10.1016/j.comnet.2010.12.016. Special Issue on Architectures and Protocols for the Future Internet. http://www.sciencedirect.com/science/article/pii/S138912861000383X.

Touch, Joe, Yu-Shun Wang, and Venkata Pingali. 2006. A Recursive Network Architecture. ISI Technical Report No. ISI-TR-2006-626. Information Sciences Institute. https://www.isi.edu/touch/pubs/isi-tr-2006-626/.

Toure, Hamadoun I. 2012. Remarks to ITU Staff on World Conference on International Telecommunications (WCIT-12). http://www.itu.int/en/osg/speeches/Pages/2012-06-06-2.aspx.

Trossen, D., and G. Parisis. 2012. Designing and realizing an information-centric internet. *IEEE Communications Magazine* 50 (7): 60–67. doi:10.1109/MCOM.2012.6231280.

Trossen, Dirk, Janne Tuononen, George Xylomenos, Mikko Sarela, Andras Zahemszky, Pekka Nikander, and Teemu Rinta-aho. 2008. From Design for Tussle to Tussle Networking: PSIRP Vision and Use Cases. http://www.psirp.org/files/Deliverables/PSIRP-TR08-0001_Vision.pdf.

Tsuchiya, Paul F., and Tony Eng. 1993. Extending the IP internet through address reuse. *SIGCOMM Computer Communication Review* 23 (1): 16–33. doi:10.1145/173942.173944. http://doi.acm.org.libproxy.mit.edu/10.1145/173942.173944.

Turányi, Zoltán, András Valkó, and Andrew T. Campbell. 2003. 4 + 4: An architecture for evolving the internet address space back toward transparency. *SIGCOMM Computer Communication Review* 33 (5): 43–54. doi:10.1145/963985.963990. http://doi.acm.org/10.1145/963985.963990.

United Nations. 1948. The Universal Declaration of Human Rights. http://www.un.org/en/documents/udhr/index.shtml.

Untz, Vincent, Martin Heusse, Franck Rousseau, and Andrzej Duda. 2004. On demand label switching for spontaneous edge networks. In *Proceedings of the ACM*

SIGCOMM workshop on future directions in network architecture. FDNA '04, 35–42. New York: ACM. doi:10.1145/1016707.1016714. http://doi.acm.org.libproxy.mit.edu 10.1145/1016707.1016714.

U. S. Energy Information Administration. 2013. Annual Energy Outlook 2013. http://www.eia.gov/forecasts/aeo/MT_electric.cfm.

van der Merwe, J. E., S. Rooney, L. Leslie, and S. Crosby. 1998. The tempest—A practical framework for network programmability. *IEEE Network* 12 (3): 20–28. doi:10.1109/65.690958.

van Schewick, Barbara. 2012. *Internet architecture and innovation*. Cambridge,MA: MIT Press.

Walfish, Michael, Jeremy Stribling, Maxwell Krohn, Hari Balakrishnan, Robert Morris, and Scott Shenker. 2004. Middleboxes no longer considered harmful. In *Proceedings of the 6th conference on symposium on operating systems design & implementation: OSDI'04*, Vol. 6, 15-15. Berkeley, CA: USENIX Association. http://dl.acm.org /citation.cfm?id=1251254.1251269.

Wang, X., and Y. Xiao. 2009. IP traceback based on deterministic packet marking and logging. In *Scalable Computing and Communications: Eighth International Conference on Embedded Computing: bedded computing: Scalcom embeddedcom'09*, 178–182. doi:10.1109/EmbeddedCom-ScalCom.2009.40.

Wetherall, David. 1999. Active network vision and reality: Lessons from a capsule-based system. In *Proceedings of the seventeenth ACM symposium on operating systems principles*: *SOSP '99*, 64–79. New York: ACM. doi:10.1145/319151.319156. http://doi.acm.org/10.1145/319151.319156.

Wilkes, M. V., and D. J. Wheeler. 1979. The Cambridge digital communication ring. In *Local Area Communication Networks Symposium*. Boston: Mitre Corporation and the National Bureau of Standards.

Wing, D., S. Cheshire, M. Boucadair, R. Penno, and P. Selkirk. 2013. Port Control Protocol (PCP), RFC 6887. http://www.ietf.org/rfc/rfc6887.txt.

Wolf, Tilman, James Griffioen, Kenneth L. Calvert, Rudra Dutta, George N. Rouskas, Ilya Baldin, and Anna Nagurney. 2014. Choicenet: Toward an economy plane for the internet. *SIGCOMM Computer Communication Review* 44 (3): 58–65. doi:10.1145/2656877.2656886. http://doi.acm.org.libproxy.mit.edu/10.1145 /2656877.2656886.

Wright, Charles V., Lucas Ballard, Scott E. Coull, Fabian Monrose, and Gerald M. Masson. 2008. Spot me if you can: Uncovering spoken phrases in encrypted VoIP conversations. In *Proceedings of the 2008 IEEE symposium on security and privacy: SP '08*, 35–49. Washington, DC: IEEE Computer Society. doi:10.1109/SP.2008.21. http://dx.doi.org/10.1109/SP.2008.21.

Wroclawski, John. 1997. The Metanet. In *Workshop on research directions for the next generation internet.* Vienna, VA: Computing Research Association. http://archive.cra.org/Policy/NGI/papers/wroklawWP.

Yaar, A., A. Perrig, and D. Song. 2003. PI: A path identification mechanism to defend against DDoS attacks. In *Proceedings of the 2003 IEEE symposium on Security and privacy*, 93–107. doi:10.1109/SECPRI.2003.1199330.

Yaar, A., A. Perrig, and D. Song. 2004. SIFF: A stateless internet flow filter to mitigate DDOS flooding attacks. In *Proceedings of the 2004 IEEE symposium on Security and privacy*, 130–143. doi:10.1109/SECPRI.2004.1301320.

Yang, Xiaowei. 2003. NIRA A new internet routing architecture. In *Proceedings of the ACM SIGCOMM workshop on future directions in network architecture: FDNA '03*, 301–312. New York: ACM. doi:10.1145/944759.944768. http://doi.acm.org/10.1145/944759.944768.

Yang, Xiaowei, David Wetherall, and Thomas Anderson. 2005. A DOS-limiting network architecture. In *Proceedings of the 2005 conference on applications, technologies, architectures, and protocols for computer communications: SIGCOMM '05*, 241–252. New York: ACM. doi:10.1145/1080091.1080120. http://doi.acm.org/10.1145/1080091.1080120.

Yemeni, Y., and S. da Silva. 1996. Towards Programmable Networks. FIP/IEEE International Workshop on Distributed Systems, October.

Zhang, Lixia, Alexander Afanasyev, Jeffrey Burke, Van Jacobson, KC. Claffy, Patrick Crowley, Christos Papadopoulos, Lan Wang, and Beichuan Zhang. 2014. Named data networking. *SIGCOMM Computer Communication Review* 44 (3): 66–73. doi:10.1145/2656877.2656887. http://doi.acm.org/10.1145/2656877.2656887.

Index

Printed in the United States
by Baker & Taylor Publisher Services